# Hedgehog Signaling Activation in Human Cancer and its Clinical Implications

Jingwu Xie
Editor

# Hedgehog Signaling Activation in Human Cancer and its Clinical Implications

Springer

*Editor*
Jingwu Xie
Wells Center for Pediatric Research
Indiana University School of Medicine
980 W. Walnut St., Rm R3-520
Indianapolis, IN 46202, USA
jinxie@iupui.edu

ISBN 978-1-4419-8434-0 e-ISBN 978-1-4419-8435-7
DOI 10.1007/978-1-4419-8435-7
Springer New York Dordrecht Heidelberg London

Library of Congress Control Number: 2011925369



Printed on acid-free paper

Springer is part of Springer Science+Business Media (www.springer.com)

# Contents

# Contributors

**K. Ayers**
Institut Biologie du Développement & Cancer – IBDC, Université de Nice Sophia-Antipolis, UMR6543 CNRS, Centre de Biochimie, Parc Valrose 06108, Nice cedex 2, France

**K. S. Clifford Chao**
Department of Radiation Oncology, College of Physicians and surgeons, Columbia University; Division of Radiation Oncology, Weil Cornell Medical College, NY, USA

**James K. Chen, Ph.D.**
Department of Chemical and Systems Biology, Stanford University School of Medicine, 269 Campus Drive, CCSR 3155, Stanford, CA 94305, USA

**Yu-Jen Chen, M.D., Ph.D.**
School of Medicine, National Yang Ming University; Department of Radiation Oncology, Mackay Memorial Hospital; Institute of Pharmacology, Taipei Medical University, Tapei, Taiwan

**Xiaodong Cheng, Ph.D.**
Department of Pharmacology and Toxicology, The University of Texas Medical Branch, 301 University Boulevard, Galveston, TX 77555-0615, USA

**G. D'Angelo**
Institut Biologie du Développement & Cancer – IBDC, Université de Nice Sophia-Antipolis, UMR6543 CNRS, Centre de Biochimie, Parc Valrose 06108, Nice cedex 2, France

**B. Mark Evers, M.D.**
Departments of Surgery and the Lucille P. Markey Cancer Center, The University of Kentucky, CC140 Roach Building, Lexington, KY 40508, USA

**Ervin Epstein, Dr.,**
Children's Hospital Oakland Research Institute, 5700 Martin Luther King Jr Way, Oakland, 94609 California, USA

**Jason Fangusaro, M.D.**
Department of Pediatrics, Division of Hematology/Oncology/Stem Cell Transplantion, Northwestern University, Feinberg School of Medicine and Children's Memorial Hospital, Chicago, IL, USA

**Ari J. Firestone**
Department of Chemical and Systems Biology, Stanford University School of Medicine, 269 Campus Drive, CCSR 3155, Stanford, CA 94305, USA

**Chi-chung Hui, Ph.D.**
Program in Developmental and Stem Cell Biology, The Hospital for Sick Children; Department of Molecular Genetics, University of Toronto, Toronto, ON M5G 1X8, Canada

**Philip Iannaccone, M.D., Ph.D.**
Developmental Biology Program, Children's Memorial Research Center and Department of Pediatrics, Northwestern University Feinberg School of Medicine and Children's Memorial Hospital, Chicago, IL, USA

**Jianhang Jia, Ph.D.**
Department of Molecular and Cellular Biochemistry, Markey Cancer Center, University of Kentucky College of Medicine, 741 South Limestone, BBSRB Rm 373, Lexington, KY 40536-0509, USA

**Jin Jiang, Ph.D.**
Department of Developmental Biology, University of Texas Southwestern Medical Center at Dallas, 6000 Harry Hines Blvd., Dallas, TX 75390, USA; Department of Pharmacology, Southwestern Medical Center at Dallas, University of Texas, Dallas, TX 75390, USA

**Michael T. Lewis, Ph.D.**
The Lester and Sue Smith Breast Center, Departments of Molecular and Cellular Biology and Radiology, Baylor College of Medicine, One Baylor Plaza, BCM600, Room N1210, Houston, TX 77030, USA

**Jing Li, Ph.D.**
Lucille P. Markey Cancer Center, The University of Kentucky, CC140 Roach Building, Lexington, KY 40508, USA

**Hui-Fen Liao**
Department of Biochemical Science and Technology, National Chiayi University, Chiayi; School of Medicine, National Yang Ming University, Taipei, Taiwan

**Karen McGovern, Ph.D.**
Infinity Pharmaceuticals, Inc., 780 Memorial Drive, Cambridge, MA 02139, USA

**Stacey K. Ogden, Ph.D.**
Department of Biochemistry, St. Jude Children's Research Hospital, 262 Danny Thomas Place, MS 340, Memphis, TN 38105, USA

**David J. Robbins, Ph.D.**
Molecular Oncology Program, Department of Surgery, Miller School of Medicine, University of Miami, Miami, FL 33136, USA

**P. P. Thérond**
Institut Biologie du Développement & Cancer – IBDC, Université de Nice Sophia-Antipolis, UMR6543 CNRS, Centre de Biochimie, Parc Valrose, 06108 Nice cedex 2, France

**Martin Tremblay, Ph.D.**
Infinity Pharmaceuticals, Inc., 780 Memorial Drive, Cambridge, MA 02139, USA

**Adriana P. Visbal, Ph.D.**
Program in Developmental Biology, Baylor College of Medicine, One Baylor Plaza, BCM600, Room N1210, Houston, TX 77030, USA

**David Walterhouse, M.D.**
Department of Pediatrics, Developmental Biology Program, Children's Memorial Research Center and Division of Hematology/Oncology/Stem Cell Transplantion, Northwestern University, Feinberg School of Medicine and Children's Memorial Hospital, Chicago, IL, USA

**F. Wendler**
Institut Biologie du Développement & Cancer – IBDC, Université de Nice Sophia-Antipolis, UMR6543 CNRS, Centre de Biochimie, Parc Valrose, 06108, Nice cedex 2, France

**Jingwu Xie, Ph.D.**
Wells Center for Pediatric Research, Indiana University School of Medicine, 980 W. Walnut St., Rm R3-520, Indianapolis, IN 46202, USA

**Joon Won Yoon, Ph.D.**
Department of Pediatrics, Developmental Biology Program, Children's Memorial Research Center, Northwestern University, Feinberg School of Medicine and Children's Memorial Hospital, Chicago, IL 60614, USA

# Chapter 1
# Overview of Hedgehog Signaling Pathway

**Chi-chung Hui and Jin Jiang**

## Introduction

Initially discovered in *Drosophila* and later found in all vertebrate model organisms, the Hedgehog (Hh) family of secreted proteins plays critical roles in both embryonic development and adult tissue homeostasis [41, 84]. Numerous human genetic disorders and cancer have been associated with aberrant Hh signaling activity [41, 63, 84].

Hh acts through a conserved pathway to influence the balance between activator and repressor forms of the Gli family of zinc finger transcription factors ($Gli^A$ and $Gli^R$; Fig. 1.1). While *Drosophila* has only one Hh and one Gli protein, Cubitus interruptus (Ci), mammals have three Hh family members (Sonic hedgehog (Shh), Indian hedgehog and Desert hedgehog) and three Gli proteins (Gli1, Gli2 and Gli3). In mice, $Gli^R$ function is mainly derived from Gli3, whereas $Gli^A$ function is primarily contributed by Gli2. Gli1 is a transcriptional target of Hh signaling and acts in a positive feedback to reinforce $Gli^A$ activity. The reception of Hh signals is mediated by a 12-span transmembrane protein Patched (Ptc) that binds directly to Hh, and a 7-span transmembrane protein Smoothened (Smo) that transduces the signal into the cytoplasm. Ptc blocks Smo activity in the absence of Hh, allowing the production of $Gli^R/Ci^R$ that represses a subset of Hh target genes. Binding of Hh to Ptc activates Smo, which blocks $Gli^R/Ci^R$ production and promotes $Gli^A/Ci^A$ activation. The fundamentals of *Drosophila* and mammalian Hh

C.-c. Hui (✉)
Program in Developmental and Stem Cell Biology,
The Hospital for Sick Children,
Toronto, ON M5G 1X8, Canada
and
Department of Molecular Genetics, University of Toronto,
Toronto, ON M5G 1X8, Canada
e-mail: cchui@sickkids.ca

J. Xie (ed.), *Hedgehog Signaling Activation in Human Cancer and its Clinical Implications*,
DOI 10.1007/978-1-4419-8435-7_1, © Springer Science+Business Media, LLC 2011 

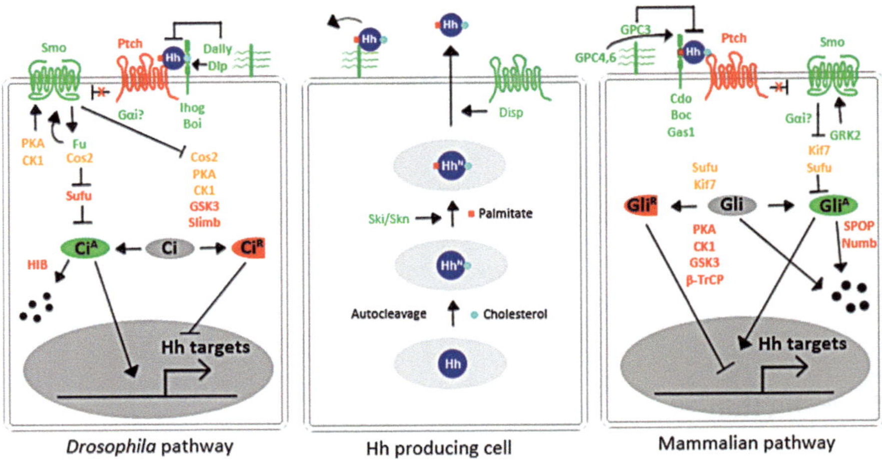


**Fig. 1.1** Sending and transducing the Hh signal. In Hh-producing cells, full-length Hh is autocatalytically cleaved to generate an N-terminal fragment (HhN) modified by cholesterol. HhN is palmitoylated by Ski/Skn. Secretion of dual lipid-modified Hh is mediated by Disp. HSPGs facilitate Hh movement. Hh signal reception is facilitated by Ihog/Boi in *Drosophila* and Cdo/Boc/Gas1 in mammals, functioning as essential coreceptors. Dally and its mammalian HSPG counterpart GPC3 inhibit Hh pathway activity, whereas Dlp and related molecules GPC4 and GPC6 promote Hh signaling. In the absence of Hh, Ptc blocks Smo and full-length Ci/Gli2/Gli3 is phosphorylated by multiple kinases and subsequently targeted to ubiquitin/proteasome-mediated proteolysis through Slimb/βTRCP to generate a truncated repressor form ($Ci^R$/$Gli^R$). In *Drosophila*, efficient phosphorylation of Ci requires the kinesin-like protein Cos2, which acts as a molecular scaffold to bridge Ci and its kinases. Hh-binding to Ptc blocks its inhibition on Smo. In *Drosophila*, Ptc inhibition triggers Smo phosphorylation by PKA and CKI, leading to the cell surface accumulation and activation of Smo. Smo then recruits Cos2-Fu to activate Fu and dissociates Cos2-Ci-kinase complexes to inhibit Ci phosphorylation and processing. Furthermore, high levels of Hh stimulate $Ci^A$ via Fu-mediated antagonism of Sufu. Hh signaling induces the expression of nuclear HIB that targets $Ci^A$ for degradation. Fu-Cos2 is also involved in a feedback regulation of Smo phosphorylation. In mammalian systems, Kif7 is the mammalian Cos2 homolog but it does not interact directly with mSmo. mSmo phosphorylation requires GRK2. In mammals, Fu homolog is not required for Hh signaling and Sufu is a key negative regulator of Hh signaling. Kif7 and Sufu seem to play dual roles in positive and negative regulation of the Hh pathway. In addition to SPOP, which targets full length Gli2 and Gli3 for degradation, Numb is involved in Gli1 degradation. This figure is adapted from [41]

signal transduction pathways are similar, though major difference can be found in several regulatory steps. There is accumulating evidence suggesting that Hh signaling can also exert Gli-independent non-transcriptional effects [94]. In this chapter, we review the basics of the Hh signaling pathway and highlight some of the recent findings in the field.

## Hh Signal Transduction

### *Lipid Modification and Multimerization of Hh*

In Hh-producing cells, full-length Hh precursor undergoes autocleavage to release an N-terminal fragment (HhN) with a cholesterol moiety covalently linked to its C-terminus (Fig. 1.1) [70]. HhN is then palmitoylated near its N-terminus by the acyltransferase Skinny Hedgehog (Ski/Skn) [9]. While cholesterol modification increases the affinity of Hh for cell membranes and restricts its free dispersal [7, 49], dual lipid modifications facilitate the formation of large multimeric Hh complexes, allowing Hh to move over a long distance ([98] and references therein). HhN forms nanoscale oligomers with heparan sulfate proteoglycans (HSPGs), and disruption of HhN oligomerization and HSPGs interaction compromises specifically long-range signaling [86]. Dispatched (Disp), a transmembrane protein structurally related to Ptc, is required for the secretion of lipidated Hh to the extracellular space [2, 7, 55]. A recent study suggested that Disp might also act with Ptc1 to mediate the transport of Shh through tissues [29].

### *Heparan Sulfate Proteoglycans Regulate Hh Signaling*

Genetic studies in *Drosophila* have shown that members of the glypican subfamily of HSPGs, Dally and Dally-like (Dlp), modulate the transport and reception of Hh signals [95]. Mutations in these genes as well as those affecting the biosynthesis of HSPGs impede the spread of Hh signals and reduce Hh pathway activity [50]. HSPGs seem to affect Hh signaling in many different ways (Fig. 1.1). In the absence of HSPGs, cell surface Hh diminishes, suggesting that HSPGs contribute to the stability of Hh. HSPGs appear to be required for Hh movement as a narrow strip of HSPG-deficient cells is sufficient to completely block Hh signaling in wild-type cells behind the mutant clone. In addition, Dlp is critical for Hh signaling activity and may act as an essential coreceptor [96]. A recent study suggested that there are two functional families of glypicans in *Drosophila* and mammals [92]: Dlp and its mammalian counterparts, including GPC4 and GPC6, constitute a group that acts positively and cell-autonomously for Hh signaling, whereas Dally and other glypicans, such as GPC3, form another group that inhibits Hh response. Consistent with this, GPC3 competes with Ptc for Hh binding in vitro and inhibits Hh signaling during mouse development [8]. It is important to note that HSPGs also regu late other signaling molecules, including Wg/Wnt and Dpp/TGF-β [95], and thus these extracellular matrix proteins likely exert differential effects on multiple signaling pathways during development and tumorigenesis.

## *Modulation of Pathway Activity by Multiple Hh-Binding Proteins*

In addition to Ptc, there are multiple Hh-binding proteins identified in *Drosophila* and mammals. Some of them might act as coreceptors of Hh (Fig. 1.1). Genetic analysis in *Drosophila* revealed that the Ihog family of immunoglobin/fibronectin repeat-containing proteins, Ihog (Interference hedgehog) and Boi (Brother of Ihog), are essential for Hh pathway activity [96, 105]. Mammalian homologs of Ihog/Boi, Cdo and Boc, are also positively involved in Shh signaling [81, 97, 102]. The Ihog/Cdo family proteins bind Hh through fibronectin domains [81, 97], and Ihog can enhance Hh binding to Ptc [97], suggesting that they act as Hh coreceptors. Indeed, Ihog promotes surface presentation of Ptc, and both Ihog and Ptc are required for high-affinity Hh binding, supporting the notion that Ihog and Ptc constitute the Hh receptor in *Drosophila* [105].

Hip1 and Gas1 are two vertebrate-specific Hh-interacting proteins. *Hip1* encodes a membrane-bound glycoprotein that acts as a negative regulator of Hh signaling by competing with Ptc for Hh binding [18]. *Hip1* expression is induced by Hh signaling and restricts Hh signaling through a negative feedback mechanism [17, 36]. On the contrary, *Gas1* encodes a GPI-anchored membrane protein that promotes Shh signaling [1, 58]. Since Gas1 acts cooperatively with Cdo in the positive regulation of Hh response [1], it might function as a coreceptor of Hh.

## *Ptc Inhibits Smo Catalytically*

Being the core Hh-binding receptor, Ptc paradoxically functions as an inhibitor of Hh signaling and blocks pathway activation in the absence of Hh. The precise mechanism by which Ptc regulates Smo remains a mystery. Recent studies suggest that Ptc and Hh reciprocally regulate Smo subcellular localization and conformation. Ptc and Smo are largely segregated in *Drosophila* imaginal discs [22] and they do not form stable protein complexes [43, 79]. Cultured cell experiments suggested that Ptc inhibits Smo at a substochiometrical concentration [79]. Ptc is homologous to the resistance-nodulation-division (RND) family of prokaryotic proton-driven transporter, and might function by transporting an endogenous small molecule Smo agonist or antagonist across membranes, as conserved residues in RND-like transporters are essential for Ptc function [79]. Indeed, Ptc regulates trafficking of lipoproteins through endosomes [44]. Several natural and synthetic small molecules can inhibit or activate Hh pathway at the level of Smo [10, 11]. In cultured cells, Ptc induces the secretion of pro-vitamin D3, and both pro-vitamin D3 and vitamin D3 inhibit Hh signaling at high concentrations [6]. Oxysterols, which lie downstream of vitamin D3 in the cholesterol biosynthetic pathway, act as positive regulators of Hh signaling at a level upstream of Smo [21, 25]. Whether oxysterols or related molecules function as physiological Smo regulators remains to be determined. A recent genetic study in *Drosophila* suggested that the phospholipid, phosphatidylinositol-4 phosphate (PI4P), is a target of Ptc action. In *Drosophila* cells, PI4P promotes Smo accumulation and Hh pathway activation, and Ptc restricts the production of PI4P by regulating its kinase/phosphatase directly or indirectly [98]. Exactly how Ptc regulates PI4P levels and whether oxysterols or lipoprotein-derived lipids are linked to the effects of PI4P on Smo await further investigation.

## *Regulation of Smo Trafficking and Conformation*

In *Drosophila*, Ptc restricts Smo cell surface expression by promoting endocytosis and degradation of Smo. Hh induces opposite changes in the subcellular distribution of Ptc and Smo, with Smo accumulating on the cell surface and Ptc entering the cytoplasm [22, 39, 106]. How Hh and Ptc reciprocally regulate Smo trafficking is not clear, but it is mediated at least in part by Smo phosphorylation. Phosphorylation-deficient Smo variants fail to accumulate on the cell surface in response to Hh, whereas phospho-mimicking Smo variants constitutively accumulate on the cell surface [39, 104].

A similar reciprocal trafficking relationship is observed for mammalian Ptc1 and Smo but this occurs in the primary cilium, a microtubule-based cell surface protrusion present in most mammalian cells (Fig. 1.2). Genetic studies in mice have implicated primary cilia as essential cellular organelles for mammalian Hh signaling.

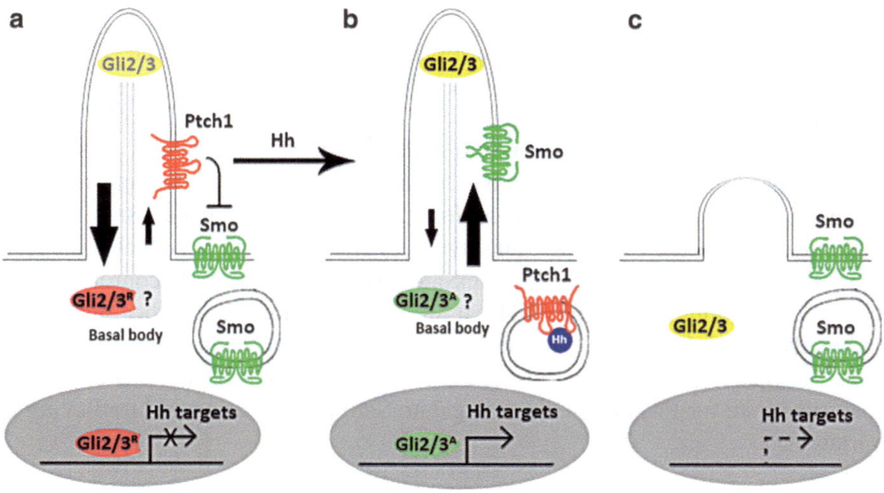


**Fig. 1.2** Hh signaling and primary cilia. (**a**) In the absence of Hh, Ptch1 localizes to the primary cilia and inhibits Smo from entering primary cilia. Due to high retrograde transport activities, little or low levels of full length Gli2 and Gli3 are detected at the ciliary tip. Gli3 and Gli2 (to a lesser extent) are processed to form truncated repressors, which enter the nucleus to inhibit a subset of Hh target genes. (**b**) Hh binding of Ptch1 leads to the elimination of Ptch1 from the primary cilia and, subsequently, the entry of Smo into primary cilia. Full length Gli2 and Gli3 are found at the ciliary tip probably due to high anterograde transport activities. By ill-defined processes, Gli2 and Gli3 are converted into active forms, which promote the transcription of Hh target genes. (**c**) Deletion of primary cilia abolishes the processing as well as activation processes of Gli2 and Gli3. In addition to regulating Smo ciliary localization, Hh also induces a conformational change of Smo essential for its activation. See text for details

Mutations affecting the intraflagellar transport (IFT) machinery or other components that are involved in the assembly and function of cilia affect Hh signaling in several developmental contexts [30]. In the absence of Hh, Ptc localizes to cilia and prevents Smo from accumulating in the cilia; binding of Hh to Ptc triggers reciprocal trafficking of Ptc and Smo, with Ptc moving out of and Smo accumulating in the cilia [20, 72]. Ciliary localization of Smo correlates with Hh pathway activation: both an oncogenic Smo mutation and Smo agonists, such as SAG and oxysterols, promoted accumulation of Smo in the cilia [20, 72], and mutation of a conserved ciliary localization motif in Smo prevented its ciliary accumulation and abolished its signaling activity [20]. How Ptc restricts Smo ciliary accumulation is not clear. Smo may constantly move in and out of the cilia in equilibrium by binding to anterograde and retrograde IFT motors and Ptc may tilt this balance. In support of this model, β-arrestins promote Smo ciliary localization by mediating its association with the anterograde IFT motor kinesin-II in response to Hh [48], and Smo is enriched in the cilia of cells defective in retrograde transport [46, 64]. However, recent studies indicated that ciliary entry of Smo does not require microtubule-dependent cytoplasmic motors [46], and that Smo moves through a lateral transport pathway from the plasma membrane to the ciliary membrane [62].

Ciliary localization of Smo is not sufficient for its activation [3, 46, 73, 89], and Smo activation at the cilia likely may involve additional steps including conformational change [104]. FRET analysis demonstrated that both *Drosophila* and mammalian Smo proteins undergo a conformational change in response to Hh [104]. In response to Hh stimulation or Ptc inhibition, *Drosophila* Smo is phosphorylated by protein kinase A (PKA) and casein kinase I (CK1) at its C-terminal tail (C-tail) [4, 22, 39, 99], which triggers a conformational switch and increased proximity of two Smo C-tails within a Smo dimer [104]. Mechanistically, these phosphorylation events activate Smo by counteracting multiple Arg clusters that maintain Smo in a closed inactive conformation [104]. Mammalian Smo (mSmo) C-tail does not harbor PKA/CK1 sites, but does contain a long stretch of basic residues that inhibits its activity; and mSmo undergoes a similar conformational change upon Shh stimulation [104]. mSmo is phosphorylated either directly or indirectly by the G protein-coupled receptor kinase GRK2, which positively regulates Hh signaling [14, 59, 69], raising the possibility that GRK2 and related kinases may substitute for PKA and CK1 to regulate Smo conformation and trafficking in vertebrates.

## *Downstream of Smo: G Protein and Cos2/Kif7-Ci/Gli Signaling Complex*

G protein $G\alpha_i$ is activated by Smo in both *Drosophila* and mammalian cultured cells [66, 71], and $G\alpha_i$ is required for the expression of Hh target gene *decapentaplegic* (*dpp*) in *Drosophila* wing imaginal discs [66]. However, whether $G\alpha_i$ plays a physiological role in Shh signaling is not clear, as inhibition of $G\alpha_i$ activity had minimal effects on Hh-dependent ventral neural tube patterning in chick embryos [53].

Smo likely signals through both $G\alpha_i$-dependent and -independent mechanisms. In *Drosophila*, Smo directly interacts with a multi-protein signaling complex containing Ci, the kinesin-like protein Costal 2 (Cos2), and the Ser/Thr protein kinase Fused (Fu) [38, 54, 65, 75]. Cos2 serves as a molecular scaffold to bring Ci and Fu together with PKA, GSK3, and CK1, leading to efficient phosphorylation and proteolytic processing of Ci [103]. Activated Smo attenuates Cos2-Ci-kinase complex formation, thus inhibiting Ci phosphorylation and processing [74, 103].

mSmo does not interact directly with the vertebrate Cos2 homologs Kif7 and Kif27 [83]. However, recent studies demonstrated that Kif7 is a functional homolog of Cos2. Kif7 forms complexes with Gli proteins and its deletion or mutation leads to aberrant regulation of Hh signaling [16, 26, 51]. Cos2 can move along the microtubules and its motor activity appears to be required for Ci processing [28]. Similarly, Kif7 function is dependent on intact IFT machinery and Hh signaling promotes ciliary localization of Kif7 [26]. Furthermore, Gli3 processing is compromised in *Kif7* null embryos [16, 26, 51]. In *Drosophila*, $G\alpha_i$ is associated with Cos2 upon Hh stimulation [66]. It remains to be determined whether $G\alpha_i$ or related proteins serve as a link between Smo and Kif7.

## *Control of Gli Protein Degradation and Processing*

Ci/Gli activity is regulated by multiple mechanisms, including phosphorylation, proteolysis, and cytoplasmic/nuclear shuttling. In the absence of Hh, full-length Ci/Gli protein can be proteolytically processed into a truncated repressor (Ci, Gli3 and, to a lesser extent, Gli2) or degraded (Gli1 and Gli2). Hh signaling blocks the production of the truncated repressor, and stimulates nuclear translocation and activation of accumulated full-length Ci/Gli. Ci/Gli processing requires the activities of PKA, GSK3, and CK1 as well as the F-box protein Slimb/β-TRCP of the SCF ubiquitin ligase complex [42]. PKA, GSK3, and CK1 sequentially phosphorylate multiple sites in the C-terminal region of Ci/Gli, resulting in the recruitment of Slimb/β-TRCP [40, 76, 80, 87]. A *p*rocessing *d*eterminant *d*omain (PDD) located between the Zn-finger DNA-binding and Slimb/β-TRCP-binding domains of Ci/Gli appears to be critical for proteasome-mediated degradation that selectively removes its C-terminal half. Deletion of this domain from Ci blocks the production of $Ci^R$ [61] and renders complete degradation of Ci [77]. Gli3 is processed more efficiently than Gli2 into a truncated repressor form probably due to a more potent PDD, and Gli1 lacks a PDD and does not exhibit repressor activity [68].

In mammalian cells, Gli2 and Gli3 are localized to the tip of primary cilia in an Hh-dependent manner (Fig. 1.2; [13, 33, 46, 90]). Upon Hh stimulation, Gli2 shifts from a predominantly cytoplasmic localization to the distal tip of the cilium and within the nucleus [51]. While $Gli3^R$ is predominantly nuclear and not found at the ciliary tip [33, 90], Hh stimulation leads to its disappearance and accumulation of full-length Gli3 ($Gli3^{FL}$) at the tip of the cilium as well as in the nucleus [103]. Importantly, Hh signaling also promotes degradation of full-length Ci/Gli2/Gli3 through an ubiquitin ligase containing HIB/SPOP [13, 45, 88, 90, 100, 101], and this mechanism serves as a negative feedback loop to tune down Hh signaling activity in *Drosophila* [45, 100]. Gli1 is not a strong substrate for SPOP [13, 101] and its degradation involves Numb, which acts in conjunction with the E3 ubiquitin ligase Itch [23]. $Gli3^R$ is also degraded by the proteasome but this likely utilizes a different ubiquitin ligase system [90]. Control of Gli protein degradation might play a central role in preventing tumorigenesis [23, 35].

The exact locations for phosphorylation and proteasomal degradation/processing of Ci/Gli proteins are not known. As the proteasome is enriched at centrosomes that give rise to the basal body underneath the primary cilia [91], Gli proteins might be phosphorylated at primary cilia and then targeted to the centrosome-associated proteasomes for proteolysis. A recent study showed that, in the presence of Shh, the inactive catalytic subunit of PKA is enriched in the cilium base of proliferative cerebellar granular neuronal precursors and that this localization of PKA is essential for Shh-induced proliferation [5]. These observations raise an intriguing possibility that the cilium base might serve as the prime site for phosphorylation and degradation/processing of Gli proteins. In the absence of Hh, the primary cilium may act as a "cAMP gun" to locally activate PKA. Smo might activate $G\alpha_i$ in the ciliary membrane, which in turn represses the adenyl cyclase in the cilium, leading to a local drop of cAMP level and PKA activity. This model is consistent with the genetic data that $Gli^A$ and $Gli^R$ levels are affected in various mutant backgrounds with defective IFT and/or ciliogenesis. How Gli proteins in the cilium are linked to the transcriptional activation of Hh target genes in the nucleus remains unknown. A recent study has highlighted the involvement of cytoplasmic microtubules in ciliary entry of Gli2, but not of Smo [46]. Full-length Gli proteins may need to be "activated" at the cilia before they translocate to the nucleus to activate Hh target genes.

## *Sufu: A Key Regulator of Mammalian Hh Signaling*

A striking difference between *Drosophila* and mammalian Hh signal transduction is the divergent roles of Fu and Sufu [41]. In *Drosophila*, *fu* is a positive regulator essential for Hh signaling, whereas *Sufu* is a genetic suppressor of the *fu* mutation, but its loss does not elicit ectopic Hh signaling and has minimal effects on development. However, in mice, Fu is not involved in Hh signaling [12, 60] and loss of Sufu has profound effects on Hh signaling with ectopic pathway activation [19, 78, 83]. Sufu may have assumed a major inhibitory function in the mammalian Hh pathway due to the existence of multiple Gli proteins. To inhibit $Gli^A$ function, Sufu could impede Gli nuclear localization [24] or suppress Gli activity by recruiting a corepressor complex [15]. Recent studies indicate that Sufu plays a major role in Gli3 processing [13, 34, 37, 47]. Furthermore, Sufu also plays a positive role in mammalian Hh signaling through stabilization of Gli2, in part through counteracting the activity of SPOP [13, 101]. Why Fu kinase is not involved in mammalian Hh signaling? One possibility is that the role of Fu kinase in *Drosophila* Hh signaling is replaced by other protein kinases in mammals. Indeed, multiple protein kinases, including DYRK1a, DYRK2, MAP3K10, ULK3 and Cdc2l1, have been identified to influence Gli activity in mammalian cultured cells [27, 56, 57, 82].

## Unresolved Questions in Mammalian Hh Signaling

Numerous studies have revealed the differential utilization of $Gli^A$ and $Gli^R$ in various developmental systems during mammalian embryogenesis [41]. While $Gli^A$ levels are central to cancer formation [35], the involvement of $Gli^R$ has been implicated by several recent reports linking primary cilia to Hh pathway-dependent tumorigenesis [32, 93]. Though deletion of primary cilia blocks the ability of an oncogenic form of Smo (SmoM2) to induce tumorigenesis, it promotes tumorigenesis induced by activated Gli2, Gli2ΔN. Since primary cilia are essential for Gli3 processing, these results suggest that reduction of $Gli^R$ levels may accelerate Gli2ΔN-induced tumorigenesis. Sufu and Kif7 have different functional requirements for IFT or primary cilia [13, 26, 37]. It is possible that they function in separate processes downstream of Smo and deletion of primary cilia may disrupt Kif7 function, leading to increased tumor incidence in the above studies. Further studies will be needed to decipher the distinct as well as potentially overlapping functions of Sufu and Kif7 in Hh signaling during development and tumorigenesis.

Several genomic scale studies on Gli target genes revealed that though many target promoters contain a consensus related to the sequence TGGGTGGTC, other target genes may not require this consensus sequence for Gli-dependent transcriptional regulation [31, 85]. Whether $Gli^A$ or $Gli^R$ regulates these genes through interactions with other transcription factors or cofactors remains to be determined. Furthermore, there is increasing evidence that Hh exerts its effects through Gli-independent non-transcriptional mechanisms [52, 67, 94]. However, the involvement of this Gli-independent Hh signaling in development and tumorigenesis has not been studied. As detailed in the rest of this book, Hh signaling plays major roles in a wide variety of tumors and it can act via both autocrine and paracrine mechanisms. Importantly, the requirement of Hh pathway activity in tumor formation and growth seems to differ largely in a context-dependent manner. Further understanding of Hh signal transduction mechanisms at different levels along the pathway will certainly be rewarding to current efforts in targeting the pathway for cancer therapy.

**Acknowledgments** C. C. H. is supported by the Canadian Cancer Society Research Institute, and J. J. is supported by grants from NIH, CPRIT, and Welch Foundation (I-1603). We thank Julie Yu for help in illustrations.

## References

1. Allen BL, Tenzen T, McMahon AP (2007) The Hedgehog-binding proteins Gas1 and Cdo cooperate to positively regulate Shh signaling during mouse development. Genes Dev 21:1244–1257
2. Amanai K, Jiang J (2001) Distinct roles of Central missing and Dispatched in sending the Hedgehog signal. Development 128:5119–5127

3. Aanstad P, Santos N, Corbit KC, Scherz PJ, le Trinh A, Salvenmoser W, Huisken J, Reiter JF, Stainier DY (2009) The extracellular domain of Smoothened regulates ciliary localization and is required for high-level Hh signaling. Curr Biol 19:1034–1039
4. Apionishev S, Katanayeva NM, Marks SA, Kalderon D, Tomlinson A (2005) Drosophila Smoothened phosphorylation sites essential for Hedgehog signal transduction. Nat Cell Biol 7:86–92
5. Barzi M, Berenguer J, Menendez A, Alvarez-Rodriguez R, Pons S (2010) Sonic hedgehog mediated proliferation requires the localization of PKA to the cilium base. J Cell Sci 123:62–69
6. Bijlsma MF, Spek CA, Zivkovic D, van de Water S, Rezaee F, Peppelenbosch MP (2006) Repression of smoothened by patched-dependent (pro-)vitamin D3 secretion. PLoS Biol 4:e232
7. Burke R, Nellen D, Bellotto M, Hafen E, Senti KA, Dickson BJ, Basler K (1999) Dispatched, a novel sterol-sensing domain protein dedicated to the release of cholesterol-modified hedgehog from signaling cells. Cell 99:803–815
8. Capurro MI, Xu P, Shi W, Li F, Jia A, Filmus J (2008) Glypican-3 inhibits Hedgehog signaling during development by competing with patched for Hedgehog binding. Dev Cell 14:700–711
9. Chamoun Z, Mann RK, Nellen D, von Kessler DP, Bellotto M, Beachy PA, Basler K (2001) Skinny hedgehog, an acyltransferase required for palmitoylation and activity of the hedgehog signal. Science 293:2080–2084
10. Chen JK, Taipale J, Cooper MK, Beachy PA (2002) Inhibition of Hedgehog signaling by direct binding of cyclopamine to Smoothened. Genes Dev 16:2743–2748
11. Chen JK, Taipale J, Young KE, Maiti T, Beachy PA (2002) Small molecule modulation of Smoothened activity. Proc Natl Acad Sci USA 99:14071–14076
12. Chen MH, Gao N, Kawakami T, Chuang PT (2005) Mice deficient in the fused homolog do not exhibit phenotypes indicative of perturbed hedgehog signaling during embryonic development. Mol Cell Biol 25:7042–7053
13. Chen MH, Wilson CW, Li JY, Law KK, Lu CS, Gacayan R, Zhang X, Hui C-c, Chuang PT (2009) Cilium-independent regulation of Gli proteins by Sufu in Hedgehog signaling is evolutionarily conserved. Genes Dev 23:1910–1928
14. Chen W, Ren XR, Nelson CD, Barak LS, Chen JK, Beachy PA, de Sauvage F, Lefkowitz RJ (2004) Activity-dependent internalization of smoothened mediated by beta-arrestin 2 and GRK2. Science 306:2257–2260
15. Cheng SY, Bishop JM (2002) Suppressor of Fused represses Gli-mediated transcription by recruiting the SAP18-mSin3 corepressor complex. Proc Natl Acad Sci USA 99:5442–5447
16. Cheung HO-L, Zhang X, Ribeiro A, Mo R, Makino S, Puviindran V, Law KKL, Briscoe J, Hui C-C (2009) The kinesin protein Kif7 is a critical regulator of Gli transcription factors in mammalian hedgehog signaling. Sci Signal 2:1–7
17. Chuang PT, Kawcak T, McMahon AP (2003) Feedback control of mammalian Hedgehog signaling by the Hedgehog-binding protein, Hip1, modulates Fgf signaling during branching morphogenesis of the lung. Genes Dev 17:342–347
18. Chuang PT, McMahon AP (1999) Vertebrate Hedgehog signalling modulated by induction of a Hedgehog-binding protein. Nature 397:617–621
19. Cooper AF, Yu KP, Brueckner M, Brailey LL, Johnson L, McGrath JM, Bale AE (2005) Cardiac and CNS defects in a mouse with targeted disruption of suppressor of fused. Development 132:4407–4417
20. Corbit KC, Aanstad P, Singla V, Norman AR, Stainier DY, Reiter JF (2005) Vertebrate Smoothened functions at the primary cilium. Nature 437:1018–1021
21. Corcoran RB, Scott MP (2006) Oxysterols stimulate Sonic hedgehog signal transduction and proliferation of medulloblastoma cells. Proc Natl Acad Sci USA 103:8408–8413
22. Denef N, Neubuser D, Perez L, Cohen SM (2000) Hedgehog induces opposite changes in turnover and subcellular localization of patched and smoothened. Cell 102:521–531

23. Di Marcotullio L, Ferretti E, Greco A, De Smaele E, Po A, Sico MA, Alimandi M, Giannini G, Maroder M, Screpanti I et al (2006) Numb is a suppressor of Hedgehog signalling and targets Gli1 for Itch-dependent ubiquitination. Nat Cell Biol 8:1415–1423
24. Ding Q, Fukami S, Meng X, Nishizaki Y, Zhang X, Sasaki H, Dlugosz A, Nakafuku M, Hui C-C (1999) Mouse suppressor of fused is a negative regulator of sonic hedgehog signaling and alters the subcellular distribution of Gli1. Curr Biol 9:1119–1122
25. Dwyer JR, Sever N, Carlson M, Nelson SF, Beachy PA, Parhami F (2007) Oxysterols are novel activators of the hedgehog signaling pathway in pluripotent mesenchymal cells. J Biol Chem 282:8959–8968
26. Endoh-Yamagami S, Evangelista M, Wilson D, Wen X, Theunissen J-W, Phamluong K, Davis M, Scales SJ, Solloway MJ, de Sauvage FJ, Peterson AS (2009) The mammalian Cos2 homolog Kif7 plays an essential role in modulating Hh signal transduction during development. Curr Biol 19:1320–1326
27. Evangelista M, Lim TY, Lee J, Parker L, Ashique A, Peterson AS, Ye W, Davis DP, de Sauvage FJ (2008) Kinome siRNA screen identifies regulators of ciliogenesis and Hedgehog signal transduction. Sci Signal 1:ra7
28. Farzan SF, Ascano M Jr, Ogden SK, Sanial M, Brigui A, Plessis A, Robbins DJ (2008) Costal2 functions as a kinesin-like protein in the hedgehog signal transduction pathway. Curr Biol 18:1215–1220
29. Etheridge LA, Crawford TQ, Zhang S, Roelink H (2010) Evidence for a role of vertebrate Disp1 in long-range signaling. Development 137:133–140
30. Goetz SC, Anderson KV (2010) The primary cilium: a signaling centre during vertebrate development. Nat Rev Genet 11:331–344
31. Hallikas O, Palin K, Sinjushina N, Rautiainen R, Partanen J, Ukkonen E, Taipale J (2006) Genome-wide prediction of mammalian enhancers based on analysis of transcription-factor binding affinity. Cell 124:47–59
32. Han Y-G, Kim HJ, Dlugosz AA, Ellison DW, Gilbertson RJ, Alvarez-Buylla A (2009) Dual and opposing roles of primary cilia in medulloblastoma development. Nat Med 15:1062–1065
33. Haycraft CJ, Banizs B, Aydin-Son Y, Zhang Q, Michaud EJ, Yoder BK (2005) Gli2 and Gli3 localize to cilia and require the intraflagellar transport protein polaris for processing and function. PLoS Genet 1:e53
34. Humke EW, Dorn KV, Milenkovic L, Scott MP, Rohatgi R (2010) The output of Hedgehog signaling is controlled by the dynamic association between Suppressor of Fused and the Gli proteins. Genes Dev 24:670–682
35. Huntzicker EG, Estay IS, Zhen H, Lokteva LA, Jackson PK, Oro AE (2006) Dual degradation signals control Gli protein stability and tumor formation. Genes Dev 20:276–281
36. Jeong J, McMahon AP (2005) Growth and pattern of the mammalian neural tube are governed by partially overlapping feedback activities of the hedgehog antagonists patched 1 and Hhip1. Development 132:143–154
37. Jia J, Kolterud A, Zeng H, Hoover A, Teglund S, Toftgård R, Liu A (2009) Suppressor of fused inhibits mammalian Hedgehog signaling in the absence of cilia. Dev Biol 330: 452–460
38. Jia J, Tong C, Jiang J (2003) Smoothened transduces Hedgehog signal by physically interacting with Costal2/Fused complex through its C-terminal tail. Genes Dev 17:2709–2720
39. Jia J, Tong C, Wang B, Luo L, Jiang J (2004) Hedgehog signalling activity of Smoothened requires phosphorylation by protein kinase A and casein kinase I. Nature 432:1045–1050
40. Jia J, Zhang L, Zhang Q, Tong C, Wang B, Hou F, Amanai K, Jiang J (2005) Phosphorylation by double-time/CKIepsilon and CKIalpha targets cubitus interruptus for Slimb/beta-TRCP-mediated proteolytic processing. Dev Cell 9:819–830
41. Jiang J, Hui C-C (2008) Hedgehog signaling in development and cancer. Dev Cell 15:801–812
42. Jiang J, Struhl G (1998) Regulation of the Hedgehog and Wingless signalling pathways by the F-box/WD40-repeat protein Slimb. Nature 391:493–496

43. Johnson RL, Milenkovic L, Scott MP (2000) In vivo functions of the patched protein: requirement of the C terminus for target gene inactivation but not Hedgehog sequestration. Mol Cell 6:467–478
44. Khaliullina H, Panakova D, Eugster C, Riedel F, Carvalho M, Eaton S (2009) Patched regulates Smoothened trafficking using lipoprotein-derived lipids. Development 136:4111–4121
45. Kent D, Bush EW, Hooper JE (2006) Roadkill attenuates Hedgehog responses through degradation of Cubitus interruptus. Development 133:2001–2010
46. Kim J, Kato M, Beachy PA (2009) Gli2 trafficking links Hedgehog-dependent activation of Smoothened in the primary cilium to transcriptional activation in the nucleus. Proc Natl Acad Sci USA 106:21666–21671
47. Kise Y, Morinaka A, Teglund S, Miki H (2009) Sufu recruits GSK3β for efficient processing of Gli3. Biochem Biophys Res Commun 387:569–574
48. Kovacs JJ, Whalen EJ, Liu R, Xiao K, Kim J, Chen M, Wang J, Chen W, Lefkowitz RJ (2008) Beta-arrestin-mediated localization of smoothened to the primary cilium. Science 320:1777–1781
49. Li Y, Zhang H, Litingtung Y, Chiang C (2006) Cholesterol modification restricts the spread of Shh gradient in the limb bud. Proc Natl Acad Sci USA 103:6548–6553
50. Lin X (2004) Functions of heparan sulfate proteoglycans in cell signaling during development. Development 131:6009–6021
51. Liem KF Jr, He M, Ocbina PJR, Anderson KV (2009) Mouse Kif7/Costal2 is a cilia-associated protein that regulates Sonic hedgehog signaling. Proc Natl Acad Sci USA 106:13377–13382
52. Lipinski RJ, Bijlsma MF, Gipp JJ, Podhaizer DJ, Bushman W (2008) Establishment and characterization of immortalized Gli-null mouse embryonic fibroblast cell lines. BMC Cell Biol 9:49.
53. Low W-C, Wang C, Pan Y, Huang X-Y, Chen JK, Wang B (2008) The decoupling of Smoothened from G$\alpha_i$ has little effect on Gli3 protein processing and Hedgehog-regulated chick neural tube patterning. Dev Biol 321:188–196
54. Lum L, Zhang C, Oh S, Mann RK, von Kessler DP, Taipale J, Weis-Garcia F, Gong R, Wang B, Beachy PA (2003) Hedgehog signal transduction via Smoothened association with a cytoplasmic complex scaffolded by the atypical kinesin, Costal-2. Mol Cell 12:1261–1274
55. Ma Y, Erkner A, Gong R, Yao S, Taipale J, Basler K, Beachy PA (2002) Hedgehog-mediated patterning of the mammalian embryo requires transporter-like function of dispatched. Cell 111:63–75
56 Maloverjan A, Piirsoo M, Michelson P, Kogerman P, Osterlund T (2010) Identification of a novel serine/threonine kinase ULK3 as a positive regulator of Hedgehog pathway. Exp Cell Res 316:627–637.
57 Mao J, Maye P, Kogerman P, Tejedor FJ, Toftgard R, Xie W, Wu G, Wu D (2002) Regulation of Gli1 transcriptional activity in the nucleus by Dyrk1. J Biol Chem 277:35156–35161.
58. Martinelli DC, Fan C-M (2007) Gas1 extends the range of Hedgehog action by facilitating its signaling. Genes Dev 21:1231–1243
59. Meloni AR, Fralish GB, Kelly P, Salahpour A, Chen JK, Wechsler-Reya RJ, Lefkowitz RJ, Caron MG (2006) Smoothened signal transduction is promoted by G protein-coupled receptor kinase 2. Mol Cell Biol 26:7550–7560
60. Merchant M, Evangelista M, Luoh SM, Frantz GD, Chalasani S, Carano RA, van Hoy M, Ramirez J, Ogasawara AK, McFarland LM et al (2005) Loss of the serine/threonine kinase fused results in postnatal growth defects and lethality due to progressive hydrocephalus. Mol Cell Biol 25:7054–7068
61. Methot N, Basler K (1999) Hedgehog controls limb development by regulating the activities of distinct transcriptional activator and repressor forms of Cubitus interruptus. Cell 96:819–831
62. Milenkovic L, Scott MP, Rohatgi R (2009) Lateral transport of smoothened from the plasma membrane to the membrane of the cilium. J Cell Biol 187:365–374
63. Nieuwenhuis E, Hui C-C (2005) Hedgehog signaling and congenital malformations. Clin Genet 67:193–208

64. Ocbina PJ, Anderson KV (2008) Intraflagellar transport, cilia, and mammalian Hedgehog signaling: analysis in mouse embryonic fibroblasts. Dev Dyn 237:2030–2038
65. Ogden SK, Ascano M Jr, Stegman MA, Suber LM, Hooper JE, Robbins DJ (2003) Identification of a functional interaction between the transmembrane protein Smoothened and the kinesin-related protein Costal2. Curr Biol 13:1998–2003
66. Ogden SK, Fei DL, Schilling NS, Ahmed YF, Hwa J, Robbins DJ (2008) G protein $G\alpha_i$ functions immediately downstream of Smoothened in Hedgehog signaling. Nature 456:967–970
67. Okada A, Charron F, Morin S, Shin DS, Wong K, Fabre PJ, Tessier-Lavigne M, McConnell SK (2006) Boc is a receptor for sonic hedgehog in the guidance of commissural axons. Nature 444:369–373
68. Pan Y, Wang B (2007) A novel protein-processing domain in Gli2 and Gli3 differentially blocks complete protein degradation by the proteasome. J Biol Chem 282:10846–10852
69. Philipp M, Fralish GB, Meloni AR, Chen W, MacInnes AW, Barak LS, Caron MG (2008) Smoothened signaling in vertebrates is facilitated by a G protein-coupled receptor kinase. Mol Biol Cell 19:5478–5489
70. Porter JA, Young KE, Beachy PA (1996) Cholesterol modification of Hedgehog signaling proteins in animal development. Science 274:255–258
71. Riobo NA, Saucy B, DiLizio C, Manning DR (2006) Activation of heterotrimeric G proteins by Smoothened. Proc Natl Acad Sci USA 103:12607–12612
72. Rohatgi R, Milenkovic L, Scott MP (2007) Patched1 regulates hedgehog signaling at the primary cilium. Science 317:372–376
73. Rohatgi R, Milenkovic L, Corcoran RB, Scott MP (2009) Hedgehog signal transduction by Smoothened: pharmacologic evidence of a 2-step activation process. Proc Natl Acad Sci USA 106:3196–3201
74. Ruel L, Gallet A, Raisin S, Truchi A, Staccini-Lavenant L, Cervantes A, Therond PP (2007) Phosphorylation of the atypical kinesin Costal2 by the kinase Fused induces the partial disassembly of the Smoothened-Fused-Costal2-Cubitus interruptus complex in Hedgehog signalling. Development 134:3677–3689
75. Ruel L, Rodriguez R, Gallet A, Lavenant-Staccini L, Therond PP (2003) Stability and association of Smoothened, Costal2 and Fused with Cubitus interruptus are regulated by Hedgehog. Nat Cell Biol 5:907–913
76. Smelkinson MG, Kalderon D (2006) Processing of the Drosophila hedgehog signaling effector Ci-155 to the repressor Ci-75 is mediated by direct binding to the SCF component slimb. Curr Biol 16:110–116
77. Smelkinson MG, Zhou Q, Kalderon D (2007) Regulation of Ci-SCFSlimb binding, Ci proteolysis, and hedgehog pathway activity by Ci phosphorylation. Dev Cell 13:481–495
78. Svard J, Heby-Henricson K, Persson-Lek M, Rozell B, Lauth M, Bergstrom A, Ericson J, Toftgard R, Teglund S (2006) Genetic elimination of Suppressor of fused reveals an essential repressor function in the mammalian Hedgehog signaling pathway. Dev Cell 10:187–197
79. Taipale J, Cooper MK, Maiti T, Beachy PA (2002) Patched acts catalytically to suppress the activity of Smoothened. Nature 418:892–897
80. Tempe D, Casas M, Karaz S, Blanchet-Tournier MF, Concordet JP (2006) Multisite protein kinase A and glycogen synthase kinase 3beta phosphorylation leads to Gli3 ubiquitination by SCFbetaTrCP. Mol Cell Biol 26:4316–4326
81. Tenzen T, Allen BL, Cole F, Kang JS, Krauss RS, McMahon AP (2006) The cell surface membrane proteins Cdo and Boc are components and targets of the Hedgehog signaling pathway and feedback network in mice. Dev Cell 10:647–656
82. Varjosalo M, Bjorklund M, Cheng F, Syvanen H, Kivioja T, Kilpinen S, Sun Z, Kallioniemi O, Stunnenberg HG, He WW et al (2008) Application of active and kinase-deficient kinome collection for identification of kinases regulating hedgehog signaling. Cell 133:537–548
83. Varjosalo M, Li SP, Taipale J (2006) Divergence of hedgehog signal transduction mechanism between Drosophila and mammals. Dev Cell 10:177–186

84. Varjosalo M, Taipale J (2008) Hedgehog: functions and mechanisms. Genes Dev 22:2454–2472
85. Vokes SA, Ji H, McCuine S, Tenzen T, Giles S, Zhong S, Longabaugh WJ, Davidson EH, Wong WH, McMahon AP (2007) Genomic characterization of Gli-activator targets in sonic hedgehog-mediated neural patterning. Development 134:1977–1989
86. Vyas N, Goswami D, Manonmani A, Sharma P, Ranganath HA, VijayRaghavan K, Shashidhara LS, Sowdhamini R, Mayor S (2008) Nanoscale organization of hedgehog is essential for long-range signaling. Cell 133:1214–1227
87. Wang B, Li Y (2006) Evidence for the direct involvement of {beta}TrCP in Gli3 protein processing. Proc Natl Acad Sci USA 103:33–38
88. Wang C, Pan Y, Wang B (2010) Suppressor of fused and Spop regulate the stability, processing and function of Gli2 and Gli3 full-length activators but not their repressors. Development 137:2001–2009
89. Wang Y, Zhou Z, Walsh CT, McMahon AP (2009) Selective translocation of intracellular Smoothened to the primary cilium in response to Hedgehog pathway modulation. Proc Natl Acad Sci USA 106:2623–2628
90. Wen X, Lai CK, Evangelista M, Hongo J-A, de Sauvage FJ, Scales SJ (2010) Kinetics of Hedgehog-induced full-length Gli3 accumulation in primary cilia and subsequent degradation. Mol Cell Biol 30:1910–1922
91. Wigley WC, Fabunmi RP, Lee MG, Marino CR, Muallem S, DeMartino GN, Thomas PJ (1999) Dynamic association of proteasomal machinery with the centrosome. J Cell Biol 145:481–490
92. Williams EH, Pappano WN, Saunders AM, Kim M-S, Leahy DJ, Beachy PA (2010) Dally-like core protein and its mammalian homologues mediate stimulatory and inhibitory effects on Hedgehog signal response. Proc Natl Acad Sci USA 107:5869–5874
93. Wong SY, Seol AD, So P-L, Ermilov AN, Bichakjian CK, Epstein EH, Dlugosz AA, Reiter JF (2009) Primary cilia can both mediate and suppress Hedgehog pathway-dependent tumorigenesis. Nat Med 15:1055–1061
94. Yam PT, Langlois SD, Morin S, Charron F (2009) Sonic hedgehog guides axons through a noncanonical, Src-family-kinase-dependent signaling pathway. Neuron 62:349–362
95. Yan D, Lin X (2009) Shaping morphogen gradients by proteoglycans. Cold Spring Harb Perspect Biol 1:a002493
96. Yan D, Wu Y, Yang Y, Belenkaya TY, Tang X, Lin X (2010) The cell-surface proteins Dally-like and Ihog differentially regulate Hedgehog signaling strength and range during development. Development 137:2033–2044
97. Yao S, Lum L, Beachy P (2006) The ihog cell-surface proteins bind Hedgehog and mediate pathway activation. Cell 125:343–357
98. Yavari A, Nagaraj R, Owusu-Ansah E, Folick A, Ngo K, Hillman T, Call G, Rohatgi R, Scott MP, Banerjee U (2010) Role of lipid metabolism in Smoothened derepression in Hedgehog signaling. Dev Cell 19:54–65
99. Zhang C, Williams EH, Guo Y, Lum L, Beachy PA (2004) Extensive phosphorylation of Smoothened in Hedgehog pathway activation. Proc Natl Acad Sci USA 101:17900–17907
100. Zhang Q, Zhang L, Wang B, Ou CY, Chien CT, Jiang J (2006) A hedgehog-induced BTB protein modulates hedgehog signaling by degrading Ci/Gli transcription factor. Dev Cell 10:719–729
101. Zhang Q, Shi S, Chen Y, Yue T, Li S, Wang B, Jiang J (2009) Multiple Ser/Thr-rich degrons mediate the degradation of Ci/Gli by the Cul3-HIB/SPOP E3 ubiquitin ligase. Proc Natl Acad Sci USA 106:21191–21196
102. Zhang W, Kang JS, Cole F, Yi MJ, Krauss RS (2006) Cdo functions at multiple points in the Sonic Hedgehog pathway, and Cdo-deficient mice accurately model human holoprosencephaly. Dev Cell 10:657–665

103. Zhang W, Zhao Y, Tong C, Wang G, Wang B, Jia J, Jiang J (2005) Hedgehog-regulated costal2-kinase complexes control phosphorylation and proteolytic processing of Cubitus interruptus. Dev Cell 8:267–278
104. Zhao Y, Tong C, Jiang J (2007) Hedgehog regulates smoothened activity by inducing a conformational switch. Nature 450:252–258
105. Zheng X, Mann RK, Sever N, Beachy PA (2010) Genetic and biochemical definition of the Hedgehog receptor. Genes Dev 24:57–71
106. Zhu AJ, Zheng L, Suyama K, Scott MP (2003) Altered localization of Drosophila Smoothened protein activates Hedgehog signal transduction. Genes Dev 17:1240–1252

# Chapter 2
# Regulation of the Hedgehog Morphogene Gradient*

**G. D'Angelo, F. Wendler, K. Ayers, and P. P. Thérond**

**Keywords** Hedgehog • Secretion • Trafficking • Morphogenetic gradient • Extracellular matrix • Heparan sulfate proteoglycans

## Introduction

The development of a multicellular organism is controlled by a genetic program that manifests itself in proliferation, cell differentiation, and apoptosis, leading to the formation of functional organs. A small number of secreted molecules work as "instructors" during these processes [1, 2]. Among them are the Hedgehog (Hh) family of proteins, which act from their source of production at short and long range. They trigger cell fate decisions by inducing a signaling cascade in the ligand-receiving tissues of invertebrates and vertebrates. Dramatic developmental abnormalities are observed in human embryos with compromised Hh signaling, and while great effort is being made to understand and manipulate the signaling cascade downstream of Hh receptor activation, less attention has been payed to the secretion and release of the Hh ligand itself. In this review, we will discuss recent progresses in the understanding of Hh ligand packaging and dispatch from producing cells and its consequences for gradient formation.

* The first two authors contributed equally to this manuscript.

P.P. Thérond (✉)
Institut Biologie du Développement & Cancer – IBDC,
Université de Nice Sophia-Antipolis, UMR6543 CNRS, Centre de Biochimie,
Parc Valrose, 06108 Nice cedex 2, France
e-mail: therond@unice.fr

J. Xie (ed.), *Hedgehog Signaling Activation in Human Cancer and its Clinical Implications*,
DOI 10.1007/978-1-4419-8435-7_2, © Springer Science+Business Media, LLC 2011

## Atypical Biosynthesis of Active Hh Signaling Peptide

An evolutionarily conserved feature of Hh family proteins resides in their hydrophobic lipid modifications. Immature Hh protein enters the secretory pathway where proteolytic autoprocessing and lipidation produce an N-terminal active Hh peptide modified on the C-terminus by a cholesterol moiety, both on *Drosophila* Hh and on its vertebrate counterpart Sonic Hedgehog (Shh) [3] (Fig. 2.1). Moreover, active Hh is also modified by the attachment of a stable amide-linked palmitic acid at the opposing N-terminal end [4, 5]. This acylation is catalyzed by the Skinny Hedgehog protein (Skn or Ski, also called Rasp, Central missing and Sightless) [5–8] that belongs to the family of membrane bound O-acyl-tranferases (MBOAT) [9]. The importance of these dual lipid modifications is underscored by the fact that removal of these modifications interferes with the biological activity of Hh in vivo. Notably, mutations that affect the human Shh autocatalytic processing are associated with holoprosencephaly [10].

## Routing of Hh to the Plasma Membrane

The presence of lipid moieties has a significant effect on Hh solubility and is, therefore, bound to affect intracellular trafficking to the plasma membrane in producing cells. Strikingly, although Hh lipidation is essential for its activity, the exact intracellular compartments in which these modifications take place during Hh secretion are not known. Nevertheless, we do know that the dependence of correct Hh intracellular trafficking on the cholesterol moiety has been demonstrated, as a mutant form of Hh lacking the cholesterol adduct shows significantly altered subcellular distribution [11]. In addition, a hydroxyl-oxygen present within the cholesterol is responsible for the cleavage of Hh [3], and as both autoproteolysis and cholesterol modification are linked, they likely occur in the same compartment. Sensitivity of the uncleaved Shh precursor to digestion with the EndoH glycosidase suggests that the autoprocessing reaction occurs before the medial Golgi [10]. In addition, two mutant forms of Shh (point mutations) that present a temperature-dependent misfolding and retention in the ER do not undergo autoprocessing cleavage. It is thus likely that cleavage and autoprocessing occur in a post-ER pre-medial Golgi compartment, such as the *cis*-Golgi. This would also be consistent with the availability of cholesterol as a critical determinant for cleavage, as cholesterol is found at only low levels in ER membranes and at higher levels in Golgi membranes [12]. This, however, must be experimentally determined, membranes and cleavage might alternatively be required for a unique or specific Hh ER-exit route.

The site of the second Hh modification, palmitoylation, is also an open question, as mouse tagged-Skn protein is mainly localized to the ER in CHO and HeLa cells [13, 14], whereas *Drosophila* Ski was found in the Golgi apparatus in insect Schneider cells [15]. It is also not known whether palmitoylation is conditional to

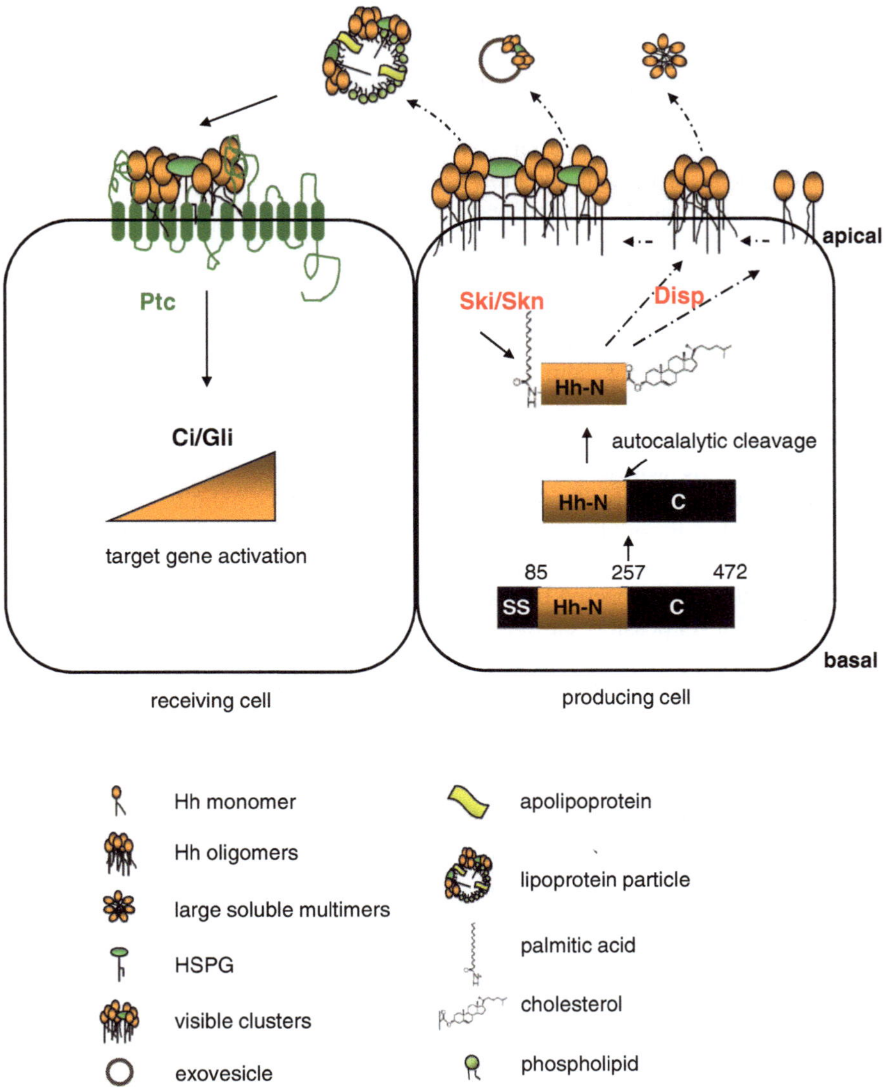


**Fig. 2.1** A schematic diagram of Hh protein biogenesis, secretion, release, and transport to target cells. In Hh-producing cells, the immature Hh protein enters the secretory pathway where proteolytic autoprocessing and lipidation produce a N-terminal active peptide modified on its C-terminus by a cholesterol moiety. Additionally, a stable amide-linked palmitic acid is attached to its N-terminus by the acyltransferase Ski/Skn. The active dually lipid-modified Hh is denoted as Hh-N. The destiny of the C-terminal fragment after cleavage is not known. The transmembrane protein Dispatched (Disp) promotes the secretion of monomers and/or oligomers of Hh-N to the apical cell membrane. Monomers of Hh-N could self-associate spontaneously to form large soluble multimers and can be released in the extracellular lumen. Conversely, Hh-N oligomers at the cell surface could be selectively enriched in visible clusters along with heparan sulfate proteoglycans (HSPGs). Upon release from producing cells, the visible clusters may be transported across several cell diameters from the Hh-producing cells either into exovesicles and/or as an integral component of lipoprotein particles. The transport of Hh-N by one or by the combination of the diverse carriers leads to transcriptional activation of Hh target genes (via the transcriptional activator Ci/Gli) in receiving cells upon its association with the transmembrane protein Patched (Ptc; Hh receptor). Activation of the Hh pathway can also be triggered by a membrane-associated Hh (not shown)

cholesterol modification, as contradictory results have been offered. Analysis of recombinant human and rat Shh produced in a cell-based assay showed that all purified forms of Shh are cholesterol-modified, with a variable percentage of dually lipidated protein as if palmitoylation was subsequent to cholesterol modification [4]. On the other hand, an uncleavable as well as a mutated form of Shh, both lacking the cholesterol modification, have been found to carry [3H]-palmitic acid or $^{125}$I-iodo palmitate when metabolically labeled in cultured cells [13, 16]. Similarly, a non-cholesterol-modified *Drosophila* Hh transgene displayed in vivo activity that depends on its palmitate, suggesting that cholesterol addition of Hh was not a prerequisite for palmitoylation [11].

Finally, one could speculate that acylation affects the routing of Hh secretion, as it was recently demonstrated that acylation of a similar morphogen, Wnt-3a protein, with a palmitoleic acid is required for appropriate trafficking from the ER [17]. However, in vivo analysis of a form of Hh lacking only the palmitic acid modification showed that it could reach and activate target cells as if Hh secretion was independent of palmitoylation [5]. Nevertheless, one must take these studies with caution, as they are hampered by the fact that this variant was expressed at a non-physiological level possibly causing misrouting within producing cells.

How the activity of the acyl-transferase Skn is regulated is also unclear, although another mammalian member of the MBOAT familly, Gup1, has been shown to negatively regulate the N-terminal palmitoylation of Shh, very likely by competing with Skn rather than catalyzing depalmitoylation [14]. This is consistent with biochemical studies showing that Shh palmitoylation does not follow cycles of palmitoylation–depalmitoylation and is thus not reversible [16]. In addition, palmitoylation or other N-terminal hydrophobic modifications of Shh greatly increase its activity in a cell-based assay that does not require secretion or transport without affecting its ability to bind Patched (Ptc) receptor [18]. This suggests that the possible role of palmitoylation is to modulate Hh activity by increasing its stability or by changing its affinity for associated cofactors.

## Secretion and Release of Lipid-Modified Hh

During embryonic development, Hh activity can exert an effect over large fields of cells, for example up to 50 μm in *Drosophila* wing imaginal disc and 300 μm in vertebrate neural tube or limb bud [2]. The long-range activity of Hh raises the question of how a dually lipidated protein can escape the membranes of producing cells to directly activate cells distant from its source of production.

Earlier studies have shown that the cholesterol moiety attached to Hh plays an important role in the protein's membrane retention as it causes strong membrane binding (reviewed in [3]). Indeed, in vitro engineered forms of Hh lacking cholesterol have been found to freely dissociate from cells after secretion, indicating that the cholesterol adduct can function as an anchor that restricts the mobility of the signal. Then how is cholesterol-modified Hh released from the lipid bilayer? A biophysical study revealed that cholesterol alone anchors proteins to membranes with significant strength and revealed a spontaneous desorption half-time of several hours [19], suggesting that a specific cellular activity or machinery is necessary for Hh release in a shorter time lapse.

Hh may be released from specialized membranes as lipidation of Hh allows its association with sterol-rich membrane microdomains, such as lipid rafts in *Drosophila* and in mammalian cells [13, 20]. Interestingly, optical imaging using tagged forms of Hh in *Drosophila* identified nanoscale oligomers at the cell surface [21]. Moreover, a higher order of Hh clustering was also observed and shown to involve ligand interactions with heparan sulfate proteoglycans (HSPG, see below), resulting in visible clusters of the Hh ligand at the cell surface. Mutants of Hh that are impaired in oligomerization lose their long-range activity, suggesting that nanoscale organization of Hh at the plasma membrane is essential for Hh to move a distance [21] (see below).

Why is this so? Within membrane microdomains, Hh may encounter proteins important for its release and efficient spreading. For example, it appears that packaging of Hh, destined for long-range movement, partially requires the cytoplasmic membrane-scaffolding proteins Reggie1/flotillin2 [22], typical members of lipid rafts [23]. Interestingly, overexpression of Reggie1 in the wing imaginal disc of *Drosophila* increased the level of extracellular Hh and its spreading, as well as the activation of the pathway at long range [22]. Reduction of Reggie in the wing disc causes a weak reduction of long-range Hh targets, without affecting short-range signaling [22], indicating that diverse routes may be taken by different pools of Hh (i.e. long vs. short range). In fact, several other proteins have been shown to be specifically required for long-range, but not short-range, bound Hh. This is discussed below.

Some insight into the process of Hh release from producing cells came from the identification of the dispatched (disp) gene that is predicted to encode a 12 transmembrane protein, containing a sterol-sensing domain (SSD), and is required in both *Drosophila* and mouse Hh-producing cells to transport the lipid-modified Hh protein [24, 25]. Importantly, autocrine activity of Shh in the notochord and juxtacrine Hh activity in imaginal discs were not affected in disp mutant animals, but no long-range signaling was observed [24, 26, 27]. These observations suggest that Hh secretion, release, and its long-range activity are closely linked to lipid modification and likely employ a novel intracellular machinery to secrete and release the membrane-anchored Hh protein.

While the exact function of Disp remains unknown, it has been suggested to act as a proton gradient-driven transporter of the Resistance-Nodulation-Division family, as mutations in conserved residues impair its activity [27]. These permeases use a proton electrochemical gradient to function as antiporters in driving out substrates such as hydrophobic drugs, heavy metals, and endogenous compounds. From this homology, it is not clear whether Disp acts by changing the membrane microenvironment around Hh favoring clustering or whether it promotes the release of lipid-modified proteins, as it has been shown for another transmembrane transporter of the bacterial ATP binding cassette family, the LolCDE complex from *Escherichia coli* [28].

It is intriguing that Disp contains a SSD, similar to the one present on proteins which are involved in cholesterol homeostasis or trafficking, including Niemann–Pick C1 (NPC1) [29]. Because only the cholesterol-modified form of Hh is sensitive to Disp activity, Disp-dependent Hh secretion could involve its SSD domain. This domain has been shown to be important for vesicular trafficking regulated by cholesterol level. Because association has been shown between the SSD domain of NPC1 and free cholesterol [30], it is possible that Hh secretion depends on the interaction between the SSD domain of Disp and the cholesterol moiety of Hh.

Based on these studies, and on the fact that in disp mutants, Hh protein production, and processing are not affected, several possible nonexclusive roles have been proposed: (1) Disp may be involved in the intracellular trafficking of Hh to the appropriate membrane microdomain dedicated to its secretion. Indeed, we have previously shown that in the absence of Disp function, cholesterol-modified Hh remains localized to the basolateral membrane of polarized embryonic epidermal secreting cells while it is apically localized in wild-type siblings [11]. However, no specific abnormal apical/basal localization of Hh was observed in disp mutant epithelial cells of the wing imaginal disc [24] (D'Angelo and Thérond unpublished observation); (2) Hh is also secreted under its monomeric form, however, its long-range activity was shown to be dependent on the clustering of Hh multimers with HSPG [21] and on the packaging into lipoprotein particles [31]. Interestingly, visible clusters of Hh are absent in disp mutant animals suggesting that it may control the oligomerization of Hh [11, 32]. It is thus possible that Disp directly promotes the clustering of Hh which is then sequentially associated with other proteins; and (3) Disp may facilitate a direct release of Hh from the plasma membrane into the extracellular space. However, secretion of an active form of Shh was observed from Disp null fibroblast, arguing against this model [25]. In summary, we believe that Disp functions in the formation of an active fraction composed of Hh oligomers and other proteins (HSPGs, lipoprotein particles, see below) dedicated to long-range transport.

## Hedgehog Spreading: A Dilemma in Motion

After release, several modes of Hh movement have been suggested including (1) transcytosis (where proteins are passed cell-to-cell by alternating endo- and exocytosis events) [33], or (2) movement on various vehicles through the extracellular matrix (ECM). However, various studies in *Drosophila* have illustrated that blocking dynamin – the "motor" involved in many cell processes, including endo- and exocytosis – has no effect on the spreading of Hh in either the wing disc [34, 35] or in the fly embryo [36]. Thus in flies, the first model has been generally rejected in favor of the second. This prediction is certainly bolstered by numerous studies which have found that secreted or membrane-attached proteins which make up the ECM can regulate the spreading or signaling of Hh. One group of these proteins attracting much interest is the HSPGs.

Proteoglycans are proteins which have long, unbranched glycosaminoglycan (GAG) chains (sugar polymers) attached to specific serine residues on their protein core. HSPGs have heparan sulfate (HS)-containing GAG chains. Originally viewed as important ECM components, recent evidence has highlighted the critical role of these sugar-modified proteins in morphogenesis and development. Accordingly, mis-regulation of the different proteins involved in HSPG biosynthesis manifests in a large variety of phenotypes [37].

A number of studies have suggested that HSPGs, more specifically the glypicans – a subtype of HSPGs anchored to the plasma membrane though a glycerophosphatidylinositol (GPI) motif – play an important role in the cells which secrete or spread Hh. Glypicans can accumulate in membrane microdomains and appear to be important for the recruitment of Hh into visible clusters (as discussed above) [21]. In addition, HSPGs at the plasma membrane may be involved in Hh stability, as loss of *tout velu* (*ttv*) or *sister of tout velu* (enzymes of the EXT family involved in the GAG chain polymerization) causes a noticeable reduction in Hh levels in the wing imaginal disc [38]. Indeed it has been demonstrated that glypicans recruit lipophorin (a Hh vehicle – see below) to *Drosophila* Hh secreting cells [39]. Thus, the GPI-anchored glypicans appear to be involved in Hh organization at the plasma membrane of secreting cells, Hh stability, and perhaps even Hh vehicle recruitment and loading.

Glypicans also have the ability to move within tissues. Eugster et al. [39] showed that glypicans are commonly shed by wing imaginal discs, moving into surrounding tissues where they are internalized. Furthermore, both fly glypicans Dally-like (Dlp) and Dally appeared to fractionate with Hh and lipoproteins in density-gradient experiments, through interactions with their GPI and the GAG chains [39], suggesting that they can move with these Hh carrying particles. Moreover, the fact that released glypicans, Hh, and lipoproteins are co-localized in endosomes in receiving cells indicates that they may be internalized together [39]. Altogether, fly glypicans do not only play a role in Hh secretion, but also appear to be associated and move with Hh carriers with which they are probably also internalized in receiving cells. This suggests that glypicans could also play an active role in the spreading of morphogens such as Hh.

## Extracellular Gradient Formation by HSPGs

Control of Hh movement and regulation of extracellular Hh gradient formation by HSPGs is an attractive hypothesis and has been researched in various labs. Hh has a high affinity for HSPGs that reside at the cell surface [40]; however, Hh is unable to enter or cross a field of mutant cells which are deficient in HSPG assembly, as is observed in mutants for proteins which are involved in HSPG biosynthesis like *ttv* and *sulfateless* (a sulfotransferase involved in GAG chain modification) [34]. In these cases, Hh is found accumulated at the edge of the mutant tissue, proximal to the source, [41–44] and cannot spread to activate signaling within or past the mutant field [44]. Furthermore, the dependence of Hh movement toward HSPGs is specific to the cholesterol modification as contrary to fully modified Hh, non-cholesterol or non-palmitoyled forms of Hh can spread through a field of cells deleted for *ttv* activity. This is mimicked in cells where glypicans themselves (Dally and Dlp) are lost in the embryo, where Hh mobility is repressed [34]. Therefore, in addition to regulating Hh stability or membrane localization in secreting cells, glypicans are thought to be involved in an active process of Hh movement from one cell to the next, perhaps by passing the morphogen from GAG chain to GAG chain.

Recent evidence from flies has suggested that one glypican, in particular, may play an important role in Hh movement. In *Drosophila*, *dally* mutant wing imaginal discs show weakly decreased Hh signaling, and data indicate that Dally is involved in posterior Hh stability and recruitment of lipophorin [39]. Indeed, Dally itself appears to be present in high levels at the apical pole, and boosts Hh levels at the apical membrane of disc cells. Moreover, Dally release may aid in Hh secretion at this pole and long-range spreading through the ECM found in the apical lumen [45]. Certainly, overexpression of secreted-Dally ectopically activates Hh signaling far from the Hh source, something which is not seen when a transmembrane-tethered form of Dally is expressed [45, 46].

The protein Notum has been studied in both flies and mammals due to its HSPG-related phenotypes. In flies, it has been suggested that Notum acts negatively on the Wnt pathway through its regulation of Dally [47, 48], and studies in mammalian cells have confirmed the ability of Notum to act as a PLC-like lipase and cleave GPI-anchor proteins, including various glypicans [49]. In flies, recent genetic studies have shown that Notum, like Dally, specifically regulates long-range signaling of Hh. Indeed, Dally is unable to augment Hh movement in the absence of Notum, suggesting a link between Notum-dependent release of Dally and Hh spreading [45].

How Dally and Dally-like release promotes long-range Hh spreading is not clear. However, it is possible that Dally augments the release of Hh on specific vehicles such as lipophorin [39] or allows it to spread in a controlled manner without being lost within the surrounding lumenal space. It is also feasible that Dally GAG chains could protect Hh from protease-mediated degradation within the ECM, and along these lines the levels of sulfatation on HSPGs were shown to be an important factor in ADAM protease-mediated Hh cleavage [50] (see below). In conclusion, in *Drosophila*, HSPGs are required for the stable retention of Hh on the cell surface, which results in the restricted and controlled movement of Hh through the surface of the epithelium. Whether glypicans mediate similar processes in vertebrates is an interesting question for future research.

## A Similar Spreading Mechanism from Flies to Mice?

In vertebrate models, like in the fly, Hh ligands can act at long range [2]. Given that the requirement of dynamin for Hh spreading has not been tested in vertebrate systems, it remains to be seen whether Hh moves through the ECM in a similar manner as is seen in *Drosophila* epithelial cells. Indeed, recent findings from the Roelink lab, showing the mammalian protein Disp may be important in Hh receiving cells for continued spreading, has led to the speculation that Hh may be recycled in receiving cells to be released for transcytosis [51]. This would, thus, mean that Hh movement in these tissues could be subject to regulation by a different set of proteins to those in invertebrates. However, the fact that several HSPGs and members of the EXT family, as well as other ECM proteins, can still regulate the non-autonomous signaling of Shh suggests at least some role of ECM in this process.

The specific roles of Hh-HSPG interaction have been tested in mice carrying mutations within a sequence essential for Shh–proteoglycan interactions (the Cardin–Weintraub sequence) [52]. These mutant mice presented growth abnormalities of the CNS, but were free of the patterning abnormalities usually associated with Shh reduction. Intriguingly, it was found that interaction of Shh with proteoglycans is essential for its role in precursor proliferation in the external granule layer of the cerebellum, and illustrated that in this model proteoglycan–Shh interaction altered the cellular response to Shh, as a different profile of target genes and transcriptional activators (Gli) was detected in cells with and without this interaction [52]. Within the developing brain, the extracellular proteoglycans worked to localize Shh to the proliferating zone and, once there, they regulate the cellular response to Shh which favors the activation of a specific subset of proliferation-activating target genes. Thus, proteoglycans, such as the glypicans, appear to be important in controlling Hh signaling by reducing spreading of the ligand. This is at odds with a role in Hh movement, but indicates that at least some forms of Hh may move through the ECM in vertebrate systems, where they encounter ECM proteins such as glypicans.

Various additional results have indicated that glypicans can adopt several different roles within the Hh pathway. For example, null-mice for the vertebrate glypican GPC3 have increased Hh signaling [40, 53], as GPC3 competes with Ptc for the Hh ligand and thus inhibits signaling. It does this by inducing Hh internalization and degradation, thereby reducing its availability to Ptc [40]. Thus in tissues such as the bones, where Indian Hh, a vertebrate member of the Hh family, stimulates the development of the endochondral skeleton, KO mice for GPC3 have increased Ihh levels and thus increased signaling, leading to bone overgrowth [54].

Consistent with the idea that Hh moves through the ECM, several other secreted glycoproteins have been implicated in extracellular Hh stability, and control the non-autonomous signaling of Hh. In *Drosophila* mutant animals for the ECM protein Shifted (Shf), most Hh target genes have reduced expression [55, 56]. Shf associates with both Hh and other ECM proteins, such as HSPGs, and its loss decreases Hh stability, and therefore, signaling [55, 56]. Shf may also play a role in

Hh movement as an Hh-GFP fusion protein does not appear to spread as far when Shf is lost, and although this could arguable be due to reduced stability, experiments where Hh levels have been restored to normal by Hh overexpression in a *shf* mutant background still show a reduced zone of target gene activation [55], indicating that both stability and movement are affected. Additionally, like many other pathway members, Shf action is specific only to cholesterol-modified Hh [55]. Akin to Shf, in Zebrafish, the vertebrate-specific Scube2 protein, an EGF and CUB domain protein, has been implicated in Hh specification of myogenic tissue [57, 58]. Scube has been described to have a cell non-autonomous requirement in the Hh receiving cells and may thus be involved in the formation of the Hh gradient, like its EGF-like domain containing fly cousin, Shf. In conclusion, several ECM proteins, whether secreted or plasma membrane tethered, control Hh stability and gradient formation in vertebrates, indicating ECM-mediated Hh movement as a common mechanism among different phyla.

## Hh Uses Multiple Carriers for Its Travels

In flies and in vertebrates, Hh acts both at short and long range [1]. When discussing Hh carriers, one should remember that, depending on the range of its activity, Hh might associate with different carriers. Whatever the carrier, Hh needs to be able to pass through the hydrophilic environment of a tissue and its lipid moieties need to be shielded or cleaved. Several mechanisms would be capable of solubilizing hydrophobic Hh protein and evidence has been produced for some of them (reviewed in [59]). Also, recent data have led to the proposition of new and interesting mechanisms.

Could Hh simply be released from the cell surface by cleavage? This theory has been tested in *Drosophila*, in which two transgenic fly lines have been generated that either contain a transmembrane anchored form of Hh, Hh-CD2, or a Hh-GPI that anchors Hh via a glycosyl-phosphatidylinositol attachment to the outer leaflead of the plasma membrane. Interestingly, absence of a long-range effect of those forms was concluded, strongly arguing that in *Drosophila* wing disc Hh is not simply released from the cell surface by cleavage. However, in mammals, the picture might be different, as Dierker et al. speculated that proteases, or lipases, could cleave and release Shh from membranes once it has been directed there. Indeed, they recently found that metalloproteases of the ADAM family function as the so-called "sheddases" in cells [50]. In particular, the action of secreted ADAM proteins, but not restricted to ADAM 17, leads to the removal of plasma membrane-bound Hh from mammalian Bosc23 cells. Interestingly, this group shows that an active form of Shh can be obtained from the medium that is devoid of both N- and C-terminal lipid modifications. Whether such mechanism also exists in intact tissues or in other species still needs to be validated.

For short-range transport, Hh could be handed over to the neighboring cell via direct cell-cell contact. ECM components such as HSPGs may be pivotal in this process.

Dally and Dally-like, as discussed above, have been widely implicated in Hh binding, and given that cholesterol-modified Hh cannot cross a field of cells depleted of HSPGs, an involvement in lateral diffusion has been suggested in *Drosophila* [48]. Although cleaved Dally participates in Hh long-range spreading [45], it certainly remains possible that noncleaved proteoglycans also mediate short-range movement of Hh along the cell surface by transferring Hh from one HSPG on one cell to another one located next to it.

Compelling experimental evidence points to the existence of multiple Hh carriers for long-range transport of Hh. These might work in parallel and/or depend on the tissue type and developmental stage of the organism as well as on the source of Hh. Four different long-range carriers have been described:

1. Micelles formation by Hh (via self-aggregation or lipid binding proteins) has been proposed to shield the ligand's lipids attachments. Indeed, the presence of multimeric Hh complexes has been known for a long time. Closer analysis of these forms has revealed large soluble multimers of 5–6 molecules of Shh-N in conditioned media of vertebrate cells [60], and multimers of Hh-N (of 160 kDa) are also present in conditioned culture media from Hh-producing *Drosophila* Schneider cells [13, 32]. What could aid the formation of these structures? It is thought that the hydrophobic cholesterol moiety plays an essential role, perhaps by favoring micelle-like structure with interactions between lipids forming a lipophilic interior. Intriguingly, in *Drosophila*, absence of palmitoylation does not prevent Hh multimer nor visible cluster formation, but Shh that lacks palmitate fails to multimerise [13]. Additionally, the formation of these palmitate-mediated aggregates is physiologically important, as abrogation of higher molecular weight complexes leads to a loss of long-range signaling in vertebrates.

   While the interaction of Hh molecules in these complexes is of a non-covalent nature, Dierker et al. additionally described covalently cross-linked Hh oligomers [50]. The formation of these so-called "undisruptable" complexes depends on transglutaminase and heparan sulfate activity [50], and whether this form exists throughout the Hh-producing species remains to be investigated.
2. Akin to micelle, Hh could be secreted as an integral component of lipoprotein particles. Lipoprotein particles seem to be involved in the transport of systemic Hh in *Drosophila*. They are made up of cholesterol, phospholipids, and scaffolding proteins – apolipoproteins, the latter two components building a monolayer membrane. It is conceivable that Hh is anchored within lipoprotein particles via its attached lipid and/or cholesterol. Lipoprotein particles are produced in the fat body (a tissue functionally related to the liver) of the fly and are secreted into the hemolymph in which they circulate and reach other tissues. The reduction of lipophorin levels in the fat body leads to a significant remote effect on Hh spreading and signaling in peripheral tissues [31]. Lipoprotein-like particles also exist in higher vertebrates, however, whether these particles have a physiological function in the transport of Hh protein remains to be investigated.

3. A third possibility is that Hh could be directed to secretory multivesicular bodies (MVBs). External vesicles, the so-called exosomes, have recently been suggested as carriers for Hh-related peptides in *Caenorhabditis elegans*, although a direct prove has not yet been demonstrated [61]. Nevertheless, the membrane-bound V0 sector of the vacuolar $H^+$-ATPase (V-ATPase), an integral membrane component of MVBs, acts along this pathway leading to the apical secretion of exosomes containing Hedgehog-related peptides [61]. It is an intriguing possibility that similar mechanisms exist in other organisms in order to release Hh. For instance, Shh has been observed in extracellular vesicular structures called nodal vesicular particles (NVPs) located at the surface of mouse ventral node during embryonic development [62]. It has been speculated that NVPs, which consist of a membrane sheath and a lipid core, enable transportation of Shh over vast distances without solubilization [62]. Further investigation into the nature and origin of these vehicles is merited.
4. Membrane-associated Hh may also provide another active form of the ligand. Although it is generally accepted that Hh is released in a soluble form, up to 90% is found in *Drosophila* in a membrane-containing fraction including the plasma membrane [31]. This cell surface associated Hh in producing cells could feasibly be extracted by adjacent receiving cells. The Hh receptor Ptc on those cells might internalize Hh which then results in Hh signaling pathway activation [63]. Interestingly, in *Drosophila* imaginal discs, long cellular protrusions, called cytonemes, are found to extend toward certain morphogen-expressing boundaries [64]. Although no direct evidence for their involvement in Hh signaling has been demonstrated, the morphogen Decapentaplectic (Dpp) is found in cytonemes through which it is directed to its receptor Thickveins in Dpp-responsive cells [65]. Whether Hh uses this mode of transport as a bridge to reach its receptors over several cell diameters is still elusive, but has recently become under intense scrutiny.

From these data, we can conclude that Hh is transmitted by a variety of diverse carriers, each of which may provide unique properties to Hh, allowing differential stability, signal duration, and/or range of activity, through dissimilar biophysical characteristics and composition of these carriers. However, one must reiterate that controversies still exist about how these numerous Hh forms exert a combined signaling activity on ligand-receiving cells, and that where and how these vectors are assembled also presents a "black-box" in our understanding. Inventive and highly technical investigations are needed to shed light on these intriguing questions.

While it has been well established that constitutive activation of the Hh pathway (by loss of function of the Hh receptor Ptc) is responsible for certain kind of cancers (medulloblastomas, rhabdomyosarcomas and basal cell carcinomas), the growth of other tumor types has been shown to be regulated by an autocrine or juxtacrine manner involving secretion of Hh from these tumors. More recently, a new paracrine model has been proposed in which tumor cells produce Hh that stimulates surrounding stromal cells to produce growth factors that in turn support tumor growth [66]. This model conforms much more to Hh developmental activities, which in most cases are typically paracrine [67]. In addition, the level of Hh and/or its carriers secreted from these tumor cells may correlate with their severity, as it has been shown in tumor biopsies from prostate cancer [68]. Thus, research focused on Hh secretion and spreading, and proteins involved in these processes is essential in the battle against Hh-related tumors.

## Conclusions

While great effort is being made to understand and manipulate the signaling cascade triggered by the Hh ligand, less attention has directed toward understanding the role of aberrant amounts of secreted Hh might have on surrounding responsive tissue. This issue is of great importance as locally concentrated Hh triggers a strong and direct gene response. Although very little is known about (1) where within the producing cells Hh processing and the addition of its unique lipid modifications occur, (2) what intracellular route Hh takes to reach the plasma membrane, (3) what constitutes the cellular machinery that leads to secretion, and (4) what kind of vehicle Hh uses to exert particular activities, emerging evidence strongly suggests that these may occur in a very specific way. It is these unique features of this molecule that might provide the Achilles tendon for therapeutic targets in the fight for a wide variety of syndromes caused by abnormal endogenous secretion of Hh ligand.

## References


1. Varjosalo M, Taipale J (2008) Hedgehog: functions and mechanisms. Genes Dev 22:2454–2472
2. Ashe HL, Briscoe J (2006) The interpretation of morphogen gradients. Development 133:385–394
3. Mann RK, Beachy PA (2004) Novel lipid modifications of secreted protein signals. Annu Rev Biochem 73:891–923
4. Pepinsky RB et al (1998) Identification of a palmitic acid-modified form of human Sonic hedgehog. J Biol Chem 273:14037–14045
5. Chamoun Z et al (2001) Skinny hedgehog, an acyltransferase required for palmitoylation and activity of the hedgehog signal. Science 293:2080–2084
6. Micchelli CA, The I, Selva E, Mogila V, Perrimon N (2002) Rasp, a putative transmembrane acyltransferase, is required for Hedgehog signaling. Development 129:843–851
7. Amanai K, Jiang J (2001) Distinct roles of Central missing and Dispatched in sending the Hedgehog signal. Development 128:5119–5127
8. Lee JD, Treisman JE (2001) Sightless has homology to transmembrane acyltransferases and is required to generate active Hedgehog protein. Curr Biol 11:1147–1152
9. Hofmann K (2000) A superfamily of membrane-bound O-acyltransferases with implications for wnt signaling. Trends Biochem Sci 25:111–112
10. Maity T, Fuse N, Beachy PA (2005) Molecular mechanisms of Sonic hedgehog mutant effects in holoprosencephaly. Proc Natl Acad Sci USA 102:17026–17031
11. Gallet A, Rodriguez R, Ruel L, Therond PP (2003) Cholesterol modification of hedgehog is required for trafficking and movement, revealing an asymmetric cellular response to hedgehog. Dev Cell 4:191–204

12. Mitra K, Ubarretxena-Belandia I, Taguchi T, Warren G, Engelman DM (2004) Modulation of the bilayer thickness of exocytic pathway membranes by membrane proteins rather than cholesterol. Proc Natl Acad Sci USA 101:4083–4088
13. Chen MH, Li YJ, Kawakami T, Xu SM, Chuang PT (2004) Palmitoylation is required for the production of a soluble multimeric Hedgehog protein complex and long-range signaling in vertebrates. Genes Dev 18:641–659
14. Abe Y, Kita Y, Niikura T (2008) Mammalian Gup1, a homolog of Saccharomyces cerevisiae glycerol uptake/transporter 1, acts as a negative regulator for N-terminal palmitoylation of Sonic hedgehog. FEBS J 275:318–331
15. Miura GI, Treisman JE (2006) Lipid modification of secreted signaling proteins. Cell Cycle 5:1184–1188
16. Buglino JA, Resh MD (2008) Hhat is a palmitoylacyltransferase with specificity for N-palmitoylation of Sonic Hedgehog. J Biol Chem 283:22076–22088
17. Takada R et al (2006) Monounsaturated fatty acid modification of Wnt protein: its role in Wnt secretion. Dev Cell 11:791–801
18. Taylor FR et al (2001) Enhanced potency of human Sonic hedgehog by hydrophobic modification. Biochemistry 40:4359–4371
19. Peters C, Wolf A, Wagner M, Kuhlmann J, Waldmann H (2004) The cholesterol membrane anchor of the Hedgehog protein confers stable membrane association to lipid-modified proteins. Proc Natl Acad Sci USA 101:8531–8536
20. Rietveld A, Neutz S, Simons K, Eaton S (1999) Association of sterol- and glycosylphosphatidylinositol-linked proteins with Drosophila raft lipid microdomains. J Biol Chem 274:12049–12054
21. Vyas N et al (2008) Nanoscale organization of hedgehog is essential for long-range signaling. Cell 133:1214–1227
22. Katanaev VL et al (2008) Reggie-1/flotillin-2 promotes secretion of the long-range signalling forms of Wingless and Hedgehog in Drosophila. EMBO J 27:509–521
23. Langhorst MF, Reuter A, Stuermer CA (2005) Scaffolding microdomains and beyond: the function of reggie/flotillin proteins. Cell Mol Life Sci 62:2228–2240
24. Burke R et al (1999) Dispatched, a novel sterol-sensing domain protein dedicated to the release of cholesterol-modified hedgehog from signaling cells. Cell 99:803–815
25. Tian H, Jeong J, Harfe BD, Tabin CJ, McMahon AP (2005) Mouse Disp1 is required in sonic hedgehog-expressing cells for paracrine activity of the cholesterol-modified ligand. Development 132:133–142
26. Caspary T et al (2002) Mouse Dispatched homolog1 is required for long-range, but not juxtacrine Hh signaling. Curr Biol 12:1628–1632
27. Ma Y et al (2002) Hedgehog-mediated patterning of the mammalian embryo requires transporter-like function of dispatched. Cell 111:63–75
28. Yakushi T, Masuda K, Narita S, Matsuyama S, Tokuda H (2000) A new ABC transporter mediating the detachment of lipid-modified proteins from membranes. Nat Cell Biol 2:212–218
29. Kuwabara PE, Labouesse M (2002) The sterol-sensing domain: multiple families, a unique role? Trends Genet 18:193–201
30. Ohgami N et al (2004) Binding between the Niemann-Pick C1 protein and a photoactivatable cholesterol analog requires a functional sterol-sensing domain. Proc Natl Acad Sci USA 101:12473–12478
31. Panakova D, Sprong H, Marois E, Thiele C, Eaton S (2005) Lipoprotein particles are required for Hedgehog and Wingless signalling. Nature 435:58–65
32. Gallet A, Ruel L, Staccini-Lavenant L, Therond PP (2006) Cholesterol modification is necessary for controlled planar long-range activity of Hedgehog in Drosophila epithelia. Development 133:407–418
33. Vincent JP, Dubois L (2002) Morphogen transport along epithelia, an integrated trafficking problem. Dev Cell 3:615–623

34. Han C, Belenkaya TY, Wang B, Lin X (2004) Drosophila glypicans control the cell-to-cell movement of Hedgehog by a dynamin-independent process. Development 131:601–611
35. Torroja C, Gorfinkiel N, Guerrero I (2004) Patched controls the Hedgehog gradient by endocytosis in a dynamin-dependent manner, but this internalization does not play a major role in signal transduction. Development 131:2395–2408
36. Gallet A, Therond PP (2005) Temporal modulation of the Hedgehog morphogen gradient by a patched-dependent targeting to lysosomal compartment. Dev Biol 277:51–62
37. Bulow HE, Hobert O (2006) The molecular diversity of glycosaminoglycans shapes animal development. Annu Rev Cell Dev Biol 22:375–407
38. Bornemann DJ, Duncan JE, Staatz W, Selleck S, Warrior R (2004) Abrogation of heparan sulfate synthesis in Drosophila disrupts the Wingless, Hedgehog and Decapentaplegic signaling pathways. Development 131:1927–1938
39. Eugster C, Panakova D, Mahmoud A, Eaton S (2007) Lipoprotein-heparan sulfate interactions in the Hh pathway. Dev Cell 13:57–71
40. Capurro MI et al (2008) Glypican-3 inhibits Hedgehog signaling during development by competing with patched for Hedgehog binding. Dev Cell 14:700–711
41. Bellaiche Y, The I, Perrimon N (1998) Tout-velu is a Drosophila homologue of the putative tumour suppressor EXT-1 and is needed for Hh diffusion. Nature 394:85–88
42. The I, Bellaiche Y, Perrimon N (1999) Hedgehog movement is regulated through tout velu-dependent synthesis of a heparan sulfate proteoglycan. Mol Cell 4:633–639
43. Takei Y, Ozawa Y, Sato M, Watanabe A, Tabata T (2004) Three Drosophila EXT genes shape morphogen gradients through synthesis of heparan sulfate proteoglycans. Development 131:73–82
44. Han C et al (2004) Distinct and collaborative roles of Drosophila EXT family proteins in morphogen signalling and gradient formation. Development 131:1563–1575
45. Ayers KL, Gallet A, Staccini-Lavenant L, Therond PP (2010) The long-range activity of Hedgehog is regulated in the apical extracellular space by the glypican Dally and the hydrolase Notum. Dev Cell 18:605–620
46. Takeo S, Akiyama T, Firkus C, Aigaki T, Nakato H (2005) Expression of a secreted form of Dally, a Drosophila glypican, induces overgrowth phenotype by affecting action range of Hedgehog. Dev Biol 284:204–218
47. Giraldez AJ, Copley RR, Cohen SM (2002) HSPG modification by the secreted enzyme Notum shapes the Wingless morphogen gradient. Dev Cell 2:667–676
48. Han C, Yan D, Belenkaya TY, Lin X (2005) Drosophila glypicans Dally and Dally-like shape the extracellular Wingless morphogen gradient in the wing disc. Development 132:667–679
49. Traister A, Shi W, Filmus J (2008) Mammalian Notum induces the release of glypicans and other GPI-anchored proteins from the cell surface. Biochem J 410:503–511
50. Dierker T, Dreier R, Migone M, Hamer S, Grobe K (2009) Heparan sulfate and transglutaminase activity are required for the formation of covalently cross-linked hedgehog oligomers. J Biol Chem 284:32562–32571
51. Etheridge LA, Crawford TQ, Zhang S, Roelink H (2010) Evidence for a role of vertebrate Disp1 in long-range Shh signaling. Development 137:133–140
52. Chan JA et al (2009) Proteoglycan interactions with Sonic Hedgehog specify mitogenic responses. Nat Neurosci 12:409–417
53. Song HH, Shi W, Xiang YY, Filmus J (2005) The loss of glypican-3 induces alterations in Wnt signaling. J Biol Chem 280:2116–2125
54. Capurro MI, Li F, Filmus J (2009) Overgrowth of a mouse model of Simpson-Golabi-Behmel syndrome is partly mediated by Indian hedgehog. EMBO Rep 10:901–907
55. Glise B et al (2005) Shifted, the Drosophila ortholog of Wnt inhibitory factor-1, controls the distribution and movement of Hedgehog. Dev Cell 8:255–266
56. Gorfinkiel N, Sierra J, Callejo A, Ibanez C, Guerrero I (2005) The Drosophila ortholog of the human Wnt inhibitor factor Shifted controls the diffusion of lipid-modified Hedgehog. Dev Cell 8:241–253

57. Woods IG, Talbot WS (2005) The you gene encodes an EGF-CUB protein essential for Hedgehog signaling in zebrafish. PLoS Biol 3:e66
58. Kawakami A et al (2005) The zebrafish-secreted matrix protein you/scube2 is implicated in long-range regulation of hedgehog signaling. Curr Biol 15:480–488
59. Eaton S (2006) Release and trafficking of lipid-linked morphogens. Curr Opin Genet Dev 16:17–22
60. Zeng X et al (2001) A freely diffusible form of Sonic hedgehog mediates long-range signalling. Nature 411:716–720
61. Liegeois S, Benedetto A, Garnier JM, Schwab Y, Labouesse M (2006) The V0-ATPase mediates apical secretion of exosomes containing Hedgehog-related proteins in Caenorhabditis elegans. J Cell Biol 173:949–961
62. Tanaka Y, Okada Y, Hirokawa N (2005) FGF-induced vesicular release of Sonic hedgehog and retinoic acid in leftward nodal flow is critical for left-right determination. Nature 435:172–177
63. Incardona JP et al (2000) Receptor-mediated endocytosis of soluble and membrane-tethered Sonic hedgehog by Patched-1. Proc Natl Acad Sci USA 97:12044–12049
64. Ramirez-Weber FA, Kornberg TB (1999) Cytonemes: cellular processes that project to the principal signaling center in Drosophila imaginal discs. Cell 97:599–607
65. Hsiung F, Ramirez-Weber FA, Iwaki DD, Kornberg TB (2005) Dependence of Drosophila wing imaginal disc cytonemes on Decapentaplegic. Nature 437:560–563
66. Yauch RL et al (2008) A paracrine requirement for hedgehog signalling in cancer. Nature 455:406–410
67. Ingham PW, McMahon AP (2001) Hedgehog signaling in animal development: paradigms and principles. Genes Dev 15:3059–3087
68. Karhadkar SS et al (2004) Hedgehog signalling in prostate regeneration, neoplasia and metastasis. Nature 431:707–712

# Chapter 3
# Smoothened Signaling Through a G-Protein Effector Network

**David J. Robbins and Stacey K. Ogden**

**Keywords** Hedgehog • Cancer • Smoothened • G-protein

## Introduction

The important role the Hedgehog (HH) signaling pathway plays in cancer was first revealed by patients diagnosed with the familial disorder known as Gorlin Syndrome, who harbor loss-of-function mutations in the HH receptor Patched (PTCH) [1–3]. Besides numerous developmental abnormalities, consistent with disruption of this important developmental signaling pathway, individuals afflicted with this disorder have an inherited predisposition to medulloblastoma, basal cell carcinoma, and rhabdomyosarcoma [2]. Similar mutations found in sporadic cases of these same tumor types implicated *PTCH* as an important tumor suppressor in human cancer [4]. Other components of the HH pathway, such as the gene encoding the seven-transmembrane (7TM) protein Smoothened (SMO), are also found mutated in sporadic forms of these same malignancies [5, 6]. More recently, constitutive activation of the HH pathway has been implicated in other human cancers including those of the breast, prostate, pancreas, and lung, where HH is thought to play a role as a tumor-survival factor [7]. Combined, it has been estimated that approximately 25% of all human tumors harbor a constitutively active HH signaling pathway [8]. As such, considerable effort has gone into identifying novel small-molecule inhibitors of HH signaling. Consistent with the rate-limiting role SMO plays in HH signaling, the vast majority of HH inhibitors isolated from numerous small-molecule screens appear to target SMO [9]. A number of these compounds are currently in clinical trials as anti-cancer agents, targeting tumors that are dependent on HH pathway activity [10]. Thus, a

S.K. Ogden (✉)
Department of Biochemistry, St. Jude Children's Research Hospital,
262 Danny Thomas Place, MS 340, Memphis, TN 38105, USA
e-mail: Stacey.Ogden@stjude.org

J. Xie (ed.), *Hedgehog Signaling Activation in Human Cancer and its Clinical Implications*,
DOI 10.1007/978-1-4419-8435-7_3, © Springer Science+Business Media, LLC 2011 

clear understanding of the mechanisms by which SMO communicates with downstream pathway components, and how such inhibitors affect these processes, will directly impact human health.

## Hedgehog Signal Transduction

Much of what is understood about HH signaling originates from studies of this signaling pathway during the development of the fruit fly *Drosophila melanogaster* [11]. It is now well accepted that the major components, and how they communicate with each other, are highly conserved from Drosophila to man. Although there are significant differences in the importance of some of the signaling components across phyla, it is not yet clear if these variations are due to specific contextual differences or to evolutionary divergence. Thus, in this chapter we generalize about the HH signaling pathway from work derived from numerous animal models, mentioning specific biological contexts only where necessary to illustrate a particular point.

HH is produced and secreted by discrete compartments within a developing field of cells, where it elicits both short- and long-range effects on target cells [12]. The receiving cells interpret the level of HH activation through poorly defined, indirect interactions between the HH receptor PTCH and the signal transducer SMO. PTCH inhibits the activity of SMO in a manner that appears to be catalytic, whereas SMO is constitutively active in the absence of PTCH [13, 14]. One of the mechanisms by which PTCH inhibits SMO activity involves PTCH-dependent trafficking of SMO to lysosomes [15, 16]. In response to HH, PTCH is removed from the cell surface, thereby allowing SMO phosphorylation, stabilization, and accumulation at the plasma membrane [15, 17]. Ultimately all signaling downstream of SMO coalesces to regulate the stability and activity of the GLI/CI family of transcription factors [18]. In the absence of HH, these proteins exist as proteolyzed transcriptional repressors. HH blocks this proteolytic conversion, and stabilizes full-length transcriptional activators. The degree of HH a cell is exposed to ultimately determines the ratio of GLI/CI repressor and activator forms to regulate a spectrum of transcriptional targets that is thought to correlate with the concentration of the initial HH signal.

While the general flow of information through the HH signaling cascade is known, the direct effectors of SMO, and the mechanism(s) by which it communicates with them are still being characterized. The first clue as to how SMO transduces the signal from the plasma membrane to the intracellular effectors came with the observations that Drosophila SMO directly associates with the kinesin-related protein Costal2 (COS2) [19–22], and that mammalian SMO binds the COS2 ortholog KIF7 [23]. The functional consequence of KIF7-SMO binding in mammalian systems is not yet clear. However, in Drosophila, COS2 serves as a scaffold upon which a complex containing CI and the protein kinases Fused (FU), cyclic-AMP (cAMP)-dependent protein kinase A (PKA) and casein kinase 1 (CK1) assembles [24, 25]. As such, a direct association between COS2 and SMO connects the membrane signaling components with the cytoplasmic effectors. Subsequent to this finding, a direct association between SMO and FU that drives a feed-forward loop to facilitate high-level signaling was described [26]. We recently demonstrated that the intracellular molar concentration of SMO is significantly lower than that of CI, COS2, or FU, suggesting that direct association between SMO, COS2, and FU is unlikely to facilitate all aspects of HH signaling [27]. It is, therefore, likely that multiple pools of intracellular effectors exist; some that are in direct contact with SMO, and some that are regulated through the use of G-proteins and/or second messengers.

## Smoothened as a G-Protein-Coupled Receptor

Much of what we know about G-protein-coupled receptor (GPCR) structure and function has resulted from studies of the prototypical GPCR rhodopsin, the first GPCR to be fully sequenced and to yield high-resolution structural data [28–30]. Sequence analysis of rhodopsin suggested the existence of several distinct functional domains, including seven predicted alpha-helical transmembrane segments, an extracellular amino-terminal domain, three extracellular loops, a carboxyl-terminal domain with multiple phosphorylation sites, and three intracellular loops [29]. Structural analysis of rhodopsin, and more recently of the $\beta_2$-adrenergic receptor and $A_{2A}$ adenosine receptor, confirm the existence and conservation of these domains, underscoring their importance in GPCR function [28]. As such, proteins possessing these well-established functional domains in their primary amino acid sequence or predicted tertiary structure are classified as members of the GPCR superfamily, which is estimated to encompass more than 1% of all human genes [30].

SMO was originally identified as a gene necessary for proper organization of the early Drosophila embryo [31]. Subsequent genetic and molecular characterization of SMO revealed it to be a requisite component of the HH signal transduction cascade [32]. Primary sequence comparisons revealed that SMO and the Frizzled (FZ) family of GPCRs are quite similar across distinct functional domains: 37% similarity across the extracellular amino-terminal domains, and 52% similarity across the seven predicted transmembrane domains [32]. As such, SMO has been classified as a member of the FZ family of GPCRs. The specific contributions of conserved GPCR functional domains to SMO-mediated regulation of HH pathway activity are discussed below (Fig. 3.1).

*Cysteine-rich domain (CRD).* A conserved CRD is situated in the extracellular amino terminus of all FZ family GPCRs [33–35]. Disulfide bonds between amino-terminal cysteine residues and/or cysteine residues in the extracellular loops of FZ drive receptor conformations that are necessary for its ligand binding and ligand-induced dimerization [34, 36]. Like FZ, SMO possesses multiple cysteines in its amino terminus and extracellular loops that are positionally conserved across species (http://www.gpcr.org). In vitro studies in mammalian cell culture suggested that the amino-terminus of SMO, which encompasses the CRD, is not required for GLI activation [37]. However, genetic analyses in both Drosophila and zebrafish

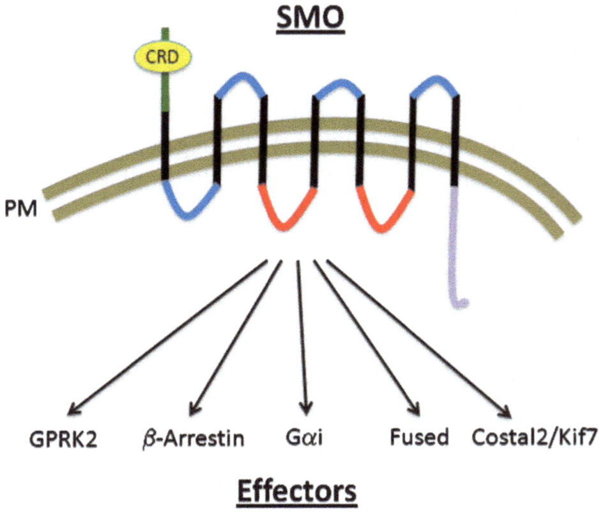


**Fig. 3.1** Domains and effectors of smoothened (SMO). A schematic depicting the predicted topology and domains of SMO in the plasma membrane is shown. The seven predicted transmembrane domains of SMO are shown in *black*. The amino-terminal domain is shown in *green* and the cysteine-rich domain (CRD) in *yellow*. Three extracellular and one intracellular loops are shown in *blue*, but intracellular loops 2 and 3 – which are thought to couple to G-proteins – are shown in *red*. The carboxyl-terminal domain is shown in *purple*. The various known direct effectors of SMO are indicated below it

support that conserved cysteine residues in the CRD are critical for SMO signaling and/or subcellular localization [38–40], suggesting that the CRD is a requisite functional domain. Further studies are needed to more clearly define contributions of the CRD to SMO signaling.

*Transmembrane domains and intracellular loops.* The topology of the 7TM regions dictates the activation state of a GPCR by directing the conformation of its intracellular loops and cytoplasmic tail [41]. In response to ligand, the TM domains of the receptor shift to allow changes in conformation of the intracellular portions of the protein that facilitate receptor phosphorylation and/or G-protein selectivity, docking, and activation [41]. The importance of SMO TM sequence/structure is underscored by known oncogenic SMO mutations, all of which are localized to predicted TM segments [6, 42]. It is likely that these mutations lock SMO TM and intracellular domains in an activated state, which is insensitive to PTCH-mediated inhibition.

The cytoplasmic tail along with intracellular loop 3 (ic3), and to a lesser extent ic2, constitute the G-protein docking site on the vast majority of GPCRs [43]. While extensive structure/function analysis of the SMO intracellular loops has not been reported, chimeric studies in cultured fibroblasts reveal a critical role for ic3 in activation of the signaling cascade [37]. These findings are supported by a loss-of-function *SMO* mutation in Drosophila of a highly conserved Arg residue localized to the carboxyl-terminal end of loop ic3 [39]. The importance of the intracellular loops is further supported by a study demonstrating that introduction of peptide analogs of either ic2 or ic3 into cultured cancer cell lines, which have an activated HH signaling pathway, attenuates their proliferation [44]. Further studies are needed to identify binding partners of SMO ic2 and ic3, and to determine whether these domains constitute a binding site for a partner G-protein.

*Carboxyl-terminal intracellular tail.* Multiple phosphorylation sites, which have been shown to be critical for pathway activation, have been identified in the SMO carboxyl-terminal tail [45–47]. Phosphorylation of such sites in response to ligand is a well-characterized event in canonical GPCR signaling, which generally serves to recruit various adaptor proteins and signaling effectors to the activated receptor [48]. Accordingly, HH-stimulated phosphorylation of SMO by PKA and G-protein regulated kinase 2 (GRK2) triggers both adaptor protein recruitment and SMO multimerization [49–51]. Mutations that prevent phosphorylation of any of these characterized phosphorylation sites compromise the ability of SMO to signal [45–47].

## SMO Signaling Through Heterotrimeric G-Proteins

We recently demonstrated that *Gαi* overexpression in Drosophila triggers activation of HH target genes and wing patterning defects consistent with excessive HH signaling [52]. These phenotypes correlated with the activation state of the expressed *Gαi* transgene, as overexpression of a *Gαi* mutant that cannot bind GTP resulted in no observable phenotype. Conversely, overexpression of wild-type Gαi triggered modest gain of function phenotypes, and overexpression of a transgene encoding constitutively active *Gαi* resulted in strong HH gain of function phenotypes.

Activation of heterotrimeric G-proteins of the Gαi family frequently serves to decrease intracellular pools of cAMP through Gαi-mediated inhibition of adenylate cyclase (AC) [53]. Accordingly, in our study, we observed a SMO- and Gαi-dependent reduction in total intracellular cAMP within 5–10 min of HH stimulation [52]. Modulation of cAMP appears to be critical for in vivo HH signaling, as a mutant allele of the cAMP phosphodiesterase *DUNCE* [54] enhanced the HH loss-of-function phenotype induced by expression of a dominant negative SMO mutant in the Drosophila wing [52]. The ability of cAMP to modulate HH pathway-dependent patterning events is further supported by studies demonstrating that overexpression of an anthrax virulence factor, which functions as a potent bacterial AC, triggers wing phenotypes similar to HH loss-of-function mutations [55]. Further, modulation of cAMP by Sonic HH (SHH) has also been demonstrated in vertebrate systems: retinal ganglion cell axons exposed to recombinant SHH reduce their intracellular pools of cytoplasmic cAMP [56], while frog melanophores exposed to SHH aggregate their melanosomes, a process favored by low concentrations of intracellular cAMP [57]. Taken together, these studies support that one mechanism by which SMO initiates HH signal transduction is to regulate cAMP production through the activation of Gαi family heterotrimeric G-proteins.

*Gαi as a context-specific modulator of HH signaling.* In vitro studies on vertebrate SMO support the ability of SMO to activate a subset of heterotrimeric G-proteins, with strongest effects on those of the Gαi family, for which SHH-induced GTP binding has been demonstrated [57–60]. Activation of Gαi in these systems fulfills the requirements of canonical HH pathway induction, as it can be inhibited by PTCH and/or small-molecule SMO inhibitors, and can be activated by SHH stimulation [57, 60]. SHH target gene induction in cultured fibroblasts is sensitive to pertussis toxin (PTX), a potent Gαi inhibitor, further supporting that Gαi is engaged by vertebrate SMO in response to ligand stimulation [60].

The above studies provide support for involvement of Gαi in HH signal transduction. However, studies performed in differing developmental or cellular contexts failed to identify a role for Gαi in the HH pathway [37, 61–63]. RNAi screens in cultured Drosophila cells did not implicate Gαi as a component of the HH signaling pathway [62, 63], while studies performed in cultured 10T1/2 cells failed to detect changes in intracellular cAMP following SHH stimulation [37]. The latter might be explained by findings that the bulk of HH signaling in vertebrate cells appears to occur in the primary cilium, a small sensory organelle that is present on most vertebrate cell types [64]. Because the volume of the primary cilium is negligible when compared to the body of the cell, localized changes in ciliary cAMP may be undetectable in whole cell lysates.

Conflicting results have also been obtained from in vivo studies examining the role of Gαi in HH signaling. Uncoupling of SMO from Gαi by expression of the PTX catalytic subunit in chick neural tube did not demonstrate compromised SHH-dependent neural cell type specification, suggesting that Gαi is not required for SHH patterning events in this developmental context [61].

Taken together, these seemingly conflicting results raise the possibility that Gαi is required only in certain cellular or tissue contexts during development. This suggestion is supported by the observation that while chick retinal ganglion axon explants are sensitive to SHH-mediated cAMP modulation and growth suppression, chick neural tube explants are not [56]. Signaling redundancy in specific tissues and/or at distinct developmental time points may also account for the apparent lack of Gαi involvement in HH signaling in some in vivo systems. This possibility is supported by work in both Drosophila and cultured vertebrate cells, which show multiple activating signals and feed-forward loops originating from SMO following ligand stimulation [26, 65, 66]. Further studies are required to ascertain if these additional SMO signals are dominant, or can compensate when Gαi is compromised.

## A SMO-Dependent G-Protein Signaling Network

Based on a series of elegant biochemical reconstitution experiments, heterotrimeric G-proteins are proposed to function as ligand-gated switches [53]. Ligand stimulation triggers the GPCR to serve as a guanine nucleotide exchange factor (GEF) for its partner heterotrimeric G-protein, allowing the GDP-bound Gα subunit to bind

GTP and become activated. Upon activation, Gα was originally thought to dissociate from its Gβγ subunits and interact with its effector(s) through random collision along the plasma membrane. Initially the only known effector of G-proteins was adenylate cyclase (AC), the enzyme that converts ATP into cAMP. As such, the Gα subunit responsible for *s*timulating AC, and driving cAMP production was named Gαs, while the Gα subunit that inhibited AC activity to lower cAMP production was named Gαi [53]. Gα subunits were originally believed to be attenuated by their own intrinsic GTPase activity to return the Gα to its inactive GDP-bound form, thereby allowing it to reassociate with its partner Gβγ subunits.

The identification of additional G-proteins, the advent of molecular biology, and the subsequent investigation of G-protein function in vivo culminated to show that more regulators of the G-protein GTPase cycle were required than initially predicted by the in vitro model [67, 68]. These regulators consist of non-receptor GEFs that promote GDP release, novel inhibitors of GDP release (GDI), *r*egulators of *G*-protein *s*ignaling (RGS) that significantly increase the GTPase activity of the Gα subunit, and GRKs that function to desensitize the activated GPCR and/or propagate receptor signaling [69]. Numerous Gα and Gγ subunits have now been described and demonstrated to be capable of signaling themselves, regulating their own spectrum of specific effectors [69].

The discovery that SMO can signal as a *bona fide* GPCR has the potential to quickly expand the number of signaling proteins regulated by SMO, to include those that act as part of a SMO regulated G-protein signaling network (SGN). Thus, we anticipate that like other GPCRs SMO might regulate a large network of signaling proteins, including other G-proteins, and modifiers and effectors of these G-proteins. We discuss below evidence for such a network of regulators, and what some of the novel components of this SGN might be.

*G-protein modulators.* Although overexpression of activated Gαi in vivo resulted in strong HH gain of function phenotypes, we found that attenuation of Gαi function triggered only mild HH loss of function phenotypes [52]. These weak phenotypes might indicate that another *Gα* gene product, of which there are five in Drosophila, functions in a redundant manner with *Gαi* during Drosophila development. A likely candidate gene for this redundant function is *Gαo*, a member of the *Gαi* family that can function redundantly with Gαi in other systems [53]. Further, the mammalian homologue of the Drosophila Concertina α subunit, Gα12/13, has been implicated in SHH-mediated regulation of the small GTP-binding protein Rho [58]. Although our survey of three *Gα* gene products did not implicate Gαs in HH pathway regulation, a genome-wide screen in cultured Drosophila cells showed that knocking down *Gαs* could enhance HH signaling activity [63]. Further work is needed to determine if Gαs might represent a feedback mechanism that resets the basal level of cAMP after HH induces a decrease in cellular cAMP concentration via Gαi.

Another group of G-proteins we anticipate will serve as novel SMO effectors are the Gβγ subunits of its partner heterotrimeric G-protein(s). At a minimum, these proteins could function as negative regulators of Gαi by acting as GDIs [69]. The Gβγ subunits might also have the capacity to regulate their own novel set of effectors in the HH signaling cascade and/or modulate effectors that they share with their Gα subunits. One such effector, AC, is stimulated by Gβγ proteins [69]. Therefore, like Gαs, Gβγ subunits activated in response to HH might be utilized to reverse Gαi-induced decreases in intracellular cAMP.

*GPCR kinases and arrestins.* As discussed above, phosphorylation of GPCRs on the cytoplasmic carboxyl-terminal tail is a common event following ligand stimulation. Phosphorylation regulates wide-ranging events including receptor subcellular localization, association with downstream pathway effectors, and commonly serves to recruit β-arrestin type adaptor proteins, which can propagate receptor signaling and/or drive receptor internalization and desensitization [70, 71]. Phosphorylation of activated receptors is driven primarily by the GRK family of kinases. GRK2 regulation of SMO follows a well-established GPCR paradigm: GRK2 phosphorylation triggers β-arrestin recruitment, which drives SMO to clathrin-coated pits where it undergoes activation-dependent internalization [72]. Interestingly, rather than desensitizing SMO to attenuate transduction of the HH signal, GRK-mediated phosphorylation and subsequent β-arrestin recruitment appear to regulate positive steps in HH signaling. Co-expression of GRK2 with SMO in cultured C3H10T1/2 cells enhances SMO-dependent activation of GLI, while GRK knockdown in cultured HEK293 cells attenuates SMO signaling in response to the SMO agonist SAG [50, 73]. An in vivo requirement for GRK2 in HH signaling was confirmed through studies analyzing zebrafish and mice lacking GRK2 function. In both cases, these animals demonstrated developmental phenotypes consistent with HH loss of function [73].

Drosophila GRK2 (dGRK2) has been demonstrated to be both a positive regulator of SMO signaling as well as a HH target gene, suggesting that it functions in a ligand-induced feed-forward loop [74]. As is the case in vertebrate systems, phosphorylation of SMO by dGRK2 results in both β-arrestin recruitment and SMO internalization in HH receiving cells [49]. dGRK2 appears to function only on activated SMO that has transduced a signal in response to ligand, as dGRK2 overexpression in wing discs triggers the removal of SMO from the plasma membrane without attenuating HH target gene induction. dGRK2-mediated internalization of SMO is independent of PTCH-driven removal of SMO from the plasma membrane in cells not receiving the HH signal, further suggesting that dGRK2 functions solely to regulate activated SMO [49]. This supports that SMO plasma membrane localization is regulated in a manner similar to numerous other GPCRs: in the absence of ligand stimulation, SMO undergoes a tonic endocytosis that is regulated by PTCH [75], while ligand stimulated SMO is internalized by the combined activity of dGRK and β-arrestin [49, 72]. As is the case with vertebrate SMO, assembly of the dGRK/β-arrestin complex on dSMO appears to be a positive regulatory event, despite it resulting in the eventual removal of SMO from the plasma membrane [49, 72]. Taken together, these studies highlight the importance of an evolutionarily conserved regulatory complex that assembles in response to ligand-induced GRK phosphorylation of SMO, of which β-arrestin appears to be paramount.

*Protein kinase A.* PKA was originally identified as a cAMP stimulated protein kinase, consisting of two regulatory subunits and two catalytic subunits [76].

The regulatory subunit inhibits the activity of the catalytic subunit, and this repression is released when the regulatory subunit binds to cAMP. PKA phosphorylates a broad spectrum of substrates, resulting in many diverse biological outputs. The various functions of PKA are thought to be spatially distinct, with PKA binding to its substrates and regulators on scaffolding proteins called A kinase anchoring proteins (AKAPs) [77]. AKAPs cluster relevant GPCRs, G-proteins, kinases, and other downstream effectors to discrete localizations within a cell [69, 77]. COS2 has been demonstrated to associate with SMO, downstream HH effectors, as well as with PKA and CK1 [20, 21, 24, 25]. As such, we hypothesized that COS2 might act as a nexus for HH signaling in a manner akin to that of AKAP proteins. We tested this hypothesis and noted that Gαi and COS2 do associate, and that this association was enhanced by HH [52]. It is, therefore, likely that COS2 acts as a scaffolding protein to recruit SMO, Gαi, and PKA, and likely, analogous to how AKAPs function, might also act to locally modulate the levels of cAMP.

PKA was initially shown to function as a negative regulator of HH signaling, phosphorylating CI in order to convert it to its repressor form [78, 79]. It was later identified as a positive regulator of HH signaling, through its ability to phosphorylate and stabilize SMO to result in SMO enriching at the plasma membrane in a highly active form [25, 46, 80]. Thus, PKA plays two seemingly opposite roles in HH signaling – in the absence of HH it acts to keep the HH pathway in its off-state and in the presence of HH functions to convert SMO into its active form. Consistent with the important role PKA plays in HH signaling, *Costal1 (COS1)* mutations, which enhance the phenotype of *COS2* mutations, were recently shown to encode mutations in both the regulatory and catalytic subunits of PKA [81].

It has been suggested that the role PKA plays in HH signaling is cAMP independent [78, 79]. This hypothesis was presented to explain the observation that a mutant mouse PKA catalytic subunit was able to rescue a *PKA* null mutation in Drosophila. These experiments assumed that the mutant PKA catalytic subunit would be unable to associate with the regulatory subunit of Drosophila PKA. However, recent demonstrations of SMO coupling to Gαi and regulating cAMP levels suggests that a cAMP independent role of PKA in HH signaling may not be correct [52, 60, 65]. Moreover, the identification of a *COS1* mutation encoding a PKA regulatory subunit, which modulates HH signaling, is consistent with a cAMP-dependent activation of PKA [81].

## Small-Molecule Modulators of SMO

Mice engineered to lack SHH die shortly after birth and exhibit a wide range of developmental defects, including cyclopia [82]. A similar phenotype was observed in offspring of livestock that ingested the corn lily *Veratrum californica*. Two groups recognized the similarities between these phenotypes and tested the hypothesis that a chemical derived from this plant, cyclopamine, functioned as an inhibitor of HH signaling [83, 84]. Cyclopamine turned out to be a potent inhibitor of HH signaling, in vitro and in vivo, and was subsequently shown to bind directly to the heptahelical bundle of SMO to functionally antagonize its signaling capability [85]. Consistent with SMO facilitating a rate-limiting step in HH pathway activation, numerous small-molecule screens for novel HH inhibitors have identified distinct SMO antagonists [9]. Many of these SMO inhibitors act in a competitive manner with cyclopamine for binding to SMO, supporting that they bind the 7TM segments. However, some of these SMO modulators bind to SMO in a non-competitive manner and/or activate SMO, suggesting that SMO may have a number of different small-molecule binding sites, as is the case with numerous GPCRs [43].

One basic tenet of pharmacology is that drugs themselves do not possess intrinsic biological properties, but rather can only act to modify existing biological processes [86]. Thus, the identification of small-molecule modulators of SMO implied the existence of endogenous SMO modulators. Furthermore, it has been known for a number of years that the HH receptor PTCH has significant homology with a family of physiological pumps in bacteria, leading to the speculation that PTCH functions to regulate the concentration of such an endogenous SMO modulator [13, 87]. Consistent with this homology, a recent study using a mixed-cell culturing system provided evidence for a lipophilic molecule being pumped into the culture medium in a PTCH-dependent manner [88]. This molecule was identified as the oxysterol, pro-vitamin D3, which was demonstrated to bind SMO in manner similar to that of cyclopamine. Purified pro-vitamin D3 inhibited HH activity, both in vitro and in vivo, with a potency similar to that of cyclopamine [88]. This was one of the first identifications of an endogenous SMO modulator, in this case an antagonist. Two other groups subsequently identified oxysterol molecules that functioned as HH activators [89, 90], suggesting that, like numerous GPCRs, SMO activity is controlled by endogenous small-molecule ligands.

## Future Directions

As the critical role(s) that HH signaling plays in tumor growth and progression continues to emerge, and the clinical use of SMO antagonists increases, the impact of on-target adverse effects is likely to become evident. For example, one SMO antagonist was recently demonstrated to have significant efficacy against medulloblastoma [91], but to induce growth defects when administered to young mice [92]. This study was performed to reveal potential problems that may be encountered by inhibiting a developmentally relevant signaling pathway in a pediatric population, the most common class of patients presenting with medulloblastoma [93]. Long bones of the limbs of animals exposed to this compound during early development were found to be significantly shortened, an effect likely due to specific inhibition of Indian HH-regulated bone growth [92]. This observation underscores the importance of delineating all the signaling events immediately downstream of SMO, as one could anticipate that various classes of small-molecule SMO inhibitors might affect distinct signaling arms. Moreover, classes of SMO antagonists that only inhibit a distinct subset of SMO effectors, such as the SGN effectors, might be used clinically for specific classes of cancer patients. Ideally, such compounds would inhibit the effectors relevant to tumor growth while having minimal impact on effectors more relevant to the role HH plays in tissue homeostasis. Such drugs would be particularly useful to medulloblastoma patients, whose ability to take at least a subset of the SMO antagonist currently in clinical trials would be severely compromised by on-target developmental defects.

**Acknowledgments** We thank C. Carroll, D. Fei, S. Marada, and S. Singh for their comments on the manuscript. This work was supported by NIH grants 1RO1CA82628 and 1RO1GM064011 (DJR), March of Dimes MOD5-FY10-6 (SKO), and by the American Lebanese Syrian Associated Charities (ALSAC) of St. Jude Children's Research Hospital.

## References

1. Johnson RL et al (1996) Human homolog of patched, a candidate gene for the basal cell nevus syndrome. Science 272(5268):1668–1671
2. Hahn H et al (1996) Mutations of the human homolog of Drosophila patched in the nevoid basal cell carcinoma syndrome. Cell 85(6):841–851
3. Unden AB et al (1996) Mutations in the human homologue of Drosophila patched (PTCH) in basal cell carcinomas and the Gorlin syndrome: different in vivo mechanisms of PTCH inactivation. Cancer Res 56(20):4562–4565
4. Wolter M, Reifenberger J, Sommer C, Ruzicka T, Reifenberger G (1997) Mutations in the human homologue of the Drosophila segment polarity gene patched (PTCH) in sporadic basal cell carcinomas of the skin and primitive neuroectodermal tumors of the central nervous system. Cancer Res 57(13):2581–2585
5. Lam CW et al (1999) A frequent activated smoothened mutation in sporadic basal cell carcinomas. Oncogene 18(3):833–836
6. Xie J et al (1998) Activating Smoothened mutations in sporadic basal-cell carcinoma. Nature 391(6662):90–92
7. Teglund S, Toftgard R (1805) Hedgehog beyond medulloblastoma and basal cell carcinoma. Biochim Biophys Acta 2:181–208
8. Lum L, Beachy PA (2004) The Hedgehog response network: sensors, switches, and routers. Science 304(5678):1755–1759
9. Robbins DJ, Goetz JA, Yuan Z, Stegman MA (2005) Inhibitors of the hedgehog signal transduction pathway. Curr Cancer Ther Rev 1:227–288
10. Xie J (2008) Hedgehog signaling pathway: development of antagonists for cancer therapy. Curr Oncol Rep 10(2):107–113
11. Jia J, Jiang J (2006) Decoding the Hedgehog signal in animal development. Cell Mol Life Sci 63(11):1249–1265
12. Ingham PW, McMahon AP (2001) Hedgehog signaling in animal development: paradigms and principles. Genes Dev 15(23):3059–3087
13. Taipale J, Cooper MK, Maiti T, Beachy PA (2002) Patched acts catalytically to suppress the activity of Smoothened. Nature 418(6900):892–897
14. Alcedo J, Noll M (1997) Hedgehog and its patched-smoothened receptor complex: a novel signalling mechanism at the cell surface. Biol Chem 378(7):583–590
15. Incardona JP, Gruenberg J, Roelink H (2002) Sonic hedgehog induces the segregation of patched and smoothened in endosomes. Curr Biol 12(12):983–995

16. Martin V, Carrillo G, Torroja C, Guerrero I (2001) The sterol-sensing domain of Patched protein seems to control Smoothened activity through Patched vesicular trafficking. Curr Biol 11(8):601–607
17. Denef N, Neubuser D, Perez L, Cohen SM (2000) Hedgehog induces opposite changes in turnover and subcellular localization of patched and smoothened. Cell 102(4):521–531
18. Robbins DJ, Hebrok M (2007) Hedgehogs: la dolce vita. Workshop on Hedgehog-Gli signaling in cancer and stem cells. EMBO Rep 8(5):451–455
19. Jia J, Tong C, Jiang J (2003) Smoothened transduces Hedgehog signal by physically interacting with Costal2/Fused complex through its C-terminal tail. Genes Dev 17(21):2709–2720
20. Lum L et al (2003) Hedgehog signal transduction via Smoothened association with a cytoplasmic complex scaffolded by the atypical kinesin, Costal-2. Mol Cell 12(5):1261–1274
21. Ogden SK et al (2003) Identification of a functional interaction between the transmembrane protein Smoothened and the kinesin-related protein Costal2. Curr Biol 13(22): 1998–2003
22. Ruel L, Rodriguez R, Gallet A, Lavenant-Staccini L, Therond PP (2003) Stability and association of Smoothened, Costal2 and Fused with Cubitus interruptus are regulated by Hedgehog. Nat Cell Biol 5(10):907–913
23. Endoh-Yamagami S et al (2009) The mammalian Cos2 homolog Kif7 plays an essential role in modulating Hh signal transduction during development. Curr Biol 19(15):1320–1326
24. Robbins DJ et al (1997) Hedgehog elicits signal transduction by means of a large complex containing the kinesin-related protein costal2. Cell 90(2):225–234
25. Zhang W et al (2005) Hedgehog-regulated Costal2-kinase complexes control phosphorylation and proteolytic processing of Cubitus interruptus. Dev Cell 8(2):267–278
26. Claret S, Sanial M, Plessis A (2007) Evidence for a novel feedback loop in the Hedgehog pathway involving Smoothened and Fused. Curr Biol 17(15):1326–1333
27. Farzan SF et al (2009) A quantification of pathway components supports a novel model of Hedgehog signal transduction. J Biol Chem 284(42):28874–28884
28. Costanzi S, Siegel J, Tikhonova IG, Jacobson KA (2009) Rhodopsin and the others: a historical perspective on structural studies of G protein-coupled receptors. Curr Pharm Des 15(35):3994–4002
29. Hofmann KP et al (2009) A G protein-coupled receptor at work: the rhodopsin model. Trends Biochem Sci 34(11):540–552
30. Morris MB, Dastmalchi S, Church WB (2009) Rhodopsin: structure, signal transduction and oligomerisation. Int J Biochem Cell Biol 41(4):721–724
31. Nusslein-Volhard C, Wieschaus E, Kluding H (1984) Mutations affecting the pattern of the larval cuticle in Drosophila melanogaster: zygotic loci on the second chromosome. Roux's Arch Dev Biol 192:267–282
32. Alcedo J, Ayzenzon M, Von Ohlen T, Noll M, Hooper JE (1996) The Drosophila smoothened gene encodes a seven-pass membrane protein, a putative receptor for the hedgehog signal. Cell 86(2):221–232
33. Bhanot P et al (1996) A new member of the frizzled family from Drosophila functions as a Wingless receptor. Nature 382(6588):225–230
34. Dann CE et al (2001) Insights into Wnt binding and signalling from the structures of two Frizzled cysteine-rich domains. Nature 412(6842):86–90
35. Wang Y et al (1996) A large family of putative transmembrane receptors homologous to the product of the Drosophila tissue polarity gene frizzled. J Biol Chem 271(8):4468–4476
36. Moro S, Hoffmann C, Jacobson KA (1999) Role of the extracellular loops of G protein-coupled receptors in ligand recognition: a molecular modeling study of the human P2Y1 receptor. Biochemistry 38(12):3498–3507
37. Murone M, Rosenthal A, de Sauvage FJ (1999) Sonic hedgehog signaling by the patched-smoothened receptor complex. Curr Biol 9(2):76–84
38. Aanstad P et al (2009) The extracellular domain of Smoothened regulates ciliary localization and is required for high-level Hh signaling. Curr Biol 19(12):1034–1039
39. Nakano Y et al (2004) Functional domains and sub-cellular distribution of the Hedgehog transducing protein Smoothened in Drosophila. Mech Dev 121(6):507–518

40. van den Heuvel M, Ingham PW (1996) Smoothened encodes a receptor-like serpentine protein required for hedgehog signalling. Nature 382(6591):547–551
41. Yeagle PL, Albert AD (2003) A conformational trigger for activation of a G protein by a G protein-coupled receptor. Biochemistry 42(6):1365–1368
42. Reifenberger J et al (1998) Missense mutations in SMOH in sporadic basal cell carcinomas of the skin and primitive neuroectodermal tumors of the central nervous system. Cancer Res 58(9):1798–1803
43. Bockaert J, Pin JP (1999) Molecular tinkering of G protein-coupled receptors: an evolutionary success. EMBO J 18(7):1723–1729
44. Remsberg JR, Lou H, Tarasov SG, Dean M, Tarasova NI (2007) Structural analogues of smoothened intracellular loops as potent inhibitors of Hedgehog pathway and cancer cell growth. J Med Chem 50(18):4534–4538
45. Apionishev S, Katanayeva NM, Marks SA, Kalderon D, Tomlinson A (2005) Drosophila Smoothened phosphorylation sites essential for Hedgehog signal transduction. Nat Cell Biol 7(1):86–92
46. Jia J, Tong C, Wang B, Luo L, Jiang J (2004) Hedgehog signalling activity of Smoothened requires phosphorylation by protein kinase A and casein kinase I. Nature 432(7020): 1045–1050
47. Zhang C, Williams EH, Guo Y, Lum L, Beachy PA (2004) Extensive phosphorylation of Smoothened in Hedgehog pathway activation. Proc Natl Acad Sci USA 101(52):17900–17907
48. Pierce KL, Lefkowitz RJ (2001) Classical and new roles of beta-arrestins in the regulation of G-protein-coupled receptors. Nat Rev Neurosci 2(10):727–733
49. Cheng S, Maier D, Neubueser D, Hipfner DR (2010) Regulation of smoothened by Drosophila G-protein-coupled receptor kinases. Dev Biol 337(1):99–109
50. Meloni AR et al (2006) Smoothened signal transduction is promoted by G protein-coupled receptor kinase 2. Mol Cell Biol 26(20):7550–7560
51. Zhao Y, Tong C, Jiang J (2007) Hedgehog regulates smoothened activity by inducing a conformational switch. Nature 450(7167):252–258
52. Ogden SK et al (2008) G protein Galphai functions immediately downstream of Smoothened in Hedgehog signalling. Nature 456(7224):967–970
53. Gilman AG (1987) G proteins: transducers of receptor-generated signals. Annu Rev Biochem 56:615–649
54. Davis RL, Kiger JA Jr (1981) Dunce mutants of Drosophila melanogaster: mutants defective in the cyclic AMP phosphodiesterase enzyme system. J Cell Biol 90(1):101–107
55. Guichard A, Park JM, Cruz-Moreno B, Karin M, Bier E (2006) Anthrax lethal factor and edema factor act on conserved targets in Drosophila. Proc Natl Acad Sci USA 103(9):3244–3249
56. Trousse F, Marti E, Gruss P, Torres M, Bovolenta P (2001) Control of retinal ganglion cell axon growth: a new role for Sonic hedgehog. Development 128(20):3927–3936
57. DeCamp DL, Thompson TM, de Sauvage FJ, Lerner MR (2000) Smoothened activates Galphai-mediated signaling in frog melanophores. J Biol Chem 275(34):26322–26327
58. Kasai K et al (2004) The G12 family of heterotrimeric G proteins and Rho GTPase mediate Sonic hedgehog signalling. Genes Cells 9(1):49–58
59. Masdeu C et al (2006) Identification and characterization of Hedgehog modulator properties after functional coupling of Smoothened to G15. Biochem Biophys Res Commun 349(2):471–479
60. Riobo NA, Saucy B, Dilizio C, Manning DR (2006) Activation of heterotrimeric G proteins by Smoothened. Proc Natl Acad Sci USA 103(33):12607–12612
61. Low WC et al (2008) The decoupling of Smoothened from Galphai proteins has little effect on Gli3 protein processing and Hedgehog-regulated chick neural tube patterning. Dev Biol 321(1):188–196
62. Lum L et al (2003) Identification of Hedgehog pathway components by RNAi in Drosophila cultured cells. Science 299(5615):2039–2045
63. Nybakken K, Vokes SA, Lin TY, McMahon AP, Perrimon N (2005) A genome-wide RNA interference screen in Drosophila melanogaster cells for new components of the Hh signaling pathway. Nat Genet 37(12):1323–1332

64. Huangfu D, Anderson KV (2005) Cilia and Hedgehog responsiveness in the mouse. Proc Natl Acad Sci USA 102(32):11325–11330
65. Chinchilla P, Xiao L, Kazanietz MG, Riobo NA (2010) Hedgehog proteins activate pro-angiogenic responses in endothelial cells through non-canonical signaling pathways. Cell Cycle 9(3):570–579
66. Ogden SK et al (2006) Smoothened regulates activator and repressor functions of Hedgehog signaling via two distinct mechanisms. J Biol Chem 281(11):7237–7243
67. Birnbaumer L (2007) Expansion of signal transduction by G proteins. The second 15 years or so: from 3 to 16 alpha subunits plus betagamma dimers. Biochim Biophys Acta 1768(4):772–793
68. Gurevich VV, Gurevich EV (2008) Rich tapestry of G protein-coupled receptor signaling and regulatory mechanisms. Mol Pharmacol 74(2):312–316
69. Malbon CC (2005) G proteins in development. Nat Rev Mol Cell Biol 6(9):689–701
70. Lefkowitz RJ, Whalen EJ (2004) Beta-arrestins: traffic cops of cell signaling. Curr Opin Cell Biol 16(2):162–168
71. Tobin AB (2008) G-protein-coupled receptor phosphorylation: where, when and by whom. Br J Pharmacol 153(Suppl 1):S167–S176
72. Chen W et al (2004) Activity-dependent internalization of smoothened mediated by beta-arrestin 2 and GRK2. Science 306(5705):2257–2260
73. Philipp M et al (2008) Smoothened signaling in vertebrates is facilitated by a G protein-coupled receptor kinase. Mol Biol Cell 19(12):5478–5489
74. Molnar C, Holguin H, Mayor F Jr, Ruiz-Gomez A, de Celis JF (2007) The G protein-coupled receptor regulatory kinase GPRK2 participates in Hedgehog signaling in Drosophila. Proc Natl Acad Sci USA 104(19):7963–7968
75. Zhu AJ, Zheng L, Suyama K, Scott MP (2003) Altered localization of Drosophila Smoothened protein activates Hedgehog signal transduction. Genes Dev 17(10):1240–1252
76. Krebs EG, Beavo JA (1979) Phosphorylation-dephosphorylation of enzymes. Annu Rev Biochem 48:923–959
77. Dessauer CW (2009) Adenylyl cyclase – A-kinase anchoring protein complexes: the next dimension in cAMP signaling. Mol Pharmacol 76(5):935–941
78. Jiang J, Struhl G (1995) Protein kinase A and hedgehog signaling in Drosophila limb development. Cell 80(4):563–572
79. Li W, Ohlmeyer JT, Lane ME, Kalderon D (1995) Function of protein kinase A in hedgehog signal transduction and Drosophila imaginal disc development. Cell 80(4):553–562
80. Zhou Q, Apionishev S, Kalderon D (2006) The contributions of protein kinase A and smoothened phosphorylation to hedgehog signal transduction in Drosophila melanogaster. Genetics 173(4):2049–2062
81. Collier LS, Suyama K, Anderson JH, Scott MP (2004) Drosophila Costal1 mutations are alleles of protein kinase A that modulate hedgehog signaling. Genetics 167(2):783–796
82. Chiang C et al (1996) Cyclopia and defective axial patterning in mice lacking Sonic hedgehog gene function. Nature 383(6599):407–413
83. Cooper MK, Porter JA, Young KE, Beachy PA (1998) Teratogen-mediated inhibition of target tissue response to Shh signaling. Science 280(5369):1603–1607
84. Incardona JP, Gaffield W, Kapur RP, Roelink H (1998) The teratogenic Veratrum alkaloid cyclopamine inhibits sonic hedgehog signal transduction. Development 125(18): 3553–3562
85. Chen JK, Taipale J, Cooper MK, Beachy PA (2002) Inhibition of Hedgehog signaling by direct binding of cyclopamine to Smoothened. Genes Dev 16(21):2743–2748
86. Hardman J, Limbird L, Gilman A (2001) Goodman & Gilman's the pharmacological basis of therapeutics, 10th edn. McGraw-Hill, New York
87. Ma Y et al (2002) Hedgehog-mediated patterning of the mammalian embryo requires transporter-like function of dispatched. Cell 111(1):63–75
88. Bijlsma MF et al (2006) Repression of smoothened by patched-dependent (pro-)vitamin D3 secretion. PLoS Biol 4(8):e232

89. Corcoran RB, Scott MP (2006) Oxysterols stimulate Sonic hedgehog signal transduction and proliferation of medulloblastoma cells. Proc Natl Acad Sci USA 103(22):8408–8413
90. Dwyer JR et al (2007) Oxysterols are novel activators of the hedgehog signaling pathway in pluripotent mesenchymal cells. J Biol Chem 282(12):8959–8968
91. Romer JT et al (2004) Suppression of the Shh pathway using a small molecule inhibitor eliminates medulloblastoma in Ptc1(+/–)p53(–/–) mice. Cancer Cell 6(3):229–240
92. Kimura H, Ng JM, Curran T (2008) Transient inhibition of the Hedgehog pathway in young mice causes permanent defects in bone structure. Cancer Cell 13(3):249–260
93. Gottardo NG, Gajjar A (2008) Chemotherapy for malignant brain tumors of childhood. J Child Neurol 23(10):1149–1159

# Chapter 4
# Kinases and Phosphatases in Hedgehog Signaling

**Jianhang Jia**

**Keywords** Smo • Ci/Gli • PKA • CK1 • CK2 • GSK3 • Fu • PP4 • PP2A

Since the nineteenth century, phosphates have been known to be bound to proteins. In 1954, an enzyme responsible for the transfer of a phosphate to another protein was discovered; this biological reaction is referred to as phosphorylation [1]. The enzyme catalyzing phosphorylation was known as a protein kinase. A year later, it was demonstrated that a protein involved in glycogen metabolism was regulated by the addition or removal of a phosphate, suggesting that reversible phosphorylation could control the activity of a specific protein [2, 3]. The study of phosphorylation was then brought into the limelight of medical research. In 1992, Fischer and Krebs received the Nobel Prize in medicine for their pioneering work. Today, it is well-known that phosphorylation/dephosphorylation is one of the most important posttranslational events in cell biology. It has been shown to be involved in many aspects of regulation such as increasing or decreasing the biological activity of a protein, facilitating the sorting of intracellular proteins, mediating protein–protein interactions, as well as labeling proteins for degradation. Importantly, the reception of a signal on the surface of a cell often results in a change in cellular phosphorylation patterns with various proteins being phosphorylated or dephosphorylated, which might eventually cause changes in cellular behavior.

Protein phosphorylation in Hedgehog (Hh) signaling has been extensively studied. Like the proteins in other signaling pathways such as Wnt and NFkB, various Hh pathway components are phosphorylated at multiple residues and most of these phosphorylation events are regulated by the upstream signal [4]. Understanding how the phosphorylation events in Hh signaling are normally controlled may provide new avenues for developing therapeutics aimed at preventing and treating human cancers

J. Jia (✉)
Department of Molecular and Cellular Biochemistry, Markey Cancer Center,
University of Kentucky College of Medicine, 741 South Limestone, BBSRB Rm 373,
Lexington, KY 40536-0509, USA
e-mail: jianhang.jia@uky.edu

J. Xie (ed.), *Hedgehog Signaling Activation in Human Cancer and its Clinical Implications*,
DOI 10.1007/978-1-4419-8435-7_4, 

due to aberrant Hh signaling. We begin this chapter with a discussion of the functions of protein phosphorylation in various situations of Hh signal transduction. We will then review the balance between phosphorylation and dephosphorylation of the critical components that are involved in transducing the upstream signal. After that, we will discuss the idea of differential phosphorylation mediating the Hh signaling activity gradient in animal development.

## Phosphorylation Events in Hh Signaling

Hh exerts its biological influence through a highly conserved signal transduction cascade. The mammalian Sonic Hedgehog (Shh), as well as the *Drosophila* Hh, can function as morphogens and regulate cell proliferation, embryonic patterning, and organ development. As discussed in the previous chapters, the Hh signal is transduced through a reception system that includes the 12-span transmembrane protein, Patched (Ptc), and the 7-span transmembrane protein, Smoothened (Smo). In the absence of Hh, Ptc inhibits Smo signaling activity by a mechanism that is still not clear, even though studies indicate that small molecules, such as lipids, could be involved [5, 6, 7]. The presence of Hh relieves the inhibition of Smo by Ptc, activating Smo and allowing Smo to regulate the downstream signaling components. The main outcome of Hh signaling is the modulation of transcriptional responses via the cubitus interruptus (Ci)/Gli family of zinc-finger transcription factors.

The *Drosophila* system is an ideal model system with which to study the Hh pathways because of the fruit fly's well-established genetics, small genome size, minimal genetic redundancy, and ease of handling. Studies with *Drosophila* have provided invaluable information on Hh signaling. In *Drosophila*, Hh induces cell-surface accumulation and phosphorylation of Smo [8]. Recent studies identified protein kinase A (PKA) and casein kinase 1 (CK1) as two of the kinases that phosphorylate Smo and increase its cell-surface localization and signaling activity [9–11]. We discuss the regulation of Smo phosphorylation in a separate section below. In addition to Smo, other pathway components, such as Fused (Fu), Costal2 (Cos2), and Suppressor of Fused (Sufu) are phosphorylated upon Hh stimulation [12, 13]. Fu was first identified as a serine/threonine (Ser/Thr) kinase that regulates segmental polarity in *Drosophila* [14, 15]. Fu is composed of a catalytic domain and a carboxy-terminal regulatory domain and has been shown to be essential for Hh pathway activation [16]. Fu is in a non-phosphorylated, inactive state in the absence of Hh and a phosphorylated active state in the presence of Hh [17, 18]. The responsible for kinase Fu phosphorylation is unknown, although it was proposed that Fu activity is autoregulated through an intramolecular mechanism [19], suggesting that Fu could be auto-phosphorylated. In addition, mutational analysis of Fu identified Thr158 in its activation loop to be a critical residue that is phosphorylated upon Hh stimulation [20]. The identification of the kinase that phosphorylates and thus activates Fu, and the biological significance of this phosphorylation event await further investigation.

Cos2, a kinesin-like protein, exists in a large protein complex with Fu and the full-length Ci [21, 22]. Cos2 mainly acts as a negative regulator in Hh signal transduction as loss of function mutations in *cos2* results in accumulation of high levels of full-length Ci ($Ci^{FL}$) and activation of Hh target genes [22, 23]. Cos2 inhibits the transcriptional activator activity of $Ci^{FL}$ by inhibiting its nuclear translocation [24–27]. Cos2 also acts as a scaffold protein to bring PKA, CK1, and glycogen synthase kinase 3 (GSK3) to $Ci^{FL}$, thus promoting sequential phosphorylation of $Ci^{FL}$ by these kinases [28, 29], which leads to proteasome-mediated $Ci^{FL}$ processing to generate its repressor form $Ci^{REP}$ (see below section). In addition to the role in Ci processing, Cos2 exerts a negative effect on Smo, which blocks the phosphorylation and activation of Smo likely by masking its phosphorylation sites [30]. Other than its negative roles in Hh signal transduction, Cos2 has a positive role in Hh-responding cells, and this correlates to its ability to form a complex with the C-terminal intracellular tail of Smo [18, 26, 31, 32]. In addition, Fu is diminished in cells lacking Cos2 [18, 33], which might explain, at least in part, why Hh signaling activity is attenuated in *cos2* mutant cells adjacent to the A/P boundary where there is Hh stimulation [34] in *Drosophila* wing discs. Interestingly, Cos2 is also phosphorylated in response to Hh and this phosphorylation depends on Fu kinase activity [18, 35]. Phosphorylation of Cos2 at Ser572 by Fu disassociates the Cos2–Smo–Ci complex, which promotes Cos2 degradation, activates Smo, and stabilizes Ci [30, 36].

Sufu was first identified in *Drosophila* and its name came from the ability to suppress the wing vein defect of the *fu* mutant [14]. In *Drosophila* Hh signaling, Sufu forms a large protein complex with Cos2, Fu, and Ci and is reported to regulate Ci subcellular localization and transcriptional activity in the nucleus [26, 37]. However, *Drosophila* embryos lacking Sufu display intact Hh signaling and develop into viable and fertile adults [38]. The role of Sufu in mammalian Hh signaling is strikingly different. Loss of Sufu in mammals leads to ectopic Hh pathway activation, suggesting that Sufu is the major intracellular inhibitor of Gli activity [39]. It has also been speculated that Sufu might have adopted other functions during evolution [39]. Interestingly again, *Drosophila* Sufu is phosphorylated in response to Hh and this phosphorylation event depends on Fu kinase activity [18, 40], although there is no evidence to indicate that Fu is a direct kinase for Sufu; it is possible that other kinase(s) are involved. In addition, further studies are necessary to characterize the consequence of Sufu phosphorylation in response to Hh stimulation.

## Kinases Regulating the Transcriptional Factor, Ci/Gli

Hh signaling regulates the balance between the transcriptional activator and repressor forms of Ci/Gli [4, 12], which is mediated by phosphorylation. Ci plays dual roles that are performed by two distinct forms. In the absence of Hh, $Ci^{FL}$ undergoes proteolytic processing to generate a truncated form ($Ci^{REP}$). $Ci^{REP}$ functions as a repressor to block the expression of Hh responsive genes such as *decapentaplegic* (*dpp*) [41, 42], which encodes a member of the TGFβ/BMP family of secreted proteins that functions as a long-range morphogen to control the growth and patterning of the wing [43, 44].

What triggers Ci proteolytic processing? The initial discovery came from a study of *Drosophila* cAMP-dependent PKA. It was reported that loss-of-function mutations in the catalytic subunit of PKA led to constitutive Hh signaling [45–48]. Subsequent work indicated that PKA phosphorylates Ci at multiple Ser/Thr residues in its C-terminal region, which is essential for the processing [49–51]. Later, GSK3 and CK1 were identified and characterized as two other kinases that act cooperatively with PKA to promote Ci processing [52, 53]. GSK3 and CK1 phosphorylation of Ci are primed by PKA phosphorylation at nearby Ser/Thr residues; these sites form critical phosphorylation clusters within the C-terminal region of Ci and this hyperphosphorylation promotes Ci processing [54]. Thus, Ci processing requires phosphorylation by multiple kinases including PKA, GSK3, and members of CK1 family.

How does phosphorylation of Ci regulate its processing? Early in 1998, an F-box/WD40 containing protein, Slimb, was identified to act downstream of PKA to promote Ci processing [55]. The vertebrate homolog of Slimb, βTRCP, functions as a substrate recognition subunit of the so-called SCF (*S*kp1, Cullin, and *F*-box) ubiquitin ligase complex that normally targets phosphorylated substrates, such as Iκ-B and β-catenin, for ubiquitination, and subsequent proteasome-mediated proteolysis [56]. The phosphorylation of Ci at multiple phosphorylation clusters creates binding sites for the SCF complex [57, 58], leading to partial degradation of Ci thus generating the truncated repressor form $Ci^{REP}$ [5, 59].

How does Hh inhibit Ci processing? An elegant study by Zhang et al. describes a precise mechanism by which Hh inhibits Ci phosphorylation [28]. $Ci^{FL}$ exists in a large protein complex including Cos2 and Fu. Complex formation impedes nuclear translocation of $Ci^{FL}$ through microtubule-dependent and independent mechanisms [60]. Interestingly, Cos2 directly associates with PKA, CK1, and GSK3 and acts as a scaffold between these kinases and $Ci^{FL}$ to facilitate $Ci^{FL}$ phosphorylation [28]. An in vivo experiment performed in this study demonstrated that a Kinesin-Cos2 chimeric protein, in which the microtubule-binding domain of Cos2 is replaced by a canonical kinesin motor domain, carried Cos2-interacting proteins to the microtubule plus end. Accumulation of Kinesin-Cos2 at the plus end of microtubules allowed for a discrete intracellular localization that allowed the enrichment of PKA, CK1, and GSK3 at the same position. In addition, Cos2 was shown to be required for Ci phosphorylation by these kinases. Importantly, Hh signaling appears to inhibit Ci phosphorylation and processing by dissociating the Cos2-Ci-kinase complex [28].

In vertebrate species, three Gli proteins, Gli1, Gli2, and Gli3, are the transcriptional mediators of Hh signaling. Genetic and biochemical studies suggest that Gli2 and Gli3 are the primary mediators of Hh signaling. Gli1 is a transcriptional target of Hh signaling and provides positive feedback to reinforce the Hh signaling activity [61–68]. Gli2 mainly functions as an activator and Gli3 as a repressor, although in some developmental contexts the repressor activity of Gli2 and the activator activity of Gli3 have also been detected [65, 69–72]. Consistent with their ability to function as both an activator and repressor, Gli2 and Gli3 are proteolytically processed to truncated forms, and there is evidence that the processing Gli2 and Gli3 are inhibited by Hh [73–76]. The multiple clusters of phosphorylation sites found in Ci are also present in Gli proteins [4], suggesting that Gli processing could also be regulated by multiple kinases and Slimb/β-TRCP in a similar manner to the aforementioned regulatory mechanisms for Ci processing. Indeed, studies have shown that the processing of Gli3 requires the binding of βTRCP to phosphorylated Gli3 [77, 78].

## Kinases Mediating the Activation of Smo in Response to Hh

As discussed in early chapters of this book, the Hh signal is transduced through a reception system at the plasma membrane that includes the receptor complexes Ptc–Ihog and the signal transducer Smo [4, 12, 79]. Binding of Hh to Ptc–Ihog relieves the inhibition of Smo by Ptc, which allows Smo to activate Ci/Gli transcription factors. Regulation of Smo activity remained poorly understood until recently. Smo has homology to G-protein-coupled receptors (GPCRs) [80], and a recent study revealed that Gαi, a G-protein in *Drosophila*, is essential for full activation of Hh signaling [81]. However, Smo is not a typical GPCR because it does not directly bind to Hh ligand and, upon Hh stimulation, it accumulates on the cell surface (see below) and recruits downstream components [18, 31–33, 85]. In *Drosophila*, Hh induces cell-surface accumulation and phosphorylation of Smo [8]. Recent studies from several labs identified PKA and CK1 as two of the kinases that directly phosphorylate Smo and regulate its cell-surface expression and activity [9–11]. Blocking PKA or CK1 activity prevents Hh-induced Smo accumulation and attenuates pathway activity, whereas increasing PKA activity promotes Smo accumulation and pathway activation [9]. Interestingly, similar to the phosphorylation clusters in Ci, Smo is phosphorylated by PKA and CK1 at three clusters of residues in its intracellular domain both in vitro and in vivo [9–11]. Phosphorylation-deficient forms of Smo in which PKA or CK1 sites are mutated to alanine (Ala) are defective in Hh signaling. Phosphorylation appears to be sufficient to activate Smo as phosphorylation-mimicking Smo variants in which PKA and CK1 sites are converted to acidic residues, such as aspartate (Asp) and glutamate (Glu), exhibit constitutive signaling activity [9, 11].

It is interesting that the same set of kinases plays opposing roles and phosphorylates distinct substrates depending on the signaling states. In the *Drosophila* Hh pathway, PKA and CK1 phosphorylate Ci in the absence of Hh to inhibit pathway activation. In the presence of Hh, PKA and CK1 phosphorylate Smo to stimulate its signaling activity. Similarly, in the Wnt signal transduction, the absence of Wnt allows GSK3β and CK1 to phosphorylate β-catenin, thus inhibiting Wnt signaling. Upon Wnt stimulation, GSK3β and CK1 phosphorylate the Wnt coreceptor LRP5/6 to activate signaling [82, 83]. These observations raise the important question of how different substrates are chosen by the same kinases in different signaling states. One solution, as suggested by the studies on the Wnt pathway, is that each pathway may utilize two distinct pools of kinases, one in proximity to the transcription factor/effector and the other to the membrane reception system [82, 83].

What is the consequence of Smo phosphorylation? Like many GPCRs, Smo forms homodimers and undergoes a confirmational change when it is phosphorylated by PKA and CK1 upon Hh stimulation [83]. Zhao et al., using fluorescence resonance energy transfer (FRET) analysis, observed a basal level of FRET when using the donor cyan fluorescent protein (CFP) and the acceptor yellow fluorescent protein (YFP) to be fused to the C-terminal tail of Smo. They further found that this basal FRET was dramatically increased by Hh treatment. In addition, CFP- and YFP-tagged Smo N-terminal part exhibited a high level of FRET even in the absence of Hh treatment, suggesting that Smo forms a constitutive dimer/oligomer through its N-terminal region while the C-tails are within a dimer and are situated at a distance but are brought together in response to Hh through a conformational switch [84]. The dimerization model explains why the phosphorylation-deficient forms of Smo, as well as a chimeric Smo variant with its C-tail replaced with that of the Wingless (Wg) receptor Frizzled 2 (Fz2), inhibited the activity of endogenous Smo [9, 10, 85]. How does Hh induce a conformational switch in the C-tail of Smo? The degree of conformational change in the C-tail of Smo, monitored by FRET, correlates with the level of its phosphorylation that can be manipulated by substitution of various Ser/Thr residues into either Ala or Asp [84]. Thus, a series of Smo conformational states rely on the level of Hh activity and Smo phosphorylation. More interestingly, stretches of positively charged arginine (Arg) residues residing adjacent to the phosphorylation clusters keep the C-tail of Smo in the inactive conformation. Phosphorylation of Smo at adjacent PKA and CK1 sites neutralizes the positive charges conferred by the Arg motif, leading to Smo dimerization and activation [84]. In addition to inducing a conformational change, phosphorylation also promotes Smo accumulation on the cell surface [8, 9].

How does Hh regulate Smo phosphorylation and cell-surface accumulation? Although Smo phosphorylation promotes its cell-surface accumulation and signaling activity, the mechanisms leading to its cell-surface accumulation are still unknown. Fu and Cos2 exist in a large protein complex [21, 22]. Smo transduces the Hh signal by physically interacting with the Cos2-Fu complex [18, 31–33, 85]. Cell-surface recruitment of an intracellular signaling complex through accumulated Smo is thought to cause dissociation of the Cos2-Ci-Kinase complex and hence inhibition of Ci processing [28]. Thus, the intracellular signaling complex containing Fu and Cos2 was thought to transduce the Hh signal downstream of Smo. Surprisingly, Liu and colleagues identified a mechanism of feedback regulation of Smo by the Fu–Cos2 protein complex [86]. They found that Fu is essential for Hh-induced Smo phosphorylation and cell-surface accumulation because Smo is inhibited in *fu* mutant clones or by expressing a dominant-negative (DN) form of Fu, and such inhibition is alleviated by removal of Cos2. Conversely, overexpressing Cos2 blocks Smo accumulation, which is reversed by coexpressing Fu. In addition, Fu antagonizes Cos2 by phosphorylation at Ser572, which attenuates the Cos2-Smo interaction and promotes Cos2 degradation. The authors further provided evidence that Fu and Cos2 control Smo cell-surface accumulation by regulating Smo phosphorylation [86].

Consistently, the observations by Claret et al. demonstrated that Fu plays a positive role in Smo activation [87]. These data suggest that Hh signaling proceeds through an amplification step mediated by Smo and its downstream components. In addition to regulating the phosphorylation of Smo, Hh may regulate the dephosphorylation of Smo that is mediated by its phosphatase (see the section "Phosphatases in Hh signaling" below).

Is there another kinase involved in regulating Smo? The answer to this question is yes. It has been shown that the C-tail of Smo has 26 Ser/Thr residues that are phosphorylated in response to Hh [11]. Several phosphorylated sites in the C-tail of Smo match the consensus sequence for casein kinase 2 (CK2); however, the link between CK2, Smo phosphorylation, and Hh signaling is lacking. We found that RNAi of CK2 attenuates Hh-induced Smo accumulation and downregulates Hh target gene expression, whereas increasing CK2 activity by overexpressing CK2 increases Smo accumulation and induces ectopic expression of Hh target gene. Mutating the CK2 consensus residues attenuates the ability of Smo to transduce high-level Hh signaling activity (unpublished). There is also the possibility that G-protein-coupled receptor kinase 2 (GRK2) phosphorylates Smo because GRK2 plays a conserved role in modulating the Hh pathway [5, 39], and GRK2 has been shown to regulate Smo in *Drosophila* [88, 89] and in mammals [90, 91]. Besides, a genetic screen in *Drosophila* also identified a candidate kinase that could possibly regulate Smo (Holmgren, meeting abstract). Above all, we have learned that multiple kinases are involved in Smo phosphorylation that is triggered by the upstream signal.

## Non-conserved Mechanisms of Regulation by Kinases

Vertebrate Smo proteins lack PKA and CK1 phosphorylation consensus sites found in *Drosophila* Smo, which implies that vertebrate Smo is not regulated by direct PKA and CK1 phosphorylation. However, GRK2 and perhaps other kinases may regulate the conformation of Smo in vertebrates. In support of this hypothesis, in response to Hh stimulation, vertebrate Smo proteins exhibit a similar conformational switch to that utilized by *Drosophila* Smo [84]. Interestingly, a long stretch of Arg residues, which is conserved among vertebrate Smo proteins, negatively regulates mouse Smo likely by keeping it in an inactive conformation [84]. Additionally, there is evidence that mammalian Smo is phosphorylated in cultured cells by GRK2 [90]. It is also tempting to speculate that Hh-induced phosphorylation of Smo through GRK2 alters the intracellular cellular localization of Smo and/or the transport of Smo into cilia. Further investigation is needed into whether GRK2-mediated phosphorylation of Smo promotes vertebrate Smo activity by inducing a similar conformational change.

The function of the Ser/Thr kinase Fu is also divergent among species from *Drosophila* to human. As discussed above, it is critically involved in *Drosophila* Hh signaling but it has no effect on mouse Hh signal transduction [92, 93].

One possible explanation could be due to gene duplication. Understanding how Fu evolved divergent functions in distinct cellular processes is one of the topics that await further investigation. Recently, researchers have been trying to identify the kinases involved in vertebrate Hh signaling as the functions of Fu, PKA, and CK1 are divergent. So far, some kinases have been identified, such as DYRK2 [94], unc-51-like kinase 3 (ULK3) [95], and protein kinase C (see Chap. 6). However, the underlying molecular mechanisms by which these newly identified kinases are involved in Hh signaling are unclear.

## Phosphatases in Hh Signaling

Levels of cellular protein phosphorylation are often modulated by the opposing action of protein kinases and phosphatases. Phosphatases are typically classified into two main groups: Ser/Thr protein phosphatases (STPs) and protein tyrosine phosphatases (PTPs). STPs can be subdivided into the phosphoprotein phosphatase (PPP) comprising PP1, PP2A, PP2B, PP4, PP5, PP6, and PP7, and the protein phosphatase $Mg^{2+}$- or $Mn^{2+}$-dependent (PPM) family, comprised primarily of PP2C [96]. In the Hh signaling cascade, multiple Ser/Thr kinases are involved, including PKA, GSK3, and CK1 family members as discussed above. Are there any phosphatases involved in Hh signal transduction? In 2005, Nybakken et al. carried out a genome-wide RNAi screen in cultured *Drosophila* cells, in which they identified hundreds of potential new regulators of Hh signaling, including *microtubule star* (*mts*) that encodes the catalytic subunit of protein phosphatase 2A (PP2A) [97]. They showed that RNAi of Mts reduced Hh signaling. PP2A is a multimeric enzyme that consists of the catalytic subunit Mts, a regulatory A subunit (encoded by CG33297 in *Drosophila*) and a regulatory B subunit. PP2A functions are mediated largely by the regulatory B subunit that directs the phosphatase to distinct substrates. In this study, Nybakken et al. found that dsRNAs targeting Widerborst (Wdb), a B subunit, reduced Hh signaling by 50%. This indicated that Wdb is likely to be the B subunit that targets Mts to its substrate in the Hh signaling pathway. Consistently, it is found that B56e, the vertebrate ortholog of Wdb, regulates the Hh pathway during Xenopus eye field separation [98]. In another genetic screen for modifiers of Hh signaling in *Drosophila*, Casso et al. found that *mts* is necessary for full activation of Hh signaling [99].

Even though both of the above studies identified PP2A to be a positive regulator in Hh signaling, its relevant substrates remain undetermined. We know that regulated phosphorylation of Smo and Ci are critical events in mediating Hh signal transduction. As discussed earlier in this chapter, multiple Ser/Thr kinases regulate Hh signaling by phosphorylating Smo and Ci; however, Smo and Ci phosphatases remained unknown until an in vivo RNAi screen was carried out by Jia et al. [100]. In this study, protein phosphatase 4 (PP4) and PP2A were identified as phosphatases that influence Hh signaling by regulating Smo and Ci, respectively. RNAi knockdown of PP4 elevates Smo phosphorylation and accumulation, leading to increased

Hh signaling activity. Furthermore, the authors mapped the PP4-interacting domain in Smo and found that deletion of this domain in Smo promotes Smo phosphorylation and signaling activity. In addition, PP4 was found to regulate Hh-induced Smo cell-surface accumulation. Mechanistically, Hh was shown to downregulate the Smo-PP4 interaction that is mediated by Cos2. In the same in vivo RNAi screen, PP2A was identified to be a Ci phosphatase. Inactivating the regulatory subunit, Wdb, by RNAi or by a loss of function mutation downregulates, whereas overexpressing this regulatory subunit upregulates the level and thus signaling activity of $Ci^{FL}$. Furthermore, Wdb counteracts kinases to prevent Ci phosphorylation. The authors obtained evidence that Wdb attenuates Ci processing by dephosphorylating Ci. Thus, PP4 and PP2A are two phosphatases that act at different positions of the Hh signaling cascade. It is possible that other phosphatases are also involved in dephosphorylating Hh signaling components including Smo and Ci. For example, it was reported that PP1 could be an additional phosphatase for Smo (Zhu, meeting abstract), which would not be surprising since multiple kinases are involved.

## Gradient Hh Signaling Activity Is Interpreted by Differential Phosphorylation of the Intracellular Components

Mammalian Shh, as well as *Drosophila* Hh, can function as morphogens and regulate cell patterning in a concentration-dependent manner [4, 101–105]. Hh proteins play broad roles in the development of appendages, such as wings, eyes, and legs [106]. One of the best systems for studying Hh signaling is the *Drosophila* wing. In the developing wing (wing imaginal disc), posterior compartment (P-compartment) cells express and secrete Hh proteins that act upon neighboring anterior compartment (A-compartment) cells adjacent to the A/P boundary to induce the expression of *dpp*. The long-range morphogen Dpp protein then diffuses bidirectionally into both the A and P compartments and functions as a morphogen to control the growth and patterning of cells in the entire wing in a concentration-dependent manner [44, 45, 107]. Hh also specifies cell patterning at the A/P boundary by activating other genes, including the short-range morphogens *engrailed* (*en*) and *ptc* [102, 108]. Low levels of Hh activity are able to induce the expression of *dpp*, whereas higher levels of Hh activity are also able to activate *ptc*. The induction of *en* appears to require the highest doses of Hh signaling activities [102]. Thus, the Hh activity levels can be monitored by expression of different responsive genes. This phenomenon is similar to the differential signal transduction of Wingless (Wg) in which different levels of Wg activity differentially regulate specific genes [109]. An important issue in developmental biology is to understand how cells perceive and transduce signal gradients, and how this results in differential cell fates.

Hh family members play a similar role in vertebrate limb development to that in *Drosophila*. *Shh* is best characterized in vertebrates and is involved in an array of developmental processes [106]. In the limb bud, Shh is expressed in the zone of polarizing activity (ZPA), which polarizes the digits along the A/P axis [110], which is analogous to that in the *Drosophila* wing. Ectopic expression of Shh in the anterior limb bud induces additional digits in a concentration-dependent manner. In the neural tube, Shh is produced first in the notochord and then in the floor-plate; different levels of Shh activity direct the formation of multiple neuronal subtypes (V0, V1, V2, and V3 interneurons, and motorneurons) [106, 111], indicating that the protein functions to induce cell identity (see Chap. 2 for Hh morphogen gradient formation). Taken together, it is an important issue as to how different levels of the Hh morphogen generate distinct developmental outcomes.

How are the phosphorylation events modulated by gradient Hh signaling activity? Although the precise mechanism(s) of how an Hh gradient is sensed and transduced inside cells is still not well understood, it is likely that the differential phosphorylation of Smo and Ci correlate to different levels of Hh stimulation. In *Drosophila* wing discs, Hh induces cell-surface accumulation and activation of Smo through phosphorylation [4]. Interestingly, the extent of Smo phosphorylation appears to determine both the abundance of Smo at the cell surface and its signaling potency, raising an interesting possibility that different thresholds of Hh are transduced by differentially phosphorylated isoforms of Smo [10]. Low Hh only induces low levels of Smo phosphorylation and a low ratio of dimerized Smo, which suffices to activate low-threshold responses such as *dpp* expression, whereas high Hh induces high levels of Smo phosphorylation and a high ratio of Smo dimerization, which activates high-threshold responses including *ptc* and *en* expression. Indeed, it has been shown that increasing the number of phospho-mimetic mutations resulted in a gradual increase of FRET that is indicative of Smo dimerization/oligomerization [112]. It is believed that hyperphosphorylation of Smo is essential for transducing high levels of Hh signaling activity [31].

How are different thresholds of Hh activity interpreted by Ci/Gli transcription factors? In *Drosophila* wing development, threshold responses to the Hh morphogen appear to be mediated by differential regulation of the two forms of Ci. Accumulation of $Ci^{FL}$ and expression of *dpp* occur in broader domains than activation of ptc and en, suggesting that low levels of Hh suffice to block Ci processing whereas higher levels of Hh are required to stimulate the activity of $Ci^{FL}$. Hence increasing the low levels of Hh activity may cause decreasing levels of Ci phosphorylation at the clusters that mediate Ci processing, thus generating less $Ci^{REP}$ repressors. On the other hand, $Ci^{FL}$ may also need to be activated by phosphorylation/dephosphorylation. For example, blockage of Ci processing in slimb mutant clones induces *dpp* but not *ptc* [52]; $Ci^{FL}$ itself is inactive and only high levels of Hh stimulate the maturation of $Ci^{FL}$ into a labile hyperactive form [113]; phosphorylation of Ci by PKA not only promotes Ci processing but also inhibits the activity of Ci155 independent of Ci processing [52].

A similar model has been proposed for regulating Gli proteins by graded Shh signals [114, 115]. In the neural tube, loss of Gli2, a major source of Gli activator, results in loss of progenitor cells of most ventral characters whereas progenitor cells of ventral-lateral and lateral characters still form [63, 64]. Ventral-lateral and lateral neural progenitors lost in *Shh* mutants are partially restored in *Shh Gli3* double mutants [116, 117]. These results suggest that high levels of Shh specify the ventral most progenitors via stimulating the Gli2 (and perhaps Gli3) activator, whereas low levels of Shh specify ventral-lateral and lateral progenitors through inhibiting the Gli3 repressor. Hence, a gradient of Shh could generate a gradient of Gli activator activity and a reverse gradient of Gli repressor activity.

## References

1. Burnett G, Kennedy EP (1954) The enzymatic phosphorylation of proteins. J Biol Chem 211:969–980
2. Fischer EH, Krebs EG (1955) Conversion of phosphorylase b to phosphorylase a in muscle extracts. J Biol Chem 216:121–132
3. Sutherland EW Jr, Wosilait WD (1955) Inactivation and activation of liver phosphorylase. Nature 175:169–170
4. Jia J, Jiang J (2006) Decoding the Hedgehog signal in animal development. Cell Mol Life Sci 63:1249–1265
5. Ayers KL, Therond PP (2010) Evaluating Smoothened as a G-protein-coupled receptor for Hedgehog signalling. Trends Cell Biol 20:287–298
6. Eaton S (2008) Multiple roles for lipids in the Hedgehog signalling pathway. Nat Rev Mol Cell Biol 9:437–445
7. Taipale J, Cooper MK, Beachy PA (2002) Patched acts catalytically to suppress the activity of Smoothened. Nature 418:892–897
8. Denef N, Neubuser D, Perez L, Cohen SM (2000) Hedgehog induces opposite changes in turnover and subcellular localization of patched and smoothened. Cell 102:521–531
9. Jia J, Tong C, Wang B, Luo L, Jiang J (2004) Hedgehog signalling activity of Smoothened requires phosphorylation by protein kinase A and casein kinase I. Nature 432:1045–1050
10. Apionishev S, Katanayeva NM, Marks SA, Kalderon D, Tomlinson A (2005) Drosophila Smoothened phosphorylation sites essential for Hedgehog signal transduction. Nat Cell Biol 7:86–92
11. Zhang C, Williams EH, Guo Y, Lum L, Beachy PA (2004) Extensive phosphorylation of Smoothened in Hedgehog pathway activation. Proc Natl Acad Sci USA 101:17900–17907
12. Lum L, Beachy PA (2004) The Hedgehog response network: sensors, switches, and routers. Science 304:1755–1759
13. Aikin RA, Ayers KL, Therond PP (2008) The role of kinases in the Hedgehog signalling pathway. EMBO Rep 9:330–336
14. Preat T et al (1993) Segmental polarity in Drosophila melanogaster: genetic dissection of fused in a Suppressor of fused background reveals interaction with costal-2. Genetics 135:1047–1062
15. Therond P et al (1993) Molecular organisation and expression pattern of the segment polarity gene fused of Drosophila melanogaster. Mech Dev 44:65–80. doi:0925-4773(93)90017-R [pii]
16. Alves G et al (1998) Modulation of Hedgehog target gene expression by the Fused serine-threonine kinase in wing imaginal discs. Mech Dev 78:17–31
17. Therond PP, Knight JD, Kornberg TB, Bishop JM (1996) Phosphorylation of the fused protein kinase in response to signaling from hedgehog. Proc Natl Acad Sci USA 93:4224–4228
18. Lum L et al (2003) Hedgehog signal transduction via Smoothened association with a cytoplasmic complex scaffolded by the atypical kinesin, Costal-2. Mol Cell 12:1261–1274
19. Ascano M Jr, Nybakken KE, Sosinski J, Stegman MA, Robbins DJ (2002) The carboxyl-terminal domain of the protein kinase fused can function as a dominant inhibitor of hedgehog signaling. Mol Cell Biol 22:1555–1566

20. Fukumoto T, Watanabe-Fukunaga R, Fujisawa K, Nagata S, Fukunaga R (2001) The fused protein kinase regulates Hedgehog-stimulated transcriptional activation in Drosophila Schneider 2 cells. J Biol Chem 276:38441–38448
21. Robbins DJ et al (1997) Hedgehog elicits signal transduction by means of a large complex containing the kinesin-related protein costal2. Cell 90:225–234
22. Sisson JC, Ho KS, Suyama K, Scott MP (1997) Costal2, a novel kinesin-related protein in the Hedgehog signaling pathway. Cell 90:235–245
23. Wang QT, Holmgren RA (1999) The subcellular localization and activity of Drosophila cubitus interruptus are regulated at multiple levels. Development 126:5097–5106
24. Chen CH et al (1999) Nuclear trafficking of Cubitus interruptus in the transcriptional regulation of Hedgehog target gene expression. Cell 98:305–316
25. Stegman MA et al (2000) Identification of a tetrameric hedgehog signaling complex. J Biol Chem 275:21809–21812
26. Wang G, Amanai K, Wang B, Jiang J (2000) Interactions with Costal2 and suppressor of fused regulate nuclear translocation and activity of cubitus interruptus. Genes Dev 14:2893–2905
27. Monnier V, Ho KS, Sanial M, Scott MP, Plessis A (2002) Hedgehog signal transduction proteins: contacts of the Fused kinase and Ci transcription factor with the kinesin-related protein Costal2. BMC Dev Biol 2:4
28. Zhang W et al (2005) Hedgehog-regulated Costal2-kinase complexes control phosphorylation and proteolytic processing of Cubitus interruptus. Dev Cell 8:267–278
29. Price MA (2006) CKI, there's more than one: casein kinase I family members in Wnt and Hedgehog signaling. Genes Dev 20:399–410
30. Liu Y, Cao X, Jiang J, Jia J (2007) Fused-Costal2 protein complex regulates Hedgehog-induced Smo phosphorylation and cell-surface accumulation. Genes Dev 21:1949–1963
31. Ogden SK et al (2003) Identification of a functional interaction between the transmembrane protein Smoothened and the kinesin-related protein Costal2. Curr Biol 13:1998–2003
32. Jia J, Tong C, Jiang J (2003) Smoothened transduces Hedgehog signal by physically interacting with Costal2/Fused complex through its C-terminal tail. Genes Dev 17:2709–2720
33. Ruel L, Rodriguez R, Gallet A, Lavenant-Staccini L, Therond PP (2003) Stability and association of Smoothened, Costal2 and Fused with Cubitus interruptus are regulated by Hedgehog. Nat Cell Biol 5:907–913
34. Ho KS, Suyama K, Fish M, Scott MP (2005) Differential regulation of Hedgehog target gene transcription by Costal2 and Suppressor of Fused. Development 132:1401–1412
35. Nybakken KE, Turck CW, Robbins DJ, Bishop JM (2002) Hedgehog-stimulated phosphorylation of the kinesin-related protein Costal2 is mediated by the serine/threonine kinase fused. J Biol Chem 277:24638–24647
36. Ruel L et al (2007) Phosphorylation of the atypical kinesin Costal2 by the kinase Fused induces the partial disassembly of the Smoothened-Fused-Costal2-Cubitus interruptus complex in Hedgehog signalling. Development 134:3677–3689
37. Methot N, Basler K (2000) Suppressor of fused opposes hedgehog signal transduction by impeding nuclear accumulation of the activator form of Cubitus interruptus. Development 127:4001–4010
38. Preat T (1992) Characterization of Suppressor of fused, a complete suppressor of the fused segment polarity gene of Drosophila melanogaster. Genetics 132:725–736
39. Jiang J, Hui CC (2008) Hedgehog signaling in development and cancer. Dev Cell 15:801–812
40. Dussillol-Godar F et al (2006) Modulation of the Suppressor of fused protein regulates the Hedgehog signaling pathway in Drosophila embryo and imaginal discs. Dev Biol 291:53–66
41. Aza-Blanc P, Ramirez-Weber FA, Laget MP, Schwartz C, Kornberg TB (1997) Proteolysis that is inhibited by hedgehog targets Cubitus interruptus protein to the nucleus and converts it to a repressor. Cell 89:1043–1053

42. Methot N, Basler K (1999) Hedgehog controls limb development by regulating the activities of distinct transcriptional activator and repressor forms of Cubitus interruptus. Cell 96:819–831
43. Lecuit T et al (1996) Two distinct mechanisms for long-range patterning by Decapentaplegic in the Drosophila wing. Nature 381:387–393
44. Nellen D, Burke R, Struhl G, Basler K (1996) Direct and long-range action of a DPP morphogen gradient. Cell 85:357–368
45. Jiang J, Struhl G (1995) Protein kinase A and hedgehog signaling in Drosophila limb development. Cell 80:563–572
46. Lepage T, Cohen SM, Diaz-Benjumea FJ, Parkhurst SM (1995) Signal transduction by cAMP-dependent protein kinase A in Drosophila limb patterning. Nature 373:711–715
47. Li W, Ohlmeyer JT, Lane ME, Kalderon D (1995) Function of protein kinase A in hedgehog signal transduction and Drosophila imaginal disc development. Cell 80:553–562
48. Pan D, Rubin GM (1995) cAMP-dependent protein kinase and hedgehog act antagonistically in regulating decapentaplegic transcription in Drosophila imaginal discs. Cell 80:543–552
49. Chen Y, Cardinaux JR, Goodman RH, Smolik SM (1999) Mutants of cubitus interruptus that are independent of PKA regulation are independent of hedgehog signaling. Development 126:3607–3616
50. Price MA, Kalderon D (1999) Proteolysis of cubitus interruptus in Drosophila requires phosphorylation by protein kinase A. Development 126:4331–4339
51. Wang G, Wang B, Jiang J (1999) Protein kinase A antagonizes Hedgehog signaling by regulating both the activator and repressor forms of Cubitus interruptus. Genes Dev 13:2828–2837
52. Jia J et al (2002) Shaggy/GSK3 antagonizes Hedgehog signalling by regulating Cubitus interruptus. Nature 416:548–552
53. Price MA, Kalderon D (2002) Proteolysis of the Hedgehog signaling effector Cubitus interruptus requires phosphorylation by glycogen synthase kinase 3 and casein kinase 1. Cell 108:823–835
54. Jiang J (2002) Degrading Ci: who is Cul-pable? Genes Dev 16:2315–2321
55. Jiang J, Struhl G (1998) Regulation of the Hedgehog and Wingless signalling pathways by the F-box/WD40-repeat protein Slimb. Nature 391:493–496
56. Maniatis T (1999) A ubiquitin ligase complex essential for the NF-kappaB, Wnt/Wingless, and Hedgehog signaling pathways. Genes Dev 13:505–510
57. Jia J et al (2005) Phosphorylation by double-time/CKIepsilon and CKIalpha targets cubitus interruptus for Slimb/beta-TRCP-mediated proteolytic processing. Dev Cell 9:819–830
58. Smelkinson MG, Kalderon D (2006) Processing of the Drosophila hedgehog signaling effector Ci-155 to the repressor Ci-75 is mediated by direct binding to the SCF component slimb. Curr Biol 16:110–116
59. Hooper JE, Scott MP (2005) Communicating with Hedgehogs. Nat Rev Mol Cell Biol 6:306–317
60. Wang G, Jiang J (2004) Multiple Cos2/Ci interactions regulate Ci subcellular localization through microtubule dependent and independent mechanisms. Dev Biol 268:493–505
61. Lee J, Platt KA, Censullo P, Altaba AR (1997) Gli1 is a target of Sonic hedgehog that induces ventral neural tube development. Development 124:2537–2552
62. Ding Q et al (1998) Diminished Sonic hedgehog signaling and lack of floor plate differentiation in Gli2 mutant mice. Development 125:2533–2543
63. Matise MP, Epstein DJ, Park HL, Platt KA, Joyner AL (1998) Gli2 is required for induction of floor plate and adjacent cells, but not most ventral neurons in the mouse central nervous system. Development 125:2759–2770
64. Dai P et al (1999) Sonic Hedgehog-induced activation of the Gli1 promoter is mediated by GLI3. J Biol Chem 274:8143–8152
65. Bai CB, Stephen D, Joyner AL (2004) All mouse ventral spinal cord patterning by hedgehog is Gli dependent and involves an activator function of Gli3. Dev Cell 6:103–115

66. Bai CB, Auerbach W, Lee JS, Stephen D, Joyner AL (2002) Gli2, but not Gli1, is required for initial Shh signaling and ectopic activation of the Shh pathway. Development 129:4753–4761
67. Bai CB, Joyner AL (2001) Gli1 can rescue the in vivo function of Gli2. Development 128:5161–5172
68. Park HL et al (2000) Mouse Gli1 mutants are viable but have defects in SHH signaling in combination with a Gli2 mutation. Development 127:1593–1605
69. Buttitta L, Mo R, Hui CC, Fan CM (2003) Interplays of Gli2 and Gli3 and their requirement in mediating Shh-dependent sclerotome induction. Development 130:6233–6243
70. Motoyama J et al (2003) Differential requirement for Gli2 and Gli3 in ventral neural cell fate specification. Dev Biol 259:150–161
71. McDermott A et al (2005) Gli2 and Gli3 have redundant and context-dependent function in skeletal muscle formation. Development 132:345–357
72. Tyurina OV et al (2005) Zebrafish Gli3 functions as both an activator and a repressor in Hedgehog signaling. Dev Biol 277:537–556
73. Pan Y, Bai CB, Joyner AL, Wang B (2006) Sonic hedgehog signaling regulates Gli2 transcriptional activity by suppressing its processing and degradation. Mol Cell Biol 26:3365–3377
74. Aza-Blanc P, Lin HY, Ruiz i Altaba A, Kornberg TB (2000) Expression of the vertebrate Gli proteins in Drosophila reveals a distribution of activator and repressor activities. Development 127:4293–4301
75. Ruiz i Altaba A (1999) Gli proteins encode context-dependent positive and negative functions: implications for development and disease. Development 126:3205–3216
76. Wang B, Fallon JF, Beachy PA (2000) Hedgehog-regulated processing of Gli3 produces an anterior/posterior repressor gradient in the developing vertebrate limb. Cell 100:423–434
77. Wang B, Li Y (2006) Evidence for the direct involvement of {beta}TrCP in Gli3 protein processing. Proc Natl Acad Sci USA 103:33–38
78. Tempe D, Casas M, Karaz S, Blanchet-Tournier MF, Concordet JP (2006) Multisite protein kinase A and glycogen synthase kinase 3beta phosphorylation leads to Gli3 ubiquitination by SCFbetaTrCP. Mol Cell Biol 26:4316–4326
79. Zheng X, Mann RK, Sever N, Beachy PA (2010) Genetic and biochemical definition of the Hedgehog receptor. Genes Dev 24:57–71
80. Alcedo J, Ayzenzon M, Von Ohlen T, Noll M, Hooper JE (1996) The Drosophila smoothened gene encodes a seven-pass membrane protein, a putative receptor for the hedgehog signal. Cell 86:221–232
81. Ogden SK et al (2008) G protein Galphai functions immediately downstream of Smoothened in Hedgehog signalling. Nature 456:967–970
82. Davidson G et al (2005) Casein kinase 1 gamma couples Wnt receptor activation to cytoplasmic signal transduction. Nature 438:867–872
83. Zeng X et al (2005) A dual-kinase mechanism for Wnt co-receptor phosphorylation and activation. Nature 438:873–877
84. Zhao Y, Tong C, Jiang J (2007) Hedgehog regulates smoothened activity by inducing a conformational switch. Nature 450:252–258
85. Hooper JE (2003) Smoothened translates Hedgehog levels into distinct responses. Development 130:3951–3963
86. Liu Y, Cao X, Jiang J, Jia J (2007) Fused Costal2 protein complex regulates Hedgehog-induced Smo phosphorylation and cell-surface accumulation. Genes Dev 21:1949–1963
87. Claret S, Sanial M, Plessis A (2007) Evidence for a novel feedback loop in the hedgehog pathway involving smoothened and fused. Curr Biol 17:1326–1333
88. Cheng S, Maier D, Neubueser D, Hipfner DR (2009) Regulation of smoothened by Drosophila G-protein-coupled receptor kinases. Dev Biol 337:99–109
89. Molnar C, Holguin H, Mayor F Jr, Ruiz-Gomez A, de Celis JF (2007) The G protein-coupled receptor regulatory kinase GPRK2 participates in Hedgehog signaling in Drosophila. Proc Natl Acad Sci USA 104:7963–7968

90. Chen W et al (2004) Activity-dependent internalization of smoothened mediated by beta-arrestin 2 and GRK2. Science 306:2257–2260
91. Meloni AR et al (2006) Smoothened signal transduction is promoted by G protein-coupled receptor kinase 2. Mol Cell Biol 26:7550–7560
92. Chen MH, Gao N, Kawakani T, and Chuang PT (2005) Mice deficient in the fused homolog do not exhibit phenotypes indicative of perturbed Hedgehog signaling during embryonic development. Mol Cell Bio 25:7042–7053
93. Merchant M et al (2005). Loss of the serine/threonine kinase fused results in postnatal growth defects and lethality due to progressive hydrocephalus. Mol Cell Biol 25:7054–7068
94. Varjosalo M et al (2008) Application of active and kinase-deficient kinome collection for identification of kinases regulating hedgehog signaling. Cell 133:537–548
95. Maloverjan A, Piirsoo M, Michelson P, Kogerman P, Osterlund T (2010) Identification of a novel serine/threonine kinase ULK3 as a positive regulator of Hedgehog pathway. Exp Cell Res 316:627–637
96. Cohen PT (1997) Novel protein serine/threonine phosphatases: variety is the spice of life. Trends Biochem Sci 22:245–251
97. Nybakken K, Vokes SA, Lin TY, McMahon AP, Perrimon N (2005) A genome-wide RNA interference screen in Drosophila melanogaster cells for new components of the Hh signaling pathway. Nat Genet 37:1323–1332
98. Rorick AM et al (2007) PP2A:B56epsilon is required for eye induction and eye field separation. Dev Biol 302:477–493
99. Casso DJ, Liu S, Iwaki DD, Ogden SK, Kornberg TB (2008) A screen for modifiers of hedgehog signaling in Drosophila melanogaster identifies swm and mts. Genetics 178:1399–1413
100. Jia H, Liu Y, Yan W, Jia J (2009) PP4 and PP2A regulate Hedgehog signaling by controlling Smo and Ci phosphorylation. Development 136:307–316
101. Heemskerk J, DiNardo S (1994) Drosophila hedgehog acts as a morphogen in cellular patterning. Cell 76:449–460
102. Strigini M, Cohen SM (1997) A Hedgehog activity gradient contributes to AP axial patterning of the Drosophila wing. Development 124:4697–4705
103. Roelink H et al (1995) Floor plate and motor neuron induction by different concentrations of the amino-terminal cleavage product of sonic hedgehog autoproteolysis. Cell 81:445–455
104. Briscoe J, Ericson J (1999) The specification of neuronal identity by graded Sonic Hedgehog signalling. Semin Cell Dev Biol 10:353–362
105. Briscoe J, Ericson J (2001) Specification of neuronal fates in the ventral neural tube. Curr Opin Neurobiol 11:43–49
106. Ingham PW, McMahon AP (2001) Hedgehog signaling in animal development: paradigms and principles. Genes Dev 15:3059–3087
107. Campbell G, Tomlinson A (1999) Transducing the Dpp morphogen gradient in the wing of Drosophila: regulation of Dpp targets by brinker. Cell 96:553–562
108. Vervoort M, Crozatier M, Valle D, Vincent A (1999) The COE transcription factor Collier is a mediator of short-range Hedgehog-induced patterning of the Drosophila wing. Curr Biol 9:632–639
109. Tomlinson A (2003) Patterning the peripheral retina of the fly: decoding a gradient. Dev Cell 5:799–809
110. Riddle RD, Johnson RL, Laufer E, Tabin C (1993) Sonic hedgehog mediates the polarizing activity of the ZPA. Cell 75:1401–1416
111. Jessell TM (2000) Neuronal specification in the spinal cord: inductive signals and transcriptional codes. Nat Rev Genet 1:20–29
112. Zhao Y, Tong C, Jiang J (2007) Transducing the Hedgehog signal across the plasma membrane. Fly (Austin) 1:333–336. doi:5570 [pii]
113. Ohlmeyer JT, Kalderon D (1998) Hedgehog stimulates maturation of Cubitus interruptus into a labile transcriptional activator. Nature 396:749–753

114. Jacob J, Briscoe J (2003) Gli proteins and the control of spinal-cord patterning. EMBO Rep 4:761–765
115. Stamataki D, Ulloa F, Tsoni SV, Mynett A, Briscoe J (2005) A gradient of Gli activity mediates graded Sonic Hedgehog signaling in the neural tube. Genes Dev 19:626–641
116. Litingtung Y, Chiang C (2000) Specification of ventral neuron types is mediated by an antagonistic interaction between Shh and Gli3. Nat Neurosci 3:979–985
117. Persson M et al (2002) Dorsal-ventral patterning of the spinal cord requires Gli3 transcriptional repressor activity. Genes Dev 16:2865–2878

# Chapter 5
# Signaling Cross-Talk of Oncogenic KRAS and Hedgehog Pathways in Pancreatic Cancer

**Xiaodong Cheng**

## Introduction

Pancreatic tumors are highly metastatic and heterogeneous; over 90% are adenocarcinomas thought to arise from the pancreatic ducts, based on established ductal differentiation features such as cuboidal shape, ductal antigen expression, and growth into tubular structures [30]. Consistent with this idea, proliferative premalignant lesions of the ductal epithelium, termed pancreatic intraepithelial neoplasia (PanINs), are found in advanced malignant tumors with similar spatial distributions [16]. In addition, PanINs advance toward increasingly atypical histological stages with a concomitant accumulation of genetic alterations that have been identified in pancreatic ductal adenocarcinoma (PDA) [45, 49, 82]. Although it is commonly believed that PanINs are precursors of pancreatic ductal adenocarcinoma, the true cell(s) of origin for this malignancy has not been unambiguously defined. This is partially due to the enormous developmental plasticity of the pancreas, which enables phenotypic change between cell lineages (transdifferentiation). The islet-ductal, acinar-ductal, ductal-islet cell transdifferentiation has been observed both in cultures [25, 63, 84] and *in vivo* [4, 34, 57]. Therefore, ductal adenocarcinoma may arise from fully differentiated ductal epithelium, from other cell lineages such as acinar cells or islets cells, or from putative pancreatic stem cells [47, 56]. It is also possible that any of the cell lineages mentioned above is capable of giving rise to pancreatic adenocarcinoma, and specific genetic alterations determine the phenotypic endpoint of the tumor regardless of the precise cellular origin.

PDA is now one of the well-characterized neoplasms at the genetic level. Microdissection and new DNA, protein, and tissue array technologies have revealed multiple genetic alterations in premalignant lesions similar to those in PDA. There is now sufficient clinical, genetic, and pathological evidence for a tumor

X. Cheng (✉)
Department of Pharmacology and Toxicology, The University of Texas Medical Branch,
301 University Boulevard, Galveston, TX 77555-0615, USA
e-mail: xcheng@utmb.edu

J. Xie (ed.), *Hedgehog Signaling Activation in Human Cancer and its Clinical Implications*,
DOI 10.1007/978-1-4419-8435-7_5, 

progression model for PDA, in which the pancreatic ductal epithelium progresses from normal to increased grades of proliferative premalignant lesions of the ductal epithelium to invasive cancer [30]. Accompanying the progressive morphological changes is the sequential accumulation of genetic alterations in oncogene KRAS, the tumor suppressors INK4A [10, 60, 78], P53 [6, 22, 26, 60] and SMAD4/DPC4 [8, 24, 72, 79], and telomeric structure [21, 70]. In addition to these frequent genetic abnormalities, mutations in the tumor suppressors BRCA2, TGFBR1, and TGFBR2, the serine–threonine kinases AKT2 and LKB1/STK11, and certain DNA mismatch-repair genes represent other less common genetic events in PDA. A recent comprehensive genome-wide genetic analysis of 24 PDA samples reveals that pancreatic cancers contain an average of 63 genetic alterations, the majority of which are point mutations [37]. While this study confirms all the aforementioned frequently mutated genes uncovered by conventional strategies, the bulk of the genes that are genetically altered in PDA have not been reported previously. These alterations define a core set of 12 cellular signaling pathways and processes that are genetically altered in 67–100% of the tumors. Among these 12 implicated signaling pathways, KRAS signaling, apoptosis, cell cycle control of G1/S transition, transforming growth factor β (TGF-β) signaling, Wnt/Notch signaling and Hedgehog (Hh) signaling pathways are altered in all of the 24 PDA samples analyzed. Dysregulation of these core pathways and processes through mutation represents the major features of pancreatic tumorigenesis and have clear functional relevance to neoplasia [37]. The major focus of this review chapter is to discuss recent progresses in understanding the roles of signaling cross-talk between the KRAS and Hh pathways in PDA tumorigenesis and their implications for developing novel therapeutic treatments of PDA.

## KRAS Signaling and PDA

RAS family small GTPases are key signaling molecules that controls a multitude of important cellular processes, including cell growth, differentiation, and survival. At the molecular level, RAS proteins function as a binary switch via cycling between the GDP-bound inactive form and GTP-bound active form. Under physiological conditions, the cellular activity of RAS proteins is tightly controlled. In basal or resting cells, RAS proteins exist predominantly in the GDP-bound state because they possess low intrinsic GTP hydrolysis activity. External stimulations such as binding of a growth factor to its receptor at the cell membrane lead to the activation of intrinsic receptor tyrosine kinase and subsequently a cascade of signaling events that eventually recruit SOS, a guanine nucleotide exchange factor (GEF), to close proximity to RAS on the plasma membrane. The binding of SOS to RAS leads to the dissociation of GDP and allows RAS to bind GTP, which is usually in large excess over GDP in cells. Binding of GTP activates RAS by inducing major conformational changes at switch I (amino acids 30–38) and switch II (amino acids 59–67) regions that are important for interaction with downstream effectors. GTP-bound RAS activates a myriad of distinct effectors, among which the serine/threonine protein kinase RAF, phosphoinositide 3′-kinase (PI3K), and RalGDS are the three most well-characterized. These downstream effectors activate distinct signaling cascades, leading to the activation of transcription factors and/or other signaling molecules. Attenuation of RAS signaling is achieved via intrinsic GTP hydrolysis to GDP, a relatively slow process that can be dramatically accelerated by GTPase-activating proteins (GAPs), which enhance the GTPase activity by approximately $10^5$-fold. Balance between GEFs and GAPs, two opposing forces involved in RAS regulation, is critical for maintaining normal cellular homeostasis.

Given the key regulatory roles that RAS family proteins play in essential cellular functions, it is not surprising that abnormal RAS signaling is associated with various major human diseases, particularly cancer. Aberrant RAS activation plays a critical role in tumorigenesis, as activating RAS mutations in one of three closely related RAS isoforms, HRAS, KRAS, or NRAS, are found in ~30% of all human cancers [9]. Oncogenic RAS mutations, most commonly associated with codon 12, 13, or 61, result in a disruption of both intrinsic and GAP-mediated GTP hydrolysis and therefore constitutively lock the protein in the GTP-bound, active form. Whereas HRAS, KRAS, and NRAS share very similar biochemical properties and downstream effectors, association of mutant RAS isoforms with human cancer is tissue specific, meaning that usually only the activating allele of one particular *RAS* isoform, but not the others, is found in a specific type of human cancer. For example, gain-of-function KRAS mutants are frequently found in pancreatic, colorectal, and non-small-cell lung cancers; activating HRAS mutations are usually detected in bladder, kidney, and thyroid carcinoma; and NRAS mutations are associated with melanoma, hepatocellular carcinoma, and hematologic malignancies. While it is most likely related to the isoform-specific functions and expression of RAS proteins, the mechanisms of this tissue-specific association of RAS isoforms with human cancer are not clear.

PDA has the highest incidence of activating KRAS mutations among all human cancers [3]. Activating KRAS mutations, representing the earliest genetic changes associated with the transformation of normal ductal epithelium and PDA development, have been detected in pancreatic duct lesions with minimal cytological and architectural atypia, and occasionally in histologically normal pancreas [11, 40, 46, 49, 71]. The frequency of KRAS mutations correlates with disease progression, reaching ~100% in PDA [37, 76]. In addition to the ubiquitous association of oncogenic KRAS and PDA in human patients, studies using genetically engineered mouse models have also firmly established that oncogenic KRAS indeed play a critical role in the initiation and progression of PDA. For example, targeted endogenous expression of an oncogenic *KRAS* allele in the mouse pancreas is sufficient to drive the development of PanINs and subsequently at low frequency the progression to both locally invasive adenocarcinoma and metastatic disease with sites of spread exactly as found in human PDA [1, 28].

## Cross-Talk Between KRAS and Hh Signaling in Cancer

Recently, the Hh signaling pathway has been implicated as playing an important role in the progression and maintenance of PDA [7, 39, 73]. Hh signaling is essential for morphogenesis, tissue patterning, and stem cell maintenance in metazoan embryos [31]. Hh binds to its membrane receptor Patched (PTCH), releasing PTCH inhibition of a seven-transmembrane protein, Smoothened (SMO), which in turn activates downstream cytoplasmic transcription factors, the Ci protein in *Drosophila* or the mammalian homologue GLI proteins [43]. Components of the Hh signaling pathway, including the ligand and the receptors, are overexpressed in human PDA tissues and cell lines. Suppressing Hh activity using cyclopamine, a steroidal alkaloid that inhibits Hh signaling through direct interaction with SMO [13], in some PDA cells with activated Hh signaling, can inhibit cell growth *in vitro* and reduce tumor growth *in vivo* in the xenograft and orthotopic mouse model [7, 18, 39, 73].

The coincidence of uncontrolled activation of the RAS and Hh pathways in the early stages of PDA suggests that cross-talk between these two pathways may be a very important mechanism for the initiation and development of PDA. However, the causal effects between KRAS and Hh signaling in pancreatic tumorigenesis are not clear. Earlier results from *Pdx-Shh* mice had suggested that ectopic expression of Hh ligands is sufficient to activate the Ras signaling pathway by inducing a mutation in the *Kras* gene [73]. However, a recent in-depth study indicates that cell-autonomous activation of the Hh pathway is not sufficient to induce mutations in the *Kras* gene or to activate MAPK downstream of Ras [54]. In addition, while expression of endogenous level of oncogenic *Kras*, $Kras^{G12D}$, leads to PanINs identical to all three stages found in the cognate human condition and eventually PDA in mice [28], activation of Hh signaling alone is not sufficient to induce PanINs and PDA in a mouse model in which Hh signaling is activated specifically in the pancreatic epithelium [54]. Since KRAS mutation represents one of the earliest genetic alterations and occurs almost universally in PDA, we hypothesized that oncogenic KRAS promotes pancreatic tumorigenesis in part through activation of the Hh signaling pathway in PDA. To test this hypothesis, we established a KRAS oncogene-based, genetically defined human pancreatic cancer model using primary human pancreatic ductal epithelia (HPDE) immortalized by E6/E7 genes of human papilloma virus (HPV)-16 virus [20, 52]. Expression of $KRAS^{V12}$ in an immortalized, but nontumorigenic HPDE cell line, HPDE6-c7, using retroviral expression vector led to the transformation of the corresponding cell line. The resultant cell line, designated as HPDE6-c7-$KRAS^{V12}$, expresses an increased level of RAS, exhibits high RAS activity, and grows anchorage-independently in soft agar. Expression of $KRAS^{V12}$ in HPDE-c7 cells also led to an increased activation of its downstream effectors, such as MAPK and AKT. The phospho-MAPK and phospho-AKT levels were enhanced in the HPDE-c7-K-$RAS^{V12}$ cells compared with the parental cells [35]. This genetically defined human pancreatic cancer model, along with several similar and independently established KRAS-based human pancreatic cancer cell models [12, 58], provides a useful tool for probing the mechanism of KRAS-mediated oncogenic transformation of HPDE. Oncogenic KRAS transformation of human pancreatic ductal epithelial cells increases GLI transcriptional activity, an effect that is inhibited by the MEK-specific inhibitors, but not by the PI3K- and SMO-specific inhibitors. Inactivation of KRAS activity by a small-interfering RNA specific for oncogenic KRAS inhibits GLI activity and GLI1 expression in PDA cell lines with activating KRAS mutation. In addition, expression of the constitutively active form of $BRAF^{E600}$, but not myr-AKT, blocks the inhibitory effects of KRAS knockdown on Hh signaling. Suppressing GLI activity leads to a selective attenuation of the oncogenic transformation activity of mutant KRAS-expressing PDA cells [36]. These results provide direct evidence that oncogenic KRAS, through RAF/MEK/MAPK, activates Hh signaling via upregulation of GLI in PDA. Although dysregulation of Ras signaling in mice lacking p53 function in the pancreatic epithelium has recently been shown to induce Shh expression [29], our studies show that the ability of oncogenic KRAS to activate Hh signaling in the absence of Hh ligand represents another important mechanism by which oncogenic KRAS promotes tumor formation and also offers an explanation for why more than 50% of PDA cells lines with sustained Hh signaling activity are resistant to cyclopamine [73]. These findings combined with an earlier observation that sustained Hh activation activates platelet-derived growth factor receptor α and the RAS pathway [80] suggest that RAS and Hh signaling pathways can potentially form a positive feedback loop to promote tumorigenesis in pancreatic cancer.

Cooperation between the KRAS and Hh signaling pathways in pancreatic tumorigenesis has also been demonstrated *in vivo* in animal models more recently. When Shh-expressing pancreatic ductal epithelia with Ink4a/Arf and Trp53 null background were orthotopically implanted in mouse, no visible tumors developed. Transplantation of the same cell expressing KRas, or both KRas and Shh, led to the efficient formation of tumors. However, the tumors induced by the combination of Shh and KRas were significantly larger than those induced by KRas alone, indicating that KRAS and Hh cooperating in pancreatic tumorigenesis [48]. *Pdx1-Cre*; *LsL-Kras*$^{G12D}$; *Ink4a/Arf*$^{lox/lox}$ transgenic mice develop tumors resembled human PDA. Cancer cells from the tumor showed positive staining for Shh, while no staining was observed in surrounding stroma and non-neoplastic epithelial cells. Oncogenic KRAS signaling is believed to contribute to the observed Hh activation as expression of oncogenic $Kras^{G12D}$ in immortalized human pancreatic ductal cells leads to significant overexpression of SHH [19]. Furthermore, using *p48-Cre/+*; *LSL-Kras*$^{G12D/+}$; *Trp53*$^{F/+}$; *Smo*$^{F/+}$ mice and *p48-Cre/+*; *LSL-Kras*$^{G12D/+}$; *Trp53*$^{F/+}$; *Smo*$^{F}$/*Smo*$^{F}$ mice, Hanahan and colleagues demonstrate that autocrine Shh–Ptch–Smo signaling is not required in pancreatic ductal cells for PDA progression. In PDA tumor cells, activation of Gli transcription, independent of upstream ligand-mediated signaling, is regulated by TGF-β and KRAS and is required for the KRAS-mediated transformed phenotype of cultured PDA cancer cells [51]. Expressing a green fluorescent protein alone or fused to oncogenic KRAS under the regulation of ptf1a regulatory elements in developing zebrafish pancreas allows real-time visualization of both normal and oncogenic KRAS-expressing pancreatic progenitor cells in living zebrafish embryos. Unlike normal GFP-labeled pancreatic progenitor cells, pancreatic progenitor cells expressing oncogenic KRAS underwent normal specification and migration but failed to differentiate. This block in differentiation resulted in the abnormal persistence of an undifferentiated progenitor pool and was associated with the subsequent formation of invasive pancreatic cancer, which displayed several characteristics in common with the human disease, including abnormal Hh pathway activation [53].

In addition to pancreatic cancer, interaction between oncogenic RAS and Hh signaling pathways has also been reported in other systems including melanoma and gastric cancers. Study of a tyrosinase promoter-driven $NRAS^{Q61K}$; $Ink4a^{-/-}$ transgenic mouse model reveals that *Gli1* and *Ptch1,* but not *Shh,* are expressed in tumor samples, an indication of an active Shh-Gli pathway downstream of ligand. The expression of *Gli1* and *Ptch1* but not of the $NRAS^{Q61K}$ transgene was significantly higher in lymph node metastases vs. primary skin tumors. It was further shown that endogenous RAS-MEK and AKT signaling regulated the nuclear localization and transcriptional activity of GLI1 in melanoma and other cancer cells [69]. Immunohistochemical analyses of 35 gastric carcinoma samples show a significant correlation between Hh pathway activation and phopsho-ERK1/2 levels. Expression of a constitutively active $KRAS^{V12}$ mutant in five gastric cancer cell lines leads to increased GLI1-activity in all five cell lines. The effect of oncogenic KRAS/MEK1 was blocked by the suppressor of fused (SUFU) and the deletion of the N-terminal domain of GLI1. These results suggest that the KRAS/MEK/ERK cascade has a positive regulatory role in GLI transcriptional activity in gastric cancer [68]. A recent study showed that simultaneous activation of GLI1 and EGFR signaling induced anchorage-independent growth of RK3E rat kidney epithelial cells and human HaCaT keratinocytes, while neither expression of GLI1 nor activation of EGFR signaling alone was sufficient to elicit transformation. This EGFR- and GLI1-mediated synergistic transformation required the activation of the RAS/RAF/MEK/ERK but not of the PI3K/AKT pathway [64]. Taken together, these results suggest that collaboration of RAS and Hh signaling pathways represents a common scheme in the oncogenic transformation of multiple type cancers.

## Anticancer Therapeutics Targeting KRAS or Hh Pathway

Clinically, PDA is one of the most lethal human diseases, with a 5-year survival rate of less than 5% and a median survival rate of less than 6 months. Furthermore, pancreatic cancer is resistant to most treatments, including chemotherapy, radiation, and combination therapy. Even for the 15–20% of patients with resectable nonmetastatic adenocarcinoma of the pancreatic head, the 5-year survival rate is only 20%, with a median survival time of 17–20 months [27]. Thus, the development and characterization of new therapeutic agents, especially those based on molecular targeting with high specificity, are desperately needed.

Because the paramount importance of oncogenic KRAS proteins in the initiation and development of different human cancers, considerable amount of efforts has been directed to target RAS for cancer treatments. Thus far, RAS has been proven to be "undruggable" as attempts to block RAS signaling directly have met with little successes. In the past, several strategies have been developed to block activated KRAS, including both farnesyl transferase inhibitors (FTIs) and antisense oligonucleotides [2, 59]. FTIs have been intensively investigated in preclinical and clinical trials as a cancer therapy [67]. Although FTIs are capable of inhibiting RAS processing *in vitro* [33, 41] and in genetically engineered mouse models that harbors *HR*AS oncogene [42], several phase II and phase III clinical trials have shown that FTIs do not have significant single-agent activity in lung, pancreatic, colorectal, bladder, and prostate cancers [14, 67]. In addition, FTIs may be more effective in preventing the membrane translocation of HRAS but not KRAS and NRAS due to the fact that KRAS and NRAS proteins can be geranylgeranylated and translocated to the membrane in the presence of FTIs [32, 77]. This may partly explain the failure of FTIs in clinical trials which reflect the decreased effectiveness of FTIs toward the inhibition of the KRAS oncoprotein more commonly present in human cancers.

Synthetic lethality screening has lately emerged as a new strategy to identify agents that act on a target or targets whose functional inactivation is lethal only in the context of a specific cancer-causing mutation allele [17, 38]. Two general approaches involving the use of either small chemical compound libraries or RNA interference (RNAi) libraries for the identification of KRAS synthetic lethal agents have been explored [23, 35, 81]. Recently, efforts to identify small molecules as KRAS synthetic lethal inhibitors via high-throughput screening have been made by several research groups, and some significant progress has been achieved [17, 23, 35]. Using a genetically defined human cancer model based on oncogenic HRAS and human neonatal fibroblasts, Stockwell and colleagues pioneered the synthetic lethal chemical screening approach to search selective synthetic lethal inhibitors for their ability to kill $HRAS^{V12}$-transformed cells but not their isogenic nontumorigenic counterparts. From a high-throughput screening of a 22,550 compound library, erastin, along with several known anticancer agents, was identified as HRAS synthetic lethal compound [17]. Since KRAS is more frequently mutated in human cancers, selective toxicity in mutant KRAS-expressing cell lines would broaden the applicability of erastin as a therapeutic agent. It was determined that erastin also exerted selectively lethality to tumor cells harboring oncogenic KRAS, albeit to less extend than that of KRAS [81]. On the other hand, oncrasin-1 has been identified via HTS as a small molecule that can effectively kill KRAS mutant ovarian and lung cancer cells but not normal isogenic cells or HRAS/NRAS mutant cancer cells [23]. Further mechanistic studies revealed that apoptosis induction by this compound is blocked by knockdown of KRAS or protein kinase C iota ($PKC_\zeta$), suggesting that oncrasin-1 is synthetic lethal to active KRAS and $PKC_\zeta$ [23]. $PKC_\zeta$ is an atypical protein kinase C that is activated by oncogenic RAS protein and is required for KRAS-induced transformation and colonic carcinogenesis *in vivo* [50], as well as pancreatic cancer cell transformation and tumorigenesis [66]. Using the genetically defined human pancreatic cancer cell model as described above, we have also developed a high-throughput screening assay for identifying small chemical inhibitors that selectively target the oncogenic KRAS-expressing cancer cell but not its nontumorigenic counterpart [35]. While these newly identified RAS synthetic lethal inhibitors hold great promise for developing novel anticancer therapeutics targeting KRAS, further structure–activity relationship analyses and new drug discovery studies based on these chemical leads are imperative for improving the selectivity, potency, and drug-like properties.

RNAi screens using a short hairpin RNA (shRNA) library targeting protein kinases, phosphatases and cancer-related genes or a genome-wide shRNA library have led to the identification of several KRAS synthetic lethal partners, whose inactivation results in selective cell death in cancer cells contained an oncogenic KRAS allele but not in cells with wild-type KRAS status [5, 23, 44, 65]. The unveiling of these cellular targets essential for the survival of oncogenic KRAS-driven cancer cells has provided mechanistic insights into understanding the mechanism of synthetic lethality for KRAS. Moreover, these discoveries also lend opportunities for developing pharmacological inhibitors for these targets as novel cancer therapeutics targeting KRAS oncogene addiction. Interestingly, one of the identified KRAS synthetic lethal partners is PTCH2, a component of Hh signaling pathway [5]. While the significance and implication of this finding remains to be uncovered, it nevertheless suggests that cooperation between KRAS and Hh signaling may play a role in tumorigenesis.

Despite that the involvement of Hh signaling pathway in cancer is revealed much later compared to RAS, inhibitors of the Hh signaling pathway have emerged in recent years as a promising new class of potential therapeutics for cancer treatment. Numerous small molecules that target different components of the pathway have been identified and currently under clinical trials. The majority of Hh pathway inhibitors reported to date target SMO with a few other direct at upstream (Hh) or downstream (GLI). Cyclopamine, a natural alkaloid isolated from *Veratrum californicum*, was the first small-molecule inhibitor of the Hh pathway to be reported [15]. Cyclopamine inhibits Hh signaling by direct binding to the seven-transmembrane helical bundle of SMO [13]. As the first-generation Hh inhibitor, cyclopamine is of low affinity, poor oral bioavailability, and suboptimal pharmacological properties. More potent and soluble analogs with improved bioavailability have been designed and synthesized [74]. In addition, noncyclopamine-like Hh inhibitors have been developed through high-throughput screening of synthetic compound libraries [55]. For a detailed list of potential anticancer Hh inhibitors, please refer to a recent review [62]. One such compound, GDC-0449, is a new generation orally bioavailable and selective SMO inhibitor. A recent clinical trial of 33 patients with advanced basal-cell carcinoma refractory to conventional treatments has shown that GDC-0449 has excellent antitumor activity in locally advanced or metastatic basal-cell carcinoma: after a median treatment of 9.8 months, 18 patients showed objective response to the drug while 11 other patients had arrested stable diseases. Only four patients had progressive cancer during the treatment [75]. In a more dramatic case study, a 26-year-old male with metastatic medulloblastoma that was refractory to multiple therapies was reported. Genetic analyses of tumor specimens obtained before treatment suggested that abnormal activation of the hedgehog pathway is associated with the tumor, with loss of heterozygosity and somatic mutation of the *PTCH1* gene, a key negative regulator of hedgehog signaling. Treatment with GDC-0449 resulted in rapid regression of the tumor and lessening of symptoms. Unfortunately, the initial response to the treatment was transient and the cancer relapsed in a resistant form that was not responsive to the drug after 3 months of treatment [61]. To determine the mechanism of drug resistance, the mutational status of Hh signaling genes was analyzed, an amino acid substitution at a conserved aspartic acid residue 473 of SMO was revealed in the tumor after disease progression but not in the primary tumor specimens before treatment. Functional analyses showed that the mutation had no effect on Hh signaling but disrupted the ability of GDC-0449 to bind SMO and suppress this pathway. A mutation altering the same amino acid was also found in a GDC-0449-resistant mouse model of medulloblastoma [83].

## Conclusion

The recent advances in cancer genomics have provided a wealth of information for deciphering the molecular anatomy of cancer. For many types of cancers, it is conceivable that we will soon have a complete set of the individual pieces of a puzzle: a list of the genes and mutations associated with cancer development. However, our knowledge of the complex signaling networks, particularly cross-talks between cancer-related signaling molecules required to piece the puzzle together, remains sparse. The realization of signaling cross-talk between KRAS and Hh suggests that cooperation between the two universally dysregulated signaling pathways in PDA may play an important role in pancreatic tumorigenesis. This finding also has therapeutic implications for the treatment of pancreatic cancer. Because of its intrinsic heterogeneity and complexity, cancer is unlikely conquered by a single magic bullet aiming at one specific signaling molecule. Combination therapy simultaneously targeting multiple signaling pathways important for oncogenesis may represent the best hope for pinning down the elusive target and preventing the development of resistance to single agent, such as GDC-0449. Inhibiting the KRAS and Hh pathways synergistically can potentially provide an effective therapeutic strategy for treating PDA.

## References

1. Aguirre AJ et al (2003) Activated Kras and Ink4a/Arf deficiency cooperate to produce metastatic pancreatic ductal adenocarcinoma. Genes Dev 17(24):3112
2. Alberts SR (2008) KRAS inhibitors and pancreatic cancer. In: Lowy AM, Leach SD, Philip PA (eds) Pancreatic cancer. Springer, New York, pp 601–607
3. Almoguera C et al (1988) Most human carcinomas of the exocrine pancreas contain mutant c-K-ras genes. Cell 53(4):549
4. Alpert S, Hanahan D, Teitelman G (1988) Hybrid insulin genes reveal a developmental lineage for pancreatic endocrine cells and imply a relationship with neurons. Cell 53(2):295
5. Barbie DA et al (2009) Systematic RNA interference reveals that oncogenic KRAS-driven cancers require TBK1. Nature 462(7269):108
6. Bardeesy N, DePinho RA (2002) Pancreatic cancer biology and genetics. Nat Rev Cancer 2(12):897

7. Berman DM et al (2003) Widespread requirement for Hedgehog ligand stimulation in growth of digestive tract tumours. Nature 425(6960):846
8. Biankin AV et al (2002) DPC4/Smad4 expression and outcome in pancreatic ductal adenocarcinoma. J Clin Oncol 20(23):4531
9. Bos JL (1989) ras oncogenes in human cancer: a review. Cancer Res 49(17):4682
10. Caldas C et al (1994) Frequent somatic mutations and homozygous deletions of the p16 (MTS1) gene in pancreatic adenocarcinoma. Nat Genet 8(1):27
11. Caldas C et al (1994) Detection of K-ras mutations in the stool of patients with pancreatic adenocarcinoma and pancreatic ductal hyperplasia. Cancer Res 54(13):3568
12. Campbell PM et al (2008) Ras-driven transformation of human nestin-positive pancreatic epithelial cells. Meth Enzymol 439:451
13. Chen JK et al (2002) Inhibition of Hedgehog signaling by direct binding of cyclopamine to Smoothened. Genes Dev 16(21):2743
14. Cohen SJ et al (2003) Phase II and pharmacodynamic study of the farnesyltransferase inhibitor R115777 as initial therapy in patients with metastatic pancreatic adenocarcinoma. J Clin Oncol 21(7):1301
15. Cooper MK et al (1998) Teratogen-mediated inhibition of target tissue response to Shh signaling. Science 280(5369):1603
16. Cubilla AL, Fitzgerald PJ (1976) Morphological lesions associated with human primary invasive nonendocrine pancreas cancer. Cancer Res 36(7 PT 2):2690
17. Dolma S et al (2003) Identification of genotype-selective antitumor agents using synthetic lethal chemical screening in engineered human tumor cells. Cancer Cell 3(3):285
18. Feldmann G et al (2007) Blockade of hedgehog signaling inhibits pancreatic cancer invasion and metastases: a new paradigm for combination therapy in solid cancers. Cancer Res 67(5):2187
19. Feldmann G et al (2008) Hedgehog inhibition prolongs survival in a genetically engineered mouse model of pancreatic cancer. Gut 57(10):1420
20. Furukawa T et al (1996) Long-term culture and immortalization of epithelial cells from normal adult human pancreatic ducts transfected by the E6E7 gene of human papilloma virus 16. Am J Pathol 148(6):1763
21. Gisselsson D et al (2001) Telomere dysfunction triggers extensive DNA fragmentation and evolution of complex chromosome abnormalities in human malignant tumors. Proc Natl Acad Sci U S A 98(22):12683
22. Gorunova L et al (1998) Cytogenetic analysis of pancreatic carcinomas: intratumor heterogeneity and nonrandom pattern of chromosome aberrations. Genes Chromosomes Cancer 23(2):81
23. Guo W et al (2008) Identification of a small molecule with synthetic lethality for K-ras and protein kinase C iota. Cancer Res 68(18):7403
24. Hahn SA et al (1996) DPC4, a candidate tumor suppressor gene at human chromosome 18q21.1. Science 271(5247):350
25. Hall PA, Lemoine NR (1992) Rapid acinar to ductal transdifferentiation in cultured human exocrine pancreas. J Pathol 166(2):97
26. Harada T et al (2002) Interglandular cytogenetic heterogeneity detected by comparative genomic hybridization in pancreatic cancer. Cancer Res 62(3):835
27. Hidalgo M (2010) Pancreatic cancer. N Engl J Med 362(17):1605
28. Hingorani SR et al (2003) Preinvasive and invasive ductal pancreatic cancer and its early detection in the mouse. Cancer Cell 4(6):437
29. Hingorani SR et al (2005) Trp53R172H and KrasG12D cooperate to promote chromosomal instability and widely metastatic pancreatic ductal adenocarcinoma in mice. Cancer Cell 7(5):469
30. Hruban RH et al (2000) Progression model for pancreatic cancer. Clin Cancer Res 6(8):2969
31. Ingham PW, McMahon AP (2001) Hedgehog signaling in animal development: paradigms and principles. Genes Dev 15(23):3059
32. James G, Goldstein JL, Brown MS (1996) Resistance of K-RasBV12 proteins to farnesyltransferase inhibitors in Rat1 cells. Proc Natl Acad Sci U S A 93(9):4454

33. James GL et al (1993) Benzodiazepine peptidomimetics: potent inhibitors of Ras farnesylation in animal cells. Science 260(5116):1937
34. Jhappan C et al (1990) TGF alpha overexpression in transgenic mice induces liver neoplasia and abnormal development of the mammary gland and pancreas. Cell 61(6):1137
35. Ji Z et al (2009) Chemical genetic screening of KRAS-based synthetic lethal inhibitors for pancreatic cancer. Front Biosci 14:2904
36. Ji Z et al (2007) Oncogenic KRAS activates hedgehog signaling pathway in pancreatic cancer cells. J Biol Chem 282(19):14048
37. Jones S et al (2008) Core signaling pathways in human pancreatic cancers revealed by global genomic analyses. Science 321(5897):1801
38. Kaelin WG Jr (2005) The concept of synthetic lethality in the context of anticancer therapy. Nat Rev Cancer 5(9):689
39. Kayed H et al (2004) Indian hedgehog signaling pathway: expression and regulation in pancreatic cancer. Int J Cancer 110(5):668
40. Klimstra DS, Longnecker DS (1994) K-ras mutations in pancreatic ductal proliferative lesions. Am J Pathol 145(6):1547
41. Kohl NE et al (1993) Selective inhibition of ras-dependent transformation by a farnesyltransferase inhibitor. Science 260(5116):1934
42. Kohl NE et al (1995) Inhibition of farnesyltransferase induces regression of mammary and salivary carcinomas in ras transgenic mice. Nat Med 1(8):792
43. Lum L, Beachy PA (2004) The Hedgehog response network: sensors, switches, and routers. Science 304(5678):1755
44. Luo J et al (2009) A genome-wide RNAi screen identifies multiple synthetic lethal interactions with the Ras oncogene. Cell 137(5):835
45. Luttges J et al (2001) Allelic loss is often the first hit in the biallelic inactivation of the p53 and DPC4 genes during pancreatic carcinogenesis. Am J Pathol 158(5):1677
46. Luttges J et al (1999) The K-ras mutation pattern in pancreatic ductal adenocarcinoma usually is identical to that in associated normal, hyperplastic, and metaplastic ductal epithelium. Cancer 85(8):1703
47. Meszoely IM et al (2001) Developmental aspects of early pancreatic cancer. Cancer J 7(4):242
48. Morton JP et al (2007) Sonic hedgehog acts at multiple stages during pancreatic tumorigenesis. Proc Natl Acad Sci U S A 104(12):5103
49. Moskaluk CA, Hruban RH, Kern SE (1997) p16 and K-ras gene mutations in the intraductal precursors of human pancreatic adenocarcinoma. Cancer Res 57(11):2140
50. Murray NR et al (2004) Protein kinase Ciota is required for Ras transformation and colon carcinogenesis in vivo. J Cell Biol 164(6):797
51. Nolan-Stevaux O et al (2009) GLI1 is regulated through Smoothened-independent mechanisms in neoplastic pancreatic ducts and mediates PDAC cell survival and transformation. Genes Dev 23(1):24
52. Ouyang H et al (2000) Immortal human pancreatic duct epithelial cell lines with near normal genotype and phenotype. Am J Pathol 157(5):1623
53. Park SW et al (2008) Oncogenic KRAS induces progenitor cell expansion and malignant transformation in zebrafish exocrine pancreas. Gastroenterology 134(7):2080
54. di Magliano Pasca M et al (2006) Hedgehog/Ras interactions regulate early stages of pancreatic cancer. Genes Dev 20(22):3161
55. Peukert S, Miller-Moslin K (2010) Small-molecule inhibitors of the hedgehog signaling pathway as cancer therapeutics. ChemMedChem 5(4):500
56. Pour PM, Pandey KK, Batra SK (2003) What is the origin of pancreatic adenocarcinoma? Mol Cancer 2(1):13
57. Pour PM, Schmied BM (1999) One thousand faces of Langerhans islets. Int J Pancreatol 25(3):181
58. Qian J et al (2005) In vitro modeling of human pancreatic duct epithelial cell transformation defines gene expression changes induced by K-ras oncogenic activation in pancreatic carcinogenesis. Cancer Res 65(12):5045

59. Rejiba S et al (2007) K-ras oncogene silencing strategy reduces tumor growth and enhances gemcitabine chemotherapy efficacy for pancreatic cancer treatment. Cancer Sci 98(7):1128
60. Rozenblum E et al (1997) Tumor-suppressive pathways in pancreatic carcinoma. Cancer Res 57(9):1731
61. Rudin CM et al (2009) Treatment of medulloblastoma with hedgehog pathway inhibitor GDC-0449. N Engl J Med 361(12):1173
62. Scales SJ, de Sauvage FJ (2009) Mechanisms of Hedgehog pathway activation in cancer and implications for therapy. Trends Pharmacol Sci 30(6):303
63. Schmied BM et al (2001) Transdifferentiation of human islet cells in a long-term culture. Pancreas 23(2):157
64. Schnidar H et al (2009) Epidermal growth factor receptor signaling synergizes with Hedgehog/GLI in oncogenic transformation via activation of the MEK/ERK/JUN pathway. Cancer Res 69(4):1284
65. Scholl C et al (2009) Synthetic lethal interaction between oncogenic KRAS dependency and STK33 suppression in human cancer cells. Cell 137(5):821
66. Scotti ML et al (2010) Protein kinase Ciota is required for pancreatic cancer cell transformed growth and tumorigenesis. Cancer Res 70(5):2064
67. Sebti SM, Adjei AA (2004) Farnesyltransferase inhibitors. Semin Oncol 31(1 Suppl 1):28
68. Seto M et al (2009) Regulation of the hedgehog signaling by the mitogen-activated protein kinase cascade in gastric cancer. Mol Carcinog 48(8):703
69. Stecca B et al (2007) Melanomas require HEDGEHOG-GLI signaling regulated by interactions between GLI1 and the RAS-MEK/AKT pathways. Proc Natl Acad Sci U S A 104(14):5895
70. Suehara N et al (1997) Telomerase elevation in pancreatic ductal carcinoma compared to nonmalignant pathological states. Clin Cancer Res 3(6):993
71. Tada M et al (1996) Analysis of K-ras gene mutation in hyperplastic duct cells of the pancreas without pancreatic disease. Gastroenterology 110(1):227
72. Tascilar M et al (2001) The SMAD4 protein and prognosis of pancreatic ductal adenocarcinoma. Clin Cancer Res 7(12):4115
73. Thayer SP et al (2003) Hedgehog is an early and late mediator of pancreatic cancer tumorigenesis. Nature 425(6960):851
74. Tremblay MR et al (2008) Semisynthetic cyclopamine analogues as potent and orally bioavailable hedgehog pathway antagonists. J Med Chem 51(21):6646
75. Von Hoff DD et al (2009) Inhibition of the hedgehog pathway in advanced basal-cell carcinoma. N Engl J Med 361(12):1164
76. Wang HL et al (2010) KRAS mutation testing in human cancers: the pathologist's role in the era of personalized medicine. Adv Anat Pathol 17(1):23
77. Whyte DB et al (1997) K- and N-Ras are geranylgeranylated in cells treated with farnesyl protein transferase inhibitors. J Biol Chem 272(22):14459
78. Wilentz RE et al (1998) Inactivation of the p16 (INK4A) tumor-suppressor gene in pancreatic duct lesions: loss of intranuclear expression. Cancer Res 58(20):4740
79. Wilentz RE et al (2000) Loss of expression of Dpc4 in pancreatic intraepithelial neoplasia: evidence that DPC4 inactivation occurs late in neoplastic progression. Cancer Res 60(7):2002
80. Xie J et al (2001) A role of PDGFRalpha in basal cell carcinoma proliferation. Proc Natl Acad Sci U S A 98(16):9255
81. Yagoda N et al (2007) RAS-RAF-MEK-dependent oxidative cell death involving voltage-dependent anion channels. Nature 447(7146):864
82. Yamano M et al (2000) Genetic progression and divergence in pancreatic carcinoma. Am J Pathol 156(6):2123
83. Yauch RL et al (2009) Smoothened mutation confers resistance to a Hedgehog pathway inhibitor in medulloblastoma. Science 326(5952):572
84. Yuan S et al (1996) Transdifferentiation of human islets to pancreatic ductal cells in collagen matrix culture. Differentiation 61(1):67

# Chapter 6
# Hedgehog and Protein Kinase C Signaling

**Jing Li and B. Mark Evers**

**Keywords** Hedgehog • PKCα • PKCδ • MEK/ERK

## Introduction

Hedgehog (HH) signaling plays an important role in human cancers through promoting cancer cell growth and proliferation of tumor stem cells. The protein kinase C (PKC) family, which comprises at least ten isoforms, has been shown to exert multiple biological functions, including adhesion, secretion, proliferation, differentiation, and apoptosis. In this review, we will summarize recent findings demonstrating crosstalk between the HH and PKC signaling pathways during development, in stem cells and in malignant and nonmalignant cells. The role of MEK/ERK pathway in this crosstalk will also be discussed. The integration of these signaling pathways in the regulation of HH signaling provides for potentially new targets in the control of HH-dependent tumorigenesis.

## Protein Kinase C

The protein kinase C (PKC) family represents a group of widely distributed serine–threonine kinases [1]. Eleven PKC isoforms have been identified and divided into three major classes: the conventional PKCs (α, βI, βII, and γ), the novel PKCs (δ, ε, θ, and η), and the atypical PKCs (ζ and ι/λ) [2]. The conventional PKC isoforms have an intact C1 diacylglycerol/phorbol ester binding domain and C2 calcium-binding domain and thus require phospholipids and calcium

B.M. Evers (✉)
Departments of Surgery and the Lucille P. Markey Cancer Center, The University of Kentucky, CC140 Roach Building, Lexington, KY 40508, USA
e-mail: mark.evers@uky.edu

J. Xie (ed.), *Hedgehog Signaling Activation in Human Cancer and its Clinical Implications*, DOI 10.1007/978-1-4419-8435-7_6, © Springer Science+Business Media, LLC 2011

for activation [2]. The novel PKCs do not require calcium for their activation [2]. The atypical PKCs can be activated in the absence of diacylglycerol and calcium [2]. Upon activation, PKC isoforms often translocate to particular subcellular compartments, including the plasma membrane, Golgi complex, nuclear membrane and nucleus [3]. PKC isoforms play important roles in signal transduction of various physiological stimuli, including growth factors, hormones, and transmitters, thus PKCs are involved in many cellular processes [4].

## Functions of Different PKC Isoforms in Human Cancer

Each of the PKC isoforms is unique in its contribution to cancer development and progression. The conventional PKCs are generally considered to be predominantly antiapoptotic and principally involved in promoting cell survival and proliferation. PKCα regulates multiple biological processes, including cell proliferation, apoptosis, and cell motility [5]. However, the role of PKCα in regulating tumor growth and development is complex and highly tissue dependent. PKCα can either act as a tumor promoter or a tumor suppressor [6]. Overexpression of PKCα has been demonstrated in tissue samples of prostate, endometrial, high-grade urinary bladder, hepatocellular and breast cancers, suggesting a role of PKCα as a tumor promoter. In contrast, PKCα is down-regulated in basal cell carcinoma (BCC) and colon cancers, demonstrating a possible role of PKCα as a tumor suppressor in these tumor types. PKCβI and βII function in various signal-transducing pathways for proliferation, differentiation, metabolism, and more cell-type-specific functions [7, 8].

The novel PKCs generally have a tumor suppressor function and are regarded as pro-apoptotic proteins; however, the evidence is complex. PKCδ has been implicated both as a tumor suppressor and positively or negatively regulates cell proliferation and apoptosis [9]. For example, in breast cancer, PKCδ has shown both pro-survival and pro-apoptotic effects [10]. PKCε has been shown to behave as an oncoprotein [11]. Overexpression of PKCε increased proliferation, motility, and invasion of fibroblasts or immortalized epithelial cells. In addition, transgenic animal models have clearly shown that overexpression of PKCε is tumorigenic resulting in metastatic disease. PKCθ has been proposed as a key player in T-cell activation and an attractive therapeutic target in T-cell-mediated disease processes [12].

Atypical PKCs have been implicated in the malignant behavior of transformed human cells as well. Evidence over the past few years has shown that PKCι is a human oncogene and that the oncogenic PKCι signaling is a target for novel mechanism-based cancer therapy [13]. For example, PKCι is critical for transformed growth in human non-small cell lung cancer cells. PKCζ is involved in diverse physiological functions [14]. For example, PKCζ is involved in the control of glioblastoma cell migration and invasion by regulating the cytoskeleton rearrangement, cell adhesion, and matrix metalloprotease-9 expression. These findings suggest that PKCζ is a potential therapeutic target for glioblastoma.

## Crosstalk of HH and PKC

Gli proteins, including Gli1, Gli2, and Gli3, represent a family of zinc-finger transcription factors and play critical roles in the mediation and interpretation of HH signals [15]. Fused (Fu) is a serine–threonine kinase required for HH signaling and hSu(fu), a human homologue of *Drosophila* Su(fu), is a suppressor of Fu [16]. hSu(fu) negatively regulates Gli activity [16, 17]. hSu(fu) contains conserved PKC phosphorylation sites [16, 17], indicating that it may be subject to regulation by PKC. Information demonstrating the crosstalk between HH and PKC signaling pathways was derived from nontumor cells, such as mammalian 293T and NIH 3T3 fibroblasts. These studies mainly focused on the crosstalk of PKCα or PKCδ with HH signaling. Neill et al. [18] first demonstrated that PKCα and PKCδ-mediated Gli activity in the mammalian 293T cells. When 293T cells were cotransfected with constitutively active PKCα or PKCδ and a luciferase reporter construct containing Gli1-binding sites (GBS), the constitutively active PKCα decreased Gli1 activity by over 60%, suggesting that PKCα is a potent negative regulator of Gli1 transcriptional activity. In contrast to PKCα, constitutively active PKCδ increased the activity of Gli1, indicating a positive role of PKCδ in the regulation of Gli1 transcriptional activity.

The interaction of PKC with HH was further studied in LIGHT2 cells, a HH-responsive NIH 3T3 fibroblast cell line stably transfected with a Gli-regulated luciferase reporter containing eight tandem copies of GBS (8×GBS-luciferase) [19]. Treatment of LIGHT2 cells with phorbol 12-myristate 13-acetate (PMA), a phorbol ester, increased Gli-luciferase activity that was blocked by the PKC inhibitor GF109203X, suggesting PMA activation of Gli is mediated through PKC. Treatment with PMA increased mRNA levels of PTCH 1 and GLI1, two endogenous Gli-regulated genes, which was inhibited by GF109203X. The specificity of Gli-dependent transcription by PMA in NIH 3T3 fibroblasts was further confirmed by transfecting a wild-type 8×GBS-luciferase reporter or a mutated 8×GBS-luciferase reporter harboring a point mutation that abolishes the binding of Gli. PMA-stimulated Gli-luciferase activity was only detected in cells transfected with wild-type GBS-luciferase reporter but not the mutant reporter, indicating that PMA activity is mediated through activation of GLI transcriptional activity. Furthermore, stimulation of GLI-dependent transcription by PMA is mediated through a novel PKC. When LIGHT2 cells were treated with PMA in the presence of Gö6976 (inhibitor of classical PKCs) or rottlerin (inhibitor of novel PKCs), PMA-mediated GLI transcriptional activity was prevented by rottlerin, but not by Gö6976, suggesting this effect is mediated through a novel PKC isoform, which is likely PKCδ since NIH 3T3 fibroblasts express PKCα and PKCδ, two isoforms responsive to PMA. The involvement of PKCδ was confirmed by transfection of LIGHT2 cells with a dominant-negative mutant of PKCδ which blocked the GLI-luciferase activity. Taken together, this study demonstrates that PKCα plays a negative role, whereas PKCδ plays a positive role, in the regulation of HH signaling.

Our group further established the crosstalk of PKCα or PKCδ with HH signaling [20]. As noted above, Neill et al. [18] have shown that PKCα is a negative regulator of GLI1 transcriptional activity in 293T cells. We further confirmed the specific regulation of PKCα on GLI activity in NIH 3T3 fibroblasts. NIH 3T3 cells were cotransfected with a wild-type or a mutated GLI-luciferase reporter (point mutation that abolishes the binding of Gli) and expression plasmids, Gli1 and constitutively active PKCα. The constitutively active PKCα significantly increased the wild-type GLI-luciferase activity, but not the mutant, confirming that PKCα negatively regulates HH signaling. It has been demonstrated that PKCδ increased the activity of Gli1 in NIH 3T3 and 293T cells. Therefore, we cotransfected NIH 3T3 cells with Gli-luciferase reporter, Gli1, and either wild-type PKCδ, kinase-dead PKCδ K376R, or constitutively active PKCδΔNPS in which the N-terminal pseudosubstrate domain was deleted. Treatment with PMA increased Gli-luciferase activity only in cells cotransfected with control vector, which is consistent with the previous findings showing that the endogenous PKCδ positively regulates Gli activity [19]. In contrast, in cells transfected with wild-type PKCδ, Gli-luciferase activity was significantly decreased by PMA treatment; this effect was blocked by rottlerin. In the cells transfected with PKCδΔNPS, Gli-luciferase activity was further decreased either in the presence or absence of PMA, whereas Gli-luciferase activity was not altered in cells transfected with kinase-dead PKCδ K376R. In cells transfected with empty vector (pcDNA3) or kinase-dead PKCδ, PTCH1 mRNA expression was not altered either with or without PMA treatment. In contrast, PMA treatment decreased PTCH1 mRNA levels in cells transfected with wild-type PKCδ as well as PKCδΔNPS either in the presence or absence of PMA. Taken together, PKCδ appears to play a negative role in the regulation of Gli activity stimulated by PMA.

## Crosstalk of HH and PKC in Development

The role of HH signaling in development is well known [21]. PKC isoforms have been implicated in a number of key steps during gametogenesis, fertilization, and early development [22]. However, the interaction of the two signaling pathways in the regulation of development has not been studied extensively. Lu et al. [23] tested the efficacy of the competitive inhibitors chelerythrine chloride and Gö6976 (specific inhibitors of PKC) and sphingosine (inhibits PKC and other kinases) in primary limb bud mesenchyme cultures. PKC inhibition resulted in smaller buds and truncated wings and caused complete loss of sonic hedgehog (Shh) expression in the buds, suggesting the possibility that PKC may control Shh expression. Indeed, the PKC inhibition-induced phenotype and lost Shh expression were rescued by providing ectopic Shh. These experiments demonstrated that, providing exogenous Shh to wing buds in which PKC signaling had been blocked, restored limb development and that Shh is one of the primary targets of PKC signaling.

## Crosstalk of HH and PKC in Stem Cells

HH signaling has an essential role in the control of stem cell growth in embryonic tissues. Heo et al. [24] examined the effect of Shh on the self-renewal of mouse embryonic stem cells and its related mechanisms. They treated these cells with Shh and noted translocation of PKCα, δ, and ζ isoforms from the cytosol to the membrane, demonstrating the activation of these PKC isoforms by Shh stimulation. On the other hand, Shh-induced PKC activation was blocked by cyclopamine, a steroid alkaloid that blocks Shh signaling. Pretreatment with bisindolylmaleimide I (a PKC inhibitor) inhibited Shh-induced Gli1 gene expression and [$^3$H] thymidine incorporation, demonstrating that Shh stimulated mouse ES cell proliferation through Gli1 activation as well as PKC. Consistently, in mesenchymal stem cells transfected with Shh, the expression of angiogenic and pro-survival growth factors was increased and migration and tube formation were significantly improved in a PKC-dependent manner.

## Crosstalk of HH and PKC in Human Cancer

The HH signaling pathway, when mutated or dysregulated, contributes to tumorigenesis. Recent studies provide evidence demonstrating the crosstalk of HH and certain PKC isoforms in human cancer cells.

Gli1 expression is associated with the development of BCC. Gli1 is expressed in the outer root sheath (ORS) of the hair follicle which is thought to be a potential source of BCC. PKCα was expressed in the epidermis and ORS of human hair follicles and PKCδ in the inner root sheath. Neill et al. [18] found that PKCα is downregulated in BCC, suggesting that loss of PKCα expression may be relevant to tumor formation. We screened PKCδ expression in a set of hepatocellular cancer (HCC) in which the activation status of HH signaling had previously been determined by in situ hybridization using probes against Gli1 and PTCH1 [20]. Interestingly, the expression of PKCδ was not detected in any of these specimens with activated HH signaling. These results suggest that decreased expression of PKCδ may account for activation of HH signaling in certain HCC, further demonstrating a negative role of PKCδ in the regulation of HH signaling in cancer cells. Additional evidence to support these findings was provided by in vitro studies using Hep3B cells, a human hepatoma cell line [20]. By a combination of overexpression of PKCδ and knockdown with PKCδ siRNA, we demonstrated that overexpression of wild-type or active PKCδ decreased Gli-luciferase activity, mRNA levels of PTCH, and Gli and endogenous Gli protein levels, whereas knockdown by PKCδ siRNA had opposite effects on these HH target proteins. Furthermore, PKCδ knockdown with siRNA enhanced the proliferation and significantly blocked the inhibitory effects of KAAD-cyclopamine (a potent derivative of cyclopamine). Taken together, the loss of PKCδ increased HH signaling and Gli1 protein expression and rescued the inhibitory effect of KAAD-cyclopamine on cellular proliferation, demonstrating that PKCδ negatively regulates HH signaling.

## MEK/ERK Pathway in PKC-Mediated HH Signaling

The Raf/MEK/ERK signaling pathways regulate a variety of cellular activities including proliferation, differentiation, survival, and death. HH signaling exerts a positive feedback effect on these pathways; furthermore, PKC is well known as an activator of the ERK pathway. Therefore, it is very likely that the ERK pathway is involved in PKC-mediated HH signaling. Riobo et al. [19] investigated whether PKC activates Gli activity through the ERK pathway in LIGHT2 cells. They showed that PMA-induced GLI-luciferase reporter activity was blocked by the selective MEK-1 inhibitor PD98059 or the dual MEK-1/2 inhibitor U0126. These findings place MEK-1 downstream of PKC in the activation of GLI. Moreover, PKCα plays a positive role in the regulation of Gli1 activity; this effect was mediated by the MEK/ERK pathway.

## Summary

We have highlighted data demonstrating evidence of crosstalk between HH and PKC. Consistently, PKCα has been shown to negatively regulate HH signaling. However, studies demonstrated that PKCδ plays either a negative or positive role in the regulation of HH signaling. We proposed that the balance between PKCα and PKCδ is important in the regulation of Gli activity. When PKCα is dominant, the negative effect of PKCδ is weak, and PMA increases Gli activity through the PKCα/MEK/ERK pathway. However, when PKCδ is dominant, PMA treatment decreases Gli activity through the activation of PKCδ. It is clear that the HH pathway plays an important role in tumor cell growth and survival. However, several issues regarding the precise role of the HH signaling pathway in human cancer remain unresolved, including the exact mechanisms of signal transduction. We anticipate that more mechanistic studies will further illuminate the conserved and divergent aspects of HH signaling. A better understanding of HH signaling and its crosstalk with other signaling pathways is of importance for developing a more precise understanding of HH-associated diseases and, furthermore, holds great promise for developing new therapies based upon this information.

## References

1. Newton AC (2010) Protein kinase C: poised to signal. Am J Physiol Endocrinol Metab 298:E395–E402
2. Steinberg SF (2008) Structural basis of protein kinase C isoform function. Physiol Rev 88:1341–1378
3. Shirai Y, Saito N (2002) Activation mechanisms of protein kinase C: maturation, catalytic activation, and targeting. J Biochem 132:663–668
4. Rosse C et al (2010) PKC and the control of localized signal dynamics. Nat Rev Mol Cell Biol 11:103–112

5. Nakashima S (2002) Protein kinase C alpha (PKC alpha): regulation and biological function. J Biochem 132:669–675
6. Konopatskaya O, Poole AW (2010) Protein kinase Calpha: disease regulator and therapeutic target. Trends Pharmacol Sci 31:8–14
7. Kawakami T, Kawakami Y, Kitaura J (2002) Protein kinase C beta (PKC beta): normal functions and diseases. J Biochem 132:677–682
8. Sledge GW Jr, Gokmen-Polar Y (2006) Protein kinase C-beta as a therapeutic target in breast cancer. Semin Oncol 33:S15–S18
9. Jackson DN, Foster DA (2004) The enigmatic protein kinase Cdelta: complex roles in cell proliferation and survival. FASEB J 18:627–636
10. Vucenik I, Ramljak D (2006) The contradictory role of PKCdelta in cellular signaling. Breast Cancer Res Treat 97:1–2
11. Basu A, Sivaprasad U (2007) Protein kinase Cepsilon makes the life and death decision. Cell Signal 19:1633–1642
12. Chaudhary D, Kasaian M (2006) PKCtheta: a potential therapeutic target for T-cell-mediated diseases. Curr Opin Investig Drugs 7:432–437
13. Fields AP, Frederick LA, Regala RP (2007) Targeting the oncogenic protein kinase Ciota signalling pathway for the treatment of cancer. Biochem Soc Trans 35:996–1000
14. Hirai T, Chida K (2003) Protein kinase Czeta (PKCzeta): activation mechanisms and cellular functions. J Biochem 133:1–7
15. Kasper M, Regl G, Frischauf AM, Aberger F (2006) GLI transcription factors: mediators of oncogenic Hedgehog signalling. Eur J Cancer 42:437–445
16. Stone DM et al (1999) Characterization of the human suppressor of fused, a negative regulator of the zinc-finger transcription factor Gli. J Cell Sci 112(Pt 23):4437–4448
17. Merchant M et al (2004) Suppressor of fused regulates Gli activity through a dual binding mechanism. Mol Cell Biol 24:8627–8641
18. Neill GW et al (2003) Loss of protein kinase Calpha expression may enhance the tumorigenic potential of Gli1 in basal cell carcinoma. Cancer Res 63:4692–4697
19. Riobo NA, Haines GM, Emerson CP Jr (2006) Protein kinase C-delta and mitogen-activated protein/extracellular signal-regulated kinase-1 control GLI activation in hedgehog signaling. Cancer Res 66:839–845
20. Cai Q, Li J, Gao T, Xie J, Evers BM (2009) Protein kinase Cdelta negatively regulates hedgehog signaling by inhibition of Gli1 activity. J Biol Chem 284:2150–2158
21. Nagase T, Nagase M, Machida M, Yamagishi M (2007) Hedgehog signaling: a biophysical or biomechanical modulator in embryonic development? Ann N Y Acad Sci 1101:412–438
22. Kalive M, Faust JJ, Koeneman BA, Capco DG (2010) Involvement of the PKC family in regulation of early development. Mol Reprod Dev 77:95–104
23. Lu HC, Swindell EC, Sierralta WD, Eichele G, Thaller C (2001) Evidence for a role of protein kinase C in FGF signal transduction in the developing chick limb bud. Development 128:2451–2460
24. Heo JS, Lee MY, Han HJ (2007) Sonic hedgehog stimulates mouse embryonic stem cell proliferation by cooperation of $Ca^{2+}$/protein kinase C and epidermal growth factor receptor as well as Gli1 activation. Stem Cells 25:3069–3080

# Chapter 7
# Activation of Hedgehog Signaling in Human Cancer

**Jingwu Xie and Ervin Epstein**

**Keywords** Hedgehog • Smoothened • PTCH1 • Cancer • Basal cell carcinomas • Signal transduction • Clinical trials and animal model

## Introduction

Major advances in the studies of the hedgehog pathway have been made in the last 30 years. The hedgehog (Hh) gene was identified in 1980 through genetic analysis of fruit fly *Drosophila* segmentation [1]. In early 1990s, three vertebrate homologues of the Hh gene were identified [2–6]. As an essential signaling pathway in embryonic development, the Hh pathway is critical for maintaining tissue polarity and stem cell population. In 1996, inactivation of this pathway was linked to hereditary developmental disorder holoprosencephaly, whereas hyperactivation of this pathway was linked to human cancer [7–11]. More recently, one Hh signaling inhibitor has been successfully used in clinical trials of human cancer, which further indicates the feasibility of Hh signaling inhibitors for treating human cancers.

The general signaling mechanisms of the Hh pathway is conserved from fly to the humans [12]. In the absence of Hh ligands, smoothened (SMO), the seven transmembrane domain containing protein, serves as the key signal transducer, whose function is inhibited by another transmembrane protein Patched (PTC). An active Hh ligand (Shh, Ihh, Dhh, or the fly Hh homolog) binds to its receptor PTC and releases this inhibition, allowing SMO to signal downstream, eventually leading to activation of Gli transcription factors. As transcription factors, Gli molecules can associate with specific consensus sequences located in the promoter region of the target genes and regulate target gene expression [13, 14]. Figure 7.1 shows a simplified diagram of the Hh pathway.

---

J. Xie (✉)
Wells Center for Pediatric Research, Indiana University School of Medicine,
980 W. Walnut St., Rm R3-520, Indianapolis, IN 46202, USA
e-mail: jinxie@iupui.edu

J. Xie (ed.), *Hedgehog Signaling Activation in Human Cancer and its Clinical Implications*,
DOI 10.1007/978-1-4419-8435-7_7, © Springer Science+Business Media, LLC 2011

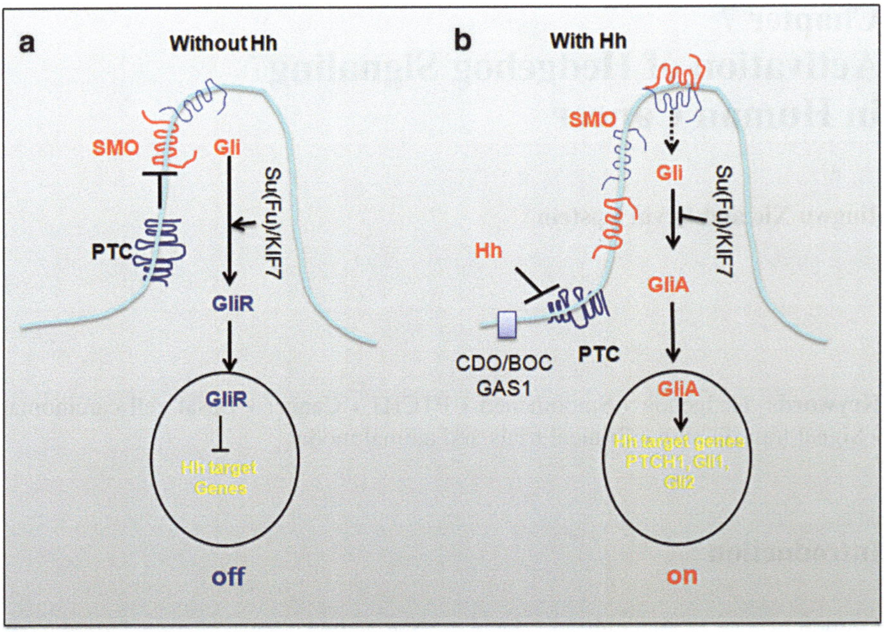


**Fig. 7.1** A simplified model for Hh signaling in mammalian cells. SMO is the key signal transducer of the Hh pathway. (**a**) In the absence of the Hh ligands, Hh receptor PTC is thought to be localized in the cilium to inhibit SMO signaling via an unknown mechanism. Gli molecules are processed with the help of Su(Fu)/KIF7 molecules into repressor forms, which turn off the Hh signaling pathway. (**b**) In the presence of Hh, PTC is thought to be shuttled out of cilium and is unable to inhibit SMO. Co-receptors of Hh ligands include CDO, BOC, and GAS1. Hh reception promotes SMO conformational changes to form dimers. Gli molecules are now processed to active forms (GliA), which will activate the Hh target genes. This process can be inhibited by KIF7 and Su(Fu). Positive regulators are in *red*, negative regulators are in *blue*, and target genes are in *yellow*. This figure was modified from Yang et al. *Oncogene* 29, 469–481 (2010)

## Signal Transduction of the Hedgehog Pathway

Hh proteins [one Hh in *Drosophila* and three Hhs in mammals – Sonic Hedgehog (Shh), Indian Hedgehog (Ihh), and Desert Hedgehog (Dhh)] are secreted molecules, functioning both at short range on nearby cells and at long range to distant cells during development [15–17]. After translation, Hh protein precursor undergoes autoprocessing to release its N-terminal fragment (HhN), which is then covalently bound to a cholesterol moiety at the C-terminal end. Palmitoylation mediated by the acyltransferase Skinny Hedgehog occurs at the N-terminus of HhN [18–21]. Several molecules are involved in the movement, extracellular transport, and release of Hh proteins, including the transmembrane transporter-like protein Dispatched (Disp) [22–24], metalloproteases [25], the heparan sulfate proteoglycans Dally-like (Dlp), and Dally [26, 27] or their regulators [28] as well as enzymes such as Sulfateless and Tout velu [29–31].

Several molecules are engaged in reception of Hh ligands, with Patched (PTC, one PTC in fly and two PTCs in vertebrates – PTCH1 and PTCH2) as the major receptor [32]. Studies from cultured cells indicate that PTC inhibits SMO at a substochiometrical concentration [33]. Hh-interacting protein (HIP) can compete with PTC on Hh binding, resulting in negative regulation of Hh signaling [34]. On the other hand, Ihog (or its vertebrate homologues CDO and BOC), GAS1, and Glypican-3 serve as co-receptors of Hh [35–42]. It is still not entirely clear how binding of Hh proteins results in the pathway activation. It is proposed that PTC limits SMO signaling via transporting endogenous small molecules specifically targeted to SMO. Candidates of these small molecules include PI4P, lipoproteins, and pro-vitamin D3 [43–46]. It is not known how these molecules regulate SMO signaling.

Significant progress has been made in our understanding of SMO signaling. Several recent reports support SMO-G protein coupling [47–50]. In particular, a study in *Drosophila* provides direct evidence for SMO-coupling to G$\alpha$i in regulation of Hh pathway activation [48]. The physiological relevance of the G-protein coupling of SMO for Hh signaling in carcinogenesis is currently not known. In *Drosophila*, SMO function is promoted through protein phosphorylation by PKA and Casein kinase I at the C-terminus [51, 52]. SMO mutants lacking these phosphorylation sites are defective in Hh signaling. However, these phosphorylation sites are not conserved in vertebrate SMO, indicative of a different mechanism for SMO signaling in higher organisms [52]. There are two important events during mammalian SMO signaling. First, SMO protein undergoes conformational change to favor SMO signaling [53] although the regulatory mechanism underlying this conformational change is not clear. Second, ciliary translocation of mammalian SMO protein is critical for Hh signaling (see below).

Accumulating evidence indicate that the primary cilia play an important role for the Hh pathway [54–59]. The function of primary cilium is regulated by protein complexes involved in intraflagellar transport (IFT), which functions in retrograde and anterograde movement of cargo within the primary cilia [60]. Mutations in IFT proteins involved in predominantly primary cilium anterograde transportation are shown to result in mice with Hh loss-of-function phenotypes [55, 61]. Gli3 processing is the most significantly affected event in IFT mutants [56, 57, 61]. Presence of several Hh components, including SMO and Gli molecules at the primary cilium upon Hh stimulation, further supports the relevance of cilium in Hh signaling [62–65]. It has been shown that a SMO mutant lacking ciliary translocation signal is unable to mediate Hh signaling [54]. However, translocation of SMO to cilium alone is not sufficient to activate Hh signaling [64, 65]. Using tissue-specific gene knockout, recent studies revealed dual roles of cilium (via knocking out cilium component Kif3a) for Hh signaling-mediated carcinogenesis in mouse models [66, 67]. While Kif3a gene is required for activated SMO-mediated tumor formation, knocking out Kif3a accelerates Gli2-mediated carcinogenesis. We still do not understand how SMO is translocated to the cilium in response to Hh signaling and how SMO activates downstream effectors. It is known that Beta-arrestin 2 can regulate ciliary localization of SMO [68]. The role of cilium for Hh signaling downstream of SMO is less clear. Not all the signaling events occur in cilium. For example, cilium is not required for Su(Fu)-mediated regulation of Gli functions [69, 70].

Several molecules are identified to be genetically downstream of SMO signaling in *Drosophila*, including COS2 and Fused. Recent in vivo studies support that a COS2 homolog KIF7 functions in the Hh pathway but no direct interaction between SMO and KIF7 is detected [71, 72], suggesting that the function of COS2 in vertebrates may be replaced by a few molecules. The phenotype from vertebrate Fused knockout is not similar to that observed as Shh null mice [73–75], and no changes of Hh signaling are observed in Fused null mice, suggesting that Fused is not critical for Hh signaling during early embryonic development of vertebrates.

In addition to the *Drosophila* homologs, mammalian cells have several novel cytoplasmic regulators of Hh signaling, including Rab23 [76] and tectonic [77]. Rab23 and tectonic are all negative regulators downstream of SMO, but the exact mechanisms of action remain to be established. Unlike many Rab proteins, we found that Rab23 is localized in the nucleus and in cytoplasm (Acta Histochem 2010, July 23 [Epub ahead of time]), suggesting that Rab23 may have other uncharacterized functions apart from membrane trafficking. Through siRNA-based screenings, several additional molecules are identified to be involved in Hh signaling in mammalian cells [78, 79], but their exact functions are not clear.

Several lines of evidence indicate that Suppressor of Fused [Su(Fu)] plays a major negative regulatory role in mammalian Hh signaling. Su(Fu) is originally identified genetically in *Drosophila* by its ability to suppress active *fused* mutations, but itself is not required for pathway activity. *Su(Fu)* null mouse mutants fail to repress the pathway [80] and have some phenotypes similar to *Ptch1* inactivation. While *Ptch1*$^{+/-}$ mice are predisposed to developing medulloblastoma, rhabdomyosarcoma, and basal cell carcinomas (BCCs) [81–83], *Su(Fu)*$^{+/-}$ mice mainly develop basaloid epidermal proliferation. Su(Fu) plays a central role in pathway repression as indicated from the data derived from *Su(Fu)* null MEFs and wild-type cells treated with Su(Fu) siRNAs [80] that loss of Su(Fu) results in Hh signaling activation. At the molecular level, Su(Fu) associates with and inhibits Gli molecule function, and is required for Gli3 processing [84, 85]. One potential molecular basis by which Hh signaling releases the suppressing activity of Su(Fu) is the enhanced Su(Fu) protein degradation upon Hh signaling activation [86].

The ultimate effect of Hh signaling activation is the activation of downstream Gli transcription factors, which can regulate target gene expression by direct binding of a consensus binding site (5′-tgggtggtc-3′) in the target gene's promoter [13, 14, 87, 88]. The activity of Gli transcription factors can be regulated at several levels. First, nuclear-cytoplasmic shuttling of Gli molecules is tightly regulated

[84, 89–91]. Protein kinase A can retain Gli1 protein in the cytoplasm via a PKA site in the nuclear localization signal domain [89], whereas activated Ras signaling induces Gli nuclear localization [91]. Second, ubiquitination, acetylation, and protein degradation of Gli molecules are regulated by several distinct mechanisms, including β-TRCP-, cul3/BTB-, and numb/Itch-mediated Gli ubiquitination [92–97]. In addition, Gli3 (Gli2 to a less extent) can be processed into transcriptional repressors, which may be mediated by the β-TRCP E3 ligase [94]. Defects in the retrograde motor for IFTs can affect Gli3 processing [98]. Furthermore, transcriptional activity of Gli molecules is tightly regulated. Su(Fu) not only prevents nuclear translocation of Gli molecules, but also inhibits Gli1-mediated transcriptional activity [99].

Several feedback regulatory loops exist in this pathway to maintain the level of Hh signaling in a given cell. PTC, HHIP, GAS1, and Gli1 are both components and the target genes of this pathway. PTC and HIP provide negative feedback regulation, whereas Gli1 forms a positive regulatory loop. On the other hand, GAS1 is down-regulated by the Hh pathway but it is a positive regulator for Hh signaling. Alterations of these loops would lead to abnormal signaling of this pathway, such as inactivation of PTCH1 in BCCs.

## The Link of Hh Signaling to Human Cancer

The initial link between Hh signaling and human cancers was made from the discovery that mutations of human *PTCH1* are associated with a rare hereditary form of BCC – Basal Cell Nevus Syndrome (also Gorlin syndrome) [100–102]. Gorlin syndrome is a rare autosomal genetic disease with two distinct sets of phenotypes: an increased risk to developing cancers such as BCCs and medulloblastomas; developmental defects such as bifid ribs, ectopic calcification. The tumor suppressor role of *PTCH1* is demonstrated in mice with knockout of one Ptch1 allele that $Ptch1^{+/-}$ mice develop tumors in addition to other features observed in Gorlin syndrome patients, such as spina bifida occulta [82, 83, 103]. Figure 7.2 shows the association of Hh signaling with human cancer.

## Activation of the Hedgehog Pathway in Human Cancer

### *BCCs and Medulloblastomas*

BCCs consistently have abnormalities of the Hh pathway with *PTCH1* mutations in 50% and mutations of *SMO* in 10% of patients [104–108]. Unlike wild-type SMO, expression of SmoM2, an activated SMO mutant molecule identified in human BCCs, in mouse skin results in the formation of BCC-like tumors [104].

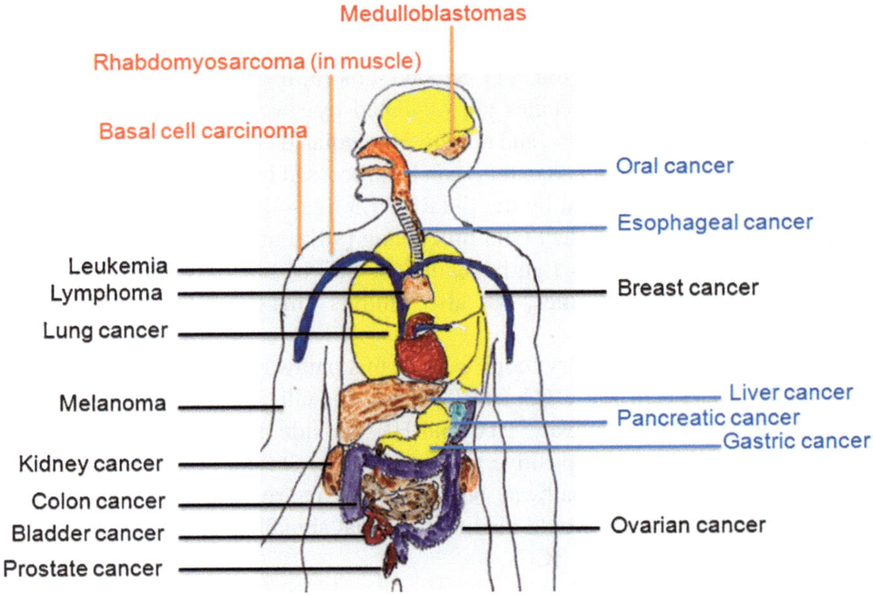


**Fig. 7.2** Activation of Hh signaling in human cancer. Since the link of Hh signaling activation in Gorlin syndrome, increasing evidence indicate that Hh signaling is frequently activated in human cancer. Based on the current data, we group these cancers into three groups. Group one is associated with Gorlin syndrome, including basal cell carcinomas, medulloblastomas, and rhabdomyosarcomas (in muscle) (in *red*). Group two includes cancer types with reproducible data of Hh signaling activation from several groups, such as oral cancer and many gastrointestinal cancers (in *blue*). Group three includes cancer types with limited reports or variable results from different groups (in *black*). Several common cancer types are in group three. Therefore, additional studies in group three cancers will provide insights as to the significance of Hh signaling in different types of cancer. This figure was from [177] with permission from the publisher

*Su(Fu)* is also mutated in a small number of BCCs [106]. Taken all the data together, the genetic alteration of the Hh pathway is detected in about 70% of BCCs. Since almost all BCCs have activated Hh signaling, we predict that alterations in other Hh signaling molecules or related molecules may be responsible for the Hh pathway activation in 30% of sporadic BCCs. At the molecular level, activated Hh signaling in BCCs leads to cell proliferation through elevated expression of PDGFR$\alpha$ [109], whereas targeted inhibition of the pathway causes apoptosis via Fas induction [110].

About one-third of medulloblastomas have activated Hh signaling. Like BCCs, loss-of-function mutations of PTCH1 are often responsible for the pathway activation. Mutations in *SMO* and *Su(Fu)* are found in a few cases. In addition, noncanonical activation of Gli2 via ATOH1 and Yap1 has been detected in medulloblastomas. Hh signaling is activated both in the desmoplastic form (more often) and in the classic form of medulloblastomas.

### *Activation of Hh Signaling in Cancers Not Associated with Gorlin Syndrome*

Increasing data indicate that Hh signaling is activated in many types of human cancers, including those associated with Gorlin syndrome as well as those not associated with this syndrome. It is estimated that over 30% of human cancers exhibit activated Hh signaling to some extent. These cancers include brain tumors, melanomas, leukemia, lymphomas, gastrointestinal, prostate, lung, and breast cancers. Unlike the situation in BCCs and meduloblastomas, which are associated with Gorlin syndrome (type I cancer), gene mutation is not primarily responsible for activated Hh signaling in those cancers not associated with Gorlin syndrome (type II cancer) [111, 112]. Current understanding is that Hh signaling activation in type II cancers is caused by ligand-dependent mechanisms or noncanonical Hh signaling activation. The association of ligand-dependence (or ligand-independence) with a specific cancer type, tumor morphology, or tumor stage has not been established.

## The Role of Hh Signaling in Cancer Initiation, Progression, and Metastasis

Increasing evidence indicate that Hh signaling is involved in different stages of carcinogenesis in different cancer types. In Barrett's esophagus, an early precursor of esophageal adenocarcinomas, both Shh and Ihh are highly expressed in the epithelium, which is associated with stromal expression of Hh target genes Ptch1 and BMP4 [113]. Sox9, as a target gene of BMP4, is highly expressed in the epithelial lesion [113]. These results indicate that Hh signaling plays an important role for the initiation of esophageal adenocarcinomas. In pancreatic cancer, activation of this pathway is found in PIN lesions as well in metastatic cancer [114–116], indicating that Hh signaling plays a significant role in pancreatic cancer. However, transgenic mice with pancreatic-specific expression of *SHH* or *GLI2* develop undifferentiated pancreatic tumors which are different from pancreatic ductal adenocarcinomas (PDAC) [116–118], suggesting that activation of Hh signaling alone is not sufficient to drive the development of PDAC. In other tumors, such as gastric and prostate cancers, Hh signaling activation is associated with cancer progression [119–122]. Consistent with these findings, inhibition of Hh signaling in prostate and gastric cancer cells reduces cell invasiveness [120, 122, 123] (our unpublished data). Reports also suggest that Hh signaling is required for development and progression of melanoma, gliomas, breast cancer, ovarian cancer, leukemia, and B-cell lymphomas [124–129]. However, the role of Hh signaling in each of the cancer types has not been completely established. It is suggested that Hh signaling plays an important role for cancer stem cells in several cancer types, such as glioma, medulloblastoma, and possibly breast cancer (see more discussion below).

Increasing evidence indicate that Hh signaling is critical for cancer stem cell maintenance and function [130–132]. For example, leukemia stem cell maintenance and expansion is dependent on Hh signaling [130, 131]. Alteration of the Hh pathway is reported to affect the hematopoietic stem cell (HSC) population in some studies but does not change HSC in other studies [131, 133–136]. Based on the cancer stem cell theory, it is anticipated that Hh signaling activation exerts resistance to cancer chemotherapy and radiotherapy [137]. Indeed, several studies have shown that Hh signaling activation is associated with chemotherapy resistance or radiotherapy resistance [138–140]. Hh signaling inhibitor IPI-926 enhances the delivery of chemotherapeutical drug Gemcitabine in a mouse model of pancreatic cancer [139]. Despite all these progress, more experiments are necessary to determine whether and how Hh signaling is involved in cancer stem cells in solid tumors.

In reviewing literatures on Hh signaling in human cancer, we notice that different results of Hh signaling activation are often reported from the same cancer type. This discrepancy may come from several reasons. First, the function of Hh signaling in human cancers may be context dependent, occurring in some tissues or cell lines but not in others. For example, accumulating data suggest that Hh signaling functions in maintaining cancer stem cell proliferation [130–132], not proliferation of all cancer cells. The percentage of cancer stem cells varies a lot from different tumor types. Second, heterogeneity in the tumor tissue often accounts for the difference in the analysis of Hh target gene expression by real-time PCR. For example, only a small portion of prostatectomy specimens (5–10%) is cancer tissue, whereas specimens of transurethral resection of the prostate (TURP) frequently have a large fraction of cancer tissue. Thus, the data from these two different types of specimens may result in different data, mainly due to the percentage of cancer cells in the tissue [122]. Even laser microscope captured tissues that will have significant amount of noncancerous cells, and the percentage may vary from operator to operator. Third, a standard is needed to define Hh signaling activation as different standards are being used. Some use increased expression of Gli1 as the read-out [91, 125], whereas others test the expression of several Hh target genes, such as Gli1, PTCH1, sFRP1, and HIP [116, 120, 141, 142]. Still others use only immunohistochemistry to detect Hh signaling activation [128, 143] while most studies use multiple approaches. Therefore, we need to read the literatures with cautions. Particular attention should be paid to the methodology used in the studies and reproducibility of the results. We believe that detection of Hh signaling pathway activation using immunohistochemistry of only one Hh target gene is not a reliable approach.

## Animal Models for Hh-Mediated Carcinogenesis

It is known that correlation of Hh target gene expression with tumor specimens is not sufficient to claim a role of Hh signaling in cancer. Establishing animal models using tissue-specific activation of Hh signaling is critical for understanding the role of Hh signaling in carcinogenesis. Currently, the mouse models for BCCs and medulloblastomas are well established, whereas mouse models for other Hh signaling-mediated carcinogenesis still need more work.

## *Mouse Models for BCCs*

Wild-type mice never develop BCCs even after carcinogen treatment, UV, or ionizing radiation. Ptch1$^{+/-}$ mice are susceptible to BCC development following UV irradiation or ionizing radiation [103]. The frequency of BCC development under these conditions is around 50% with one or two tumors per mouse [110, 144]. Due to the embryonic lethality of Ptch1$^{-/-}$, tissue-specific knockout of Ptch1 has been generated [145]. By combing conditional gene knockout and the inducible activity of the keratin 6a promoter, Krt6a-cre:Ptch1$^{neo/neo}$ mice develop BCCs following the stimulation of retinoic acid [146]. In addition to Ptch1 k/o mouse model, transgenic mice expressing Smo using Krt5 or Krt14 promoters also develop BCC-like tumors [104, 147]. However, these transgenic mice eventually lose the expression of Smo via unknown molecular mechanisms. Using conditional knock-in technology, skin-specific knock-in of SmoM2$^{YFP}$ (Krt14-creER:R26-SmoM2$^{YFP}$) mice develop multiple microscopic BCCs at very early age, which provide an easy genetic assay for putative Hh signaling downstream of Smo [148]. Su(Fu)$^{+/-}$ mice develop skin lesion resembling skin hyperplasia but not BCC-like tumors [80]. Several transgenic mice have been developed using downstream transcriptional factors Gli1 and Gli2 as well as Shh [149–151]. The inducible expression of Gli2 in the skin results in BCCs in a few weeks. These mouse models provide rich resources for further understanding of Hh signaling-mediated BCC development.

## *Mouse Models for Medulloblastomas*

A small portion of Ptch1$^{+/-}$ mice (10–30%) develops medulloblastomas and Rhabdomyosarcomas [82, 83]. The synergy between p53 pathway and Hh signaling is clearly shown in the medulloblastoma model. While p53 null mice do not develop this type of tumors, Ptch1$^{+/-}$ p53 null mice all develop medulloblastomas [152]. On the other hand, Ptch2$^{+/-}$ mice do not develop medulloblastoma per se but increase the incidence of medulloblastomas in Ptch1$^{+/-}$ mice [153, 154]. Su(Fu)$^{+/-}$ mice develop skin phenotypes similar to Gorlin syndrome but are generally not tumor prone [80]. However, Su(Fu)$^{+/-}$ mice with p53 null background frequently develop medulloblastomas with the signature of Hh signaling alterations [155]. Although Ptch1$^{+/-}$:Su(Fu)$^{+/-}$ mice are more likely to develop medulloblastoma than Ptch1$^{+/-}$ mice, the difference was not statistically significant [156]. In addition to the loss of tumor suppressor genes, transgenic mice using activated Smo mutant molecule SmoM2 under the control of neuroD2 promoter results in medulloblastomas in most mice [157]. Tissue-specific activation of Hh signaling via Ptch1 knockout or SmoM2 expression using granule neuron precursor lineage-specific promoters (Math1, GFAP, Oligo-2, and TLx3) but not the purkinje neuron-specific promoter leads to the formation of medulloblastoma [158, 159], indicating that granule neuron precursors are the source for medulloblastoma development. Further analysis shows that CD15 is the medulloblastoma stem cell marker [132, 160].

## *Mouse Models for Hh Signaling-Mediated Carcinogenesis in Other Organ Sites*

Postnatal induction of an activated allele of Smoothened (*R26-SmoM2*) using a ubiquitously expressed inducible Cre transgene (*CAGGS-CreER*) has been used to explore the role of Hh signaling-mediated carcinogenesis in mice [148]. In this model, all mice exhibited rhabdomyosarcoma and BCC; and 40% also developed medulloblastoma. In addition, pancreatic lesions resembling low-grade mucinous cystic neoplasms in humans and diverticular harmartomatous lesions in both intestine and stomach are observed. However, no other tumor types are observed in this mouse model, suggesting that activation of Hh signaling is not sufficient to drive tumor formation in prostate, lung, breast, and GI-tract.

Similar data are also observed in other studies. For example, it is shown in orthotopic mouse models that Hh signaling is necessary for pancreatic cancer tumor metastasis [161] (our unpublished data). Pancreas tissue-specific deletion of *Smo*, on the other hand, did not affect the formation of PDAC, whereas *GLI2* expression (CLEG2:Pdx1-cre mice) or Shh expression only lead to the formation of undifferentiated pancreatic tumors [116–117, 162]. These results indicate that activation of Hh signaling alone is not enough to drive PDAC formation, but it is essential for tumor progression and metastasis. In consistent with this theory, PDAC development of Kras$^{+/G12D}$:Pdx1-cre mice is not affected by the removal of Smo gene, and pdx-1 cre-driven expression of SmoM2 does not result in PIN lesions despite the fact that paracrine Hh signaling is observed in the pancreatic tissue [117, 162].

Recent study of Barrett's esophagus indicates that Shh expression in the epithelium of Barrett's esophagus can lead to stromal expression of Hh signaling target genes [113]. Using Shh transgenic mouse model, it is shown that epithelial expression of Shh can lead to stromal expression of Hh target gene BMP4, its target gene Sox9, and columnar phenotype of mouse esophageal epithelium, which resembles a feature in human Barrett's esophagus. These data suggest that Hh signaling activation can drive the formation of some features resembling Barrett's esophagus in mice.

For the role of Hh signaling in other cancer types, the major mouse models are based on xenograft in immunodeficient mice (nude or SCID mice) [163]. With potential implications of Hh signaling inhibitors for clinical cancer treatments, more established mouse models will be needed. Because modeling cancer metastasis is quite a challenge, we anticipate an increase in the use of orthotopic mouse models for studying Hh signaling in cancer progression and metastasis in the next few years.

## Small Molecule Modulators of Hedgehog Signaling

More than 50 compounds have been disclosed to have inhibitory effects on Hh signaling. Of these, four are being used in clinical trials. There are three major targeting sites for Hh signaling inhibitors identified so far: Hh molecules (Shh neutralizing antibodies and small molecule Robotnikinin); SMO protein (cyclopamine and its derivatives IPI-926, Cyc-T, and synthetic compounds GDC-0449, Cur61414, XL-139, and LDE-225); and Gli inhibitors (HPI-1, HPI-2, GANT-56, and GANT-61) [164]. We can divide Hh signaling inhibitors into three groups: natural products (cyclopamine, its derivatives, and other natural products); synthetic small molecules, and Hh signaling modulators. Table 7.1 lists the major Hh signaling inhibitors.

### *Natural Products (Cyclopamine, Its Derivatives, and Others)*

Cyclopamine, a plant-derived steroidal alkaloid, inhibits Hh signaling through direct binding to the transmembrane helices of SMO [165]. Identification of

**Table 7.1** Hedgehog signaling inhibitors[a]

| Inhibitor | 50% Maximal inhibition (IC50) | In vitro/in vivo studies |
|---|---|---|
| Cyclopamine | 300 nM | In vivo and in vitro |
| KAAD-cyclopamine | 20 nM | In vitro cultured cells |
| Jervine | 500 nM | In vitro and cultured embryos |
| Cyclopamine tartrate acid salt (Cyc-T) | 20 nM | In vitro and in vivo studies |
| Cur-61414 | 200 nM | Phase I clinical trial |
| Sant-1,2,3,4 | 20–200 nM | In vitro studies |
| Gant-58,61 | 5 μM | In vitro and in vivo studies |
| IPI-926 | <20 nM | Phase I clinical trial |
| GDC-0449 | <20 nM | Phase I/II/III clinical trials |
| BMS-833923 (XL139) | <20 nM | Phase I clinical trial |
| LDE-225 | <20 nM | Phase I clinical trial |
| Vitamin D3 | 100 μM | In vitro |
| Robotnikinin | >10 μM | In vitro |
| HPI-1,2,3,4 | <10 μmol/L | In vitro |
| Itraconazole | <1.5 μM | In vitro and xenograft |

[a]Modified from Yang et al. *Oncogene* 29, 469–481 (2010) and see text for references

specific small molecule antagonists of SMO has opened up exciting new prospects for targeted cancer therapy for human cancers associated with Hh signaling.

Specificity of cyclopamine varies depending on the concentration used. While cyclopamine at a low concentration (<10 μM) has specific inhibitory effects on Hh signaling, high doses of cyclopamine can result in cell death without effecting Hh target gene expression [166]. In several mouse models, the in vivo effect of cyclopamine on tumor shrinkage has been demonstrated. Oral delivery of cyclopamine blocks the growth of UV-induced BCCs in *Ptch1*$^{+/-}$ mice by 50% [110]. The treatment in this mouse model also prevents the development of additional microscopic BCCs, implying a cancer prevention potential of cyclopamine. Similarly, cyclopamine is shown to be effective in reducing medulloblastoma development in *Ptch1*$^{+/-}$ mice [167] and tumor growth of cancer cell lines in nu/nu mice [91, 116, 120, 168]. Additional modifications on cyclopamine aiming at increasing acid stability and aqueous solubility are now available, such as IPI-926 and Cyc-T [169, 170]. IPI-926 is now in Phase I clinical trial.

## *Synthetic Hh Signaling Antagonists*

Increasing number of synthetic Hh antagonists are reported in the literature, with most compounds targeting at SMO. Four of these compounds are now in clinical trials (Table 7.1), including GDC-0449. The successful clinical trials with GDC-0449 on human BCCs further encourage the translational studies in this area [171]. Clinical trial of the same compound in a medulloblastoma patient led to rapid tumor shrinkage but later developed drug resistance due to a SMO mutation, which disables the binding of GDC-0449 to SMO. This case report implies a need for novel alternative strategies for the treatment of Hh signaling-associated cancers. There are also a couple of small molecules targeting at Shh or Gli [171–173]. Because the wide spread existence of nonconical regulation of Gli transcriptional factors and potential resistance to Smo inhibitors, the antagonists targeting Hh downstream effectors constitute a valuable source for developing chemotherapeutical strategies for Hh pathway-related cancers.

## *Hh Signaling Modulators*

Recent studies indicate that vitamin D3, the secretion of which can be facilitated by PTCH1, can inhibit SMO signaling through direct binding to SMO. This finding raises a possibility to treat BCCs with nutrition supplements [46]. Promising data show that the effect of tazarotene, a retinoid with retinoic acid receptor (RAR) beta/gamma specificity, against BCC carcinogenesis is sustained after its withdrawal [174]. Curcumin has been shown to be able to block Hh signaling-mediated carcinogenesis. A commonly used antifungal agent Itraconazole is shown to affect Hh signaling [175]. Several natural products, genistein, curcumin, EGCG, and resveratrol [176], are also shown to affect Hh signaling in a mouse model of prostate cancer. The detailed molecular mechanisms of action for these signaling modulators remain elusive.

## Summary

In summary, link of Hh signaling activation to a variety of human cancer implies the relevance of studying Hh signaling to human health. Rapid advance in the discovery of novel Hh signaling inhibitors has provided many opportunities for developing novel cancer therapeutic strategies. Several major challenges exist for moving the Hh signaling inhibitors into clinic including the lack of basic understanding for the molecular mechanisms of Hh signaling-mediated carcinogenesis; identifying the right tumors for therapeutic application; reliable and reproducible mouse models for testing and optimization of drug dosages to minimize the side effects.

**Acknowledgments** Current research in my laboratory is supported by grants from the National Cancer Institute CA94160 and Wells Center for Pediatric Research.

## References

1. Nusslein-Volhard C, Wieschaus E (1980) Mutations affecting segment number and polarity in Drosophila. Nature 287:795–801
2. Krauss S, Concordet JP, Ingham PW (1993) A functionally conserved homolog of the Drosophila segment polarity gene hh is expressed in tissues with polarizing activity in zebrafish embryos. Cell 75:1431–1444
3. Echelard Y et al (1993) Sonic hedgehog, a member of a family of putative signaling molecules, is implicated in the regulation of CNS polarity. Cell 75:1417–1430
4. Riddle RD, Johnson RL, Laufer E, Tabin C (1993) Sonic hedgehog mediates the polarizing activity of the ZPA. Cell 75:1401–1416
5. Chang DT et al (1994) Products, genetic linkage and limb patterning activity of a murine hedgehog gene. Development 120:3339–3353
6. Roelink H et al (1994) Floor plate and motor neuron induction by vhh-1, a vertebrate homolog of hedgehog expressed by the notochord. Cell 76:761–775
7. Epstein EH (2008) Basal cell carcinomas: attack of the hedgehog. Nat Rev Cancer 8:743–754
8. Xie J (2005) Hedgehog signaling in prostate cancer. Future Oncol 1:331–338
9. Xie J (2008) Hedgehog signaling pathway: development of antagonists for cancer therapy. Curr Oncol Rep 10:107–113
10. Xie J (2008) Molecular biology of basal and squamous cell carcinomas. Adv Exp Med Biol 624:241–251
11. Jiang J, Hui CC (2008) Hedgehog signaling in development and cancer. Dev Cell 15:801–812
12. Ingham PW, Placzek M (2006) Orchestrating ontogenesis: variations on a theme by sonic hedgehog. Nat Rev Genet 7:841–850

13. Sasaki H, Hui C, Nakafuku M, Kondoh H (1997) A binding site for Gli proteins is essential for HNF-3beta floor plate enhancer activity in transgenics and can respond to Shh in vitro. Development 124:1313–1322
14. Kinzler KW, Vogelstein B (1990) The GLI gene encodes a nuclear protein which binds specific sequences in the human genome. Mol Cell Biol 10:634–642
15. McMahon AP, Ingham PW, Tabin CJ (2003) Developmental roles and clinical significance of hedgehog signaling. Curr Top Dev Biol 53:1–114
16. Ingham PW, McMahon AP (2001) Hedgehog signaling in animal development: paradigms and principles. Genes Dev 15:3059–3087, doi:10.1101/gad.938601
17. Taipale J, Beachy PA (2001) The Hedgehog and Wnt signalling pathways in cancer. Nature 411:349–354
18. Lee JJ et al (1994) Autoproteolysis in hedgehog protein biogenesis. Science 266:1528–1537
19. Porter JA, Young KE, Beachy PA (1996) Cholesterol modification of hedgehog signaling proteins in animal development. Science 274:255–259
20. Porter JA et al (1995) The product of hedgehog autoproteolytic cleavage active in local and long-range signalling. Nature 374:363–366
21. Buglino JA, Resh MD (2008) Hhat is a palmitoylacyltransferase with specificity for N-palmitoylation of Sonic Hedgehog. J Biol Chem 283:22076–22088
22. Kawakami T et al (2002) Mouse dispatched mutants fail to distribute hedgehog proteins and are defective in hedgehog signaling. Development 129:5753–5765
23. Ma Y et al (2002) Hedgehog-mediated patterning of the mammalian embryo requires transporter-like function of dispatched. Cell 111:63–75, doi:S0092867402009777 [pii]
24. Caspary T et al (2002) Mouse dispatched homolog1 is required for long-range, but not juxtacrine, Hh signaling. Curr Biol 12:1628–1632, doi:S0960982202011478 [pii]
25. Dierker T, Dreier R, Petersen A, Bordych C, Grobe K (2009) Heparan sulfate-modulated, metalloprotease-mediated sonic hedgehog release from producing cells. J Biol Chem 284:8013–8022
26. Beckett K, Franch-Marro X, Vincent JP (2008) Glypican-mediated endocytosis of Hedgehog has opposite effects in flies and mice. Trends Cell Biol 18:360–363
27. Lum L et al (2003) Identification of Hedgehog pathway components by RNAi in Drosophila cultured cells. Science 299:2039–2045
28. Baena-Lopez LA, Rodriguez I, Baonza A (2008) The tumor suppressor genes dachsous and fat modulate different signalling pathways by regulating dally and dally-like. Proc Natl Acad Sci USA 105:9645–9650
29. Toyoda H, Kinoshita-Toyoda A, Fox B, Selleck SB (2000) Structural analysis of glycosaminoglycans in animals bearing mutations in sugarless, sulfateless, and tout-velu. Drosophila homologues of vertebrate genes encoding glycosaminoglycan biosynthetic enzymes. J Biol Chem 275:21856–21861
30. Bellaiche Y, The I, Perrimon N (1998) Tout-velu is a Drosophila homologue of the putative tumour suppressor EXT-1 and is needed for Hh diffusion. Nature 394:85–88
31. Koziel L, Kunath M, Kelly OG, Vortkamp A (2004) Ext1-dependent heparan sulfate regulates the range of Ihh signaling during endochondral ossification. Dev Cell 6:801–813
32. Stone DM et al (1996) The tumour-suppressor gene patched encodes a candidate receptor for Sonic hedgehog. Nature 384:129–134
33. Taipale J, Cooper MK, Maiti T, Beachy PA (2002) Patched acts catalytically to suppress the activity of Smoothened. Nature 418:892–897
34. Chuang PT, McMahon AP (1999) Vertebrate Hedgehog signalling modulated by induction of a Hedgehog-binding protein. Nature 397:617–621
35. Martinelli DC, Fan CM (2007) Gas1 extends the range of Hedgehog action by facilitating its signaling. Genes Dev 21:1231–1243
36. Seppala M et al (2007) Gas1 is a modifier for holoprosencephaly and genetically interacts with sonic hedgehog. J Clin Invest 117:1575–1584

37. Allen BL, Tenzen T, McMahon AP (2007) The Hedgehog-binding proteins Gas1 and Cdo cooperate to positively regulate Shh signaling during mouse development. Genes Dev 21:1244–1257
38. Okada A et al (2006) Boc is a receptor for sonic hedgehog in the guidance of commissural axons. Nature 444:369–373
39. Tenzen T et al (2006) The cell surface membrane proteins Cdo and Boc are components and targets of the Hedgehog signaling pathway and feedback network in mice. Dev Cell 10:647–656
40. Zhang W, Kang JS, Cole F, Yi MJ, Krauss RS (2006) Cdo functions at multiple points in the Sonic Hedgehog pathway, and Cdo-deficient mice accurately model human holoprosencephaly. Dev Cell 10:657–665
41. Yao S, Lum L, Beachy P (2006) The ihog cell-surface proteins bind Hedgehog and mediate pathway activation. Cell 125:343–357
42. Capurro MI et al (2008) Glypican-3 inhibits Hedgehog signaling during development by competing with patched for Hedgehog binding. Dev Cell 14:700–711
43. Yavari A et al (2010) Role of lipid metabolism in smoothened derepression in hedgehog signaling. Dev Cell 19:54–65
44. Khaliullina H et al (2009) Patched regulates Smoothened trafficking using lipoprotein-derived lipids. Development 136:4111–4121
45. Callejo A, Culi J, Guerrero I (2008) Patched, the receptor of Hedgehog, is a lipoprotein receptor. Proc Natl Acad Sci USA 105:912–917
46. Bijlsma MF et al (2006) Repression of smoothened by patched-dependent (pro-)vitamin D3 secretion. PLoS Biol 4:e232
47. Philipp M et al (2008) Smoothened signaling in vertebrates is facilitated by a G protein-coupled receptor kinase. Mol Biol Cell 19:5478–5489
48. Ogden SK et al (2008) G protein Galphai functions immediately downstream of Smoothened in Hedgehog signalling. Nature 456:967–970
49. Molnar C, Holguin H, Mayor F Jr, Ruiz-Gomez A, de Celis JF (2007) The G protein-coupled receptor regulatory kinase GPRK2 participates in Hedgehog signaling in Drosophila. Proc Natl Acad Sci USA 104:7963–7968
50. Riobo NA, Saucy B, Dilizio C, Manning DR (2006) Activation of heterotrimeric G proteins by Smoothened. Proc Natl Acad Sci USA 103:12607–12612
51. Jia J, Tong C, Wang B, Luo L, Jiang J (2004) Hedgehog signalling activity of Smoothened requires phosphorylation by protein kinase A and casein kinase I. Nature 432:1045–1050
52. Zhang C, Williams EH, Guo Y, Lum L, Beachy PA (2004) Extensive phosphorylation of Smoothened in Hedgehog pathway activation. Proc Natl Acad Sci USA 101:17900–17907
53. Zhao Y, Tong C, Jiang J (2007) Hedgehog regulates smoothened activity by inducing a conformational switch. Nature 450:252–258
54. Corbit KC et al (2005) Vertebrate Smoothened functions at the primary cilium. Nature 437:1018–1021
55. Huangfu D et al (2003) Hedgehog signalling in the mouse requires intraflagellar transport proteins. Nature 426:83–87
56. May SR et al (2005) Loss of the retrograde motor for IFT disrupts localization of Smo to cilia and prevents the expression of both activator and repressor functions of Gli. Dev Biol 287:378–389
57. Huangfu D, Anderson KV (2005) Cilia and Hedgehog responsiveness in the mouse. Proc Natl Acad Sci USA 102:11325–11330
58. Zhang Q, Davenport JR, Croyle MJ, Haycraft CJ, Yoder BK (2005) Disruption of IFT results in both exocrine and endocrine abnormalities in the pancreas of Tg737(orpk) mutant mice. Lab Invest 85:45–64
59. Hoover AN et al (2008) C2cd3 is required for cilia formation and Hedgehog signaling in mouse. Development 135:4049–4058
60. Scholey JM, Anderson KV (2006) Intraflagellar transport and cilium-based signaling. Cell 125:439–442

61. Cortellino S et al (2009) Defective ciliogenesis, embryonic lethality and severe impairment of the Sonic Hedgehog pathway caused by inactivation of the mouse complex A intraflagellar transport gene Ift122/Wdr10, partially overlapping with the DNA repair gene Med1/Mbd4. Dev Biol 325:225–237
62. Haycraft CJ et al (2005) Gli2 and Gli3 localize to cilia and require the intraflagellar transport protein polaris for processing and function. PLoS Genet 1:e53
63. Wang Y, Zhou Z, Walsh CT, McMahon AP (2009) Selective translocation of intracellular Smoothened to the primary cilium in response to Hedgehog pathway modulation. Proc Natl Acad Sci USA 106:2623–2628
64. Wilson CW, Chen MH, Chuang PT (2009) Smoothened adopts multiple active and inactive conformations capable of trafficking to the primary cilium. PLoS ONE 4:e5182
65. Rohatgi R, Milenkovic L, Corcoran RB, Scott MP (2009) Hedgehog signal transduction by Smoothened: pharmacologic evidence for a 2-step activation process. Proc Natl Acad Sci USA 106:3196–3201
66. Han YG et al (2009) Dual and opposing roles of primary cilia in medulloblastoma development. Nat Med 15:1062–1065
67. Wong SY et al (2009) Primary cilia can both mediate and suppress Hedgehog pathway-dependent tumorigenesis. Nat Med 15:1055–1061
68. Kovacs JJ et al (2008) Beta-arrestin-mediated localization of smoothened to the primary cilium. Science 320:1777–1781
69. Chen MH et al (2009) Cilium-independent regulation of Gli protein function by Sufu in Hedgehog signaling is evolutionarily conserved. Genes Dev 23:1910–1928
70. Jia J et al (2009) Suppressor of fused inhibits mammalian Hedgehog signaling in the absence of cilia. Dev Biol 330:452–460
71. Cheung HO et al (2009) The kinesin protein Kif7 is a critical regulator of Gli transcription factors in mammalian hedgehog signaling. Sci Signal 2:ra29
72. Endoh-Yamagami S et al (2009) The mammalian Cos2 homolog Kif7 plays an essential role in modulating Hh signal transduction during development. Curr Biol 19:1320–1326
73. Wilson CW et al (2009) Fused has evolved divergent roles in vertebrate Hedgehog signalling and motile ciliogenesis. Nature 459:98–102
74. Merchant M et al (2005) Loss of the serine/threonine kinase fused results in postnatal growth defects and lethality due to progressive hydrocephalus. Mol Cell Biol 25:7054–7068
75. Chen MH, Gao N, Kawakami T, Chuang PT (2005) Mice deficient in the fused homolog do not exhibit phenotypes indicative of perturbed hedgehog signaling during embryonic development. Mol Cell Biol 25:7042–7053
76. Eggenschwiler JT, Espinoza E, Anderson KV (2001) Rab23 is an essential negative regulator of the mouse Sonic hedgehog signalling pathway. Nature 412:194–198
77. Reiter JF, Skarnes WC (2006) Tectonic, a novel regulator of the Hedgehog pathway required for both activation and inhibition. Genes Dev 20:22–27
78. Varjosalo M et al (2008) Application of active and kinase-deficient kinome collection for identification of kinases regulating hedgehog signaling. Cell 133:537–548
79. Evangelista M et al (2008) Kinome siRNA screen identifies regulators of ciliogenesis and hedgehog signal transduction. Sci Signal 1:ra7
80. Svard J et al (2006) Genetic elimination of suppressor of fused reveals an essential repressor function in the Mammalian hedgehog signaling pathway. Dev Cell 10:187–197
81. Aszterbaum M et al (1999) Ultraviolet and ionizing radiation enhance the growth of BCCs and trichoblastomas in patched heterozygous knockout mice. Nat Med 5:1285–1291
82. Hahn H et al (1998) Rhabdomyosarcomas and radiation hypersensitivity in a mouse model of Gorlin syndrome. Nat Med 4:619–622
83. Goodrich LV, Milenkovic L, Higgins KM, Scott MP (1997) Altered neural cell fates and medulloblastoma in mouse patched mutants. Science 277:1109–1113
84. Barnfield PC, Zhang X, Thanabalasingham V, Yoshida M, Hui CC (2005) Negative regulation of Gli1 and Gli2 activator function by Suppressor of fused through multiple mechanisms. Differentiation 73:397–405

85. Kise Y, Morinaka A, Teglund S, Miki H (2009) Sufu recruits GSK3beta for efficient processing of Gli3. Biochem Biophys Res Commun 387:569–574
86. Yue S, Chen Y, Cheng SY (2009) Hedgehog signaling promotes the degradation of tumor suppressor Sufu through the ubiquitin-proteasome pathway. Oncogene 28:492–499
87. Kinzler KW, Ruppert JM, Bigner SH, Vogelstein B (1988) The GLI gene is a member of the Kruppel family of zinc finger proteins. Nature 332:371–374
88. Ruppert JM et al (1988) The GLI-Kruppel family of human genes. Mol Cell Biol 8:3104–3113
89. Sheng T, Chi S, Zhang X, Xie J (2006) Regulation of Gli1 localization by the cAMP/protein kinase A signaling axis through a site near the nuclear localization signal. J Biol Chem 281:9–12
90. Kogerman P et al (1999) Mammalian suppressor-of-fused modulates nuclear-cytoplasmic shuttling of Gli-1. Nat Cell Biol 1:312–319
91. Stecca B et al (2007) Melanomas require HEDGEHOG-GLI signaling regulated by interactions between GLI1 and the RAS-MEK/AKT pathways. Proc Natl Acad Sci USA 104:5895–5900
92. Pan Y, Bai CB, Joyner AL, Wang B (2006) Sonic hedgehog signaling regulates Gli2 transcriptional activity by suppressing its processing and degradation. Mol Cell Biol 26:3365–3377
93. Huntzicker EG et al (2006) Dual degradation signals control Gli protein stability and tumor formation. Genes Dev 20:276–281
94. Wang B, Li Y (2006) Evidence for the direct involvement of βTrCP in Gli3 protein processing. Proc Natl Acad Sci USA 103:33–38
95. Di Marcotullio L et al (2006) Numb is a suppressor of Hedgehog signalling and targets Gli1 for Itch-dependent ubiquitination. Nat Cell Biol 8:1415–1423
96. Jiang J (2006) Regulation of Hh/Gli signaling by dual ubiquitin pathways. Cell Cycle 5:2457–2463
97. Canettieri G et al (2010) Histone deacetylase and Cullin3-REN(KCTD11) ubiquitin ligase interplay regulates Hedgehog signalling through Gli acetylation. Nat Cell Biol 12:132–142
98. Huangfu D, Anderson KV (2006) Signaling from Smo to Ci/Gli: conservation and divergence of Hedgehog pathways from Drosophila to vertebrates. Development 133:3–14
99. Cheng SY, Bishop JM (2002) Suppressor of Fused represses Gli-mediated transcription by recruiting the SAP18-mSin3 corepressor complex. Proc Natl Acad Sci USA 99:5442–5447
100. Hahn H et al (1996) Mutations of the human homolog of Drosophila patched in the nevoid basal cell carcinoma syndrome. Cell 85:841–851
101. Johnson RL et al (1996) Human homolog of patched, a candidate gene for the basal cell nevus syndrome. Science 272:1668–1671
102. Epstein E Jr (2001) Genetic determinants of basal cell carcinoma risk. Med Pediatr Oncol 36:555–558
103. Aszterbaum M, Beech J, Epstein EH Jr (1999) Ultraviolet radiation mutagenesis of hedgehog pathway genes in basal cell carcinomas. J Investig Dermatol Symp Proc 4:41–45
104. Xie J et al (1998) Activating Smoothened mutations in sporadic basal-cell carcinoma. Nature 391:90–92
105. Lam CW et al (1999) A frequent activated smoothened mutation in sporadic basal cell carcinomas. Oncogene 18:833–836
106. Reifenberger J et al (2005) Somatic mutations in the PTCH, SMOH, SUFUH and TP53 genes in sporadic basal cell carcinomas. Br J Dermatol 152:43–51
107. Reifenberger J et al (1998) Missense mutations in SMOH in sporadic basal cell carcinomas of the skin and primitive neuroectodermal tumors of the central nervous system. Cancer Res 58:1798–1803
108. Couve-Privat S, Bouadjar B, Avril MF, Sarasin A, Daya-Grosjean L (2002) Significantly high levels of ultraviolet-specific mutations in the smoothened gene in basal cell carcinomas from DNA repair-deficient xeroderma pigmentosum patients. Cancer Res 62:7186–7189
109. Xie J et al (2001) A role of PDGFRalpha in basal cell carcinoma proliferation. Proc Natl Acad Sci USA 98:9255–9259

110. Athar M et al (2004) Inhibition of smoothened signaling prevents ultraviolet B-induced basal cell carcinomas through regulation of Fas expression and apoptosis. Cancer Res 64:7545–7552
111. Berman DM et al (2003) Widespread requirement for Hedgehog ligand stimulation in growth of digestive tract tumours. Nature 425:846–851
112. Watkins DN et al (2003) Hedgehog signalling within airway epithelial progenitors and in small-cell lung cancer. Nature 422:313–317
113. Wang DH et al (2010) Aberrant epithelial-mesenchymal Hedgehog signaling characterizes Barrett's metaplasia. Gastroenterology 138:1810–1822
114. Bailey JM, Mohr AM, Hollingsworth MA (2009) Sonic hedgehog paracrine signaling regulates metastasis and lymphangiogenesis in pancreatic cancer. Oncogene 28:3513–3525
115. Feldmann G et al (2008) An orally bioavailable small-molecule inhibitor of Hedgehog signaling inhibits tumor initiation and metastasis in pancreatic cancer. Mol Cancer Ther 7:2725–2735
116. Thayer SP et al (2003) Hedgehog is an early and late mediator of pancreatic cancer tumorigenesis. Nature 425:851–856
117. Pasca di Magliano M et al (2006) Hedgehog/Ras interactions regulate early stages of pancreatic cancer. Genes Dev 20:3161–3173
118. Nolan-Stevaux O et al (2009) GLI1 is regulated through Smoothened-independent mechanisms in neoplastic pancreatic ducts and mediates PDAC cell survival and transformation. Genes Dev 23:24–36
119. Fan L et al (2004) Hedgehog signaling promotes prostate xenograft tumor growth. Endocrinology 145:3961–3970
120. Karhadkar SS et al (2004) Hedgehog signalling in prostate regeneration, neoplasia and metastasis. Nature 431:707–712
121. Sanchez P et al (2004) Inhibition of prostate cancer proliferation by interference with SONIC HEDGEHOG-GLI1 signaling. Proc Natl Acad Sci USA 101:12561–12566
122. Sheng T et al (2004) Activation of the hedgehog pathway in advanced prostate cancer. Mol Cancer 3:29
123. He J et al (2006) Suppressing Wnt signaling by the hedgehog pathway through sFRP-1. J Biol Chem 281:35598–35602
124. Ehtesham M et al (2007) Ligand-dependent activation of the hedgehog pathway in glioma progenitor cells. Oncogene 26(39):5752–5761
125. Bhattacharya R et al (2008) Role of hedgehog signaling in ovarian cancer. Clin Cancer Res 14:7659–7666
126. Fiaschi M, Rozell B, Bergstrom A, Toftgard R (2009) Development of mammary tumors by conditional expression of GLI1. Cancer Res 69:4810–4817
127. Kasper M, Jaks V, Fiaschi M, Toftgard R (2009) Hedgehog signalling in breast cancer. Carcinogenesis 30:903–911
128. Liao X et al (2009) Aberrant activation of hedgehog signaling pathway in ovarian cancers: effect on prognosis, cell invasion and differentiation. Carcinogenesis 30:131–140
129. Lindemann RK (2008) Stroma-initiated hedgehog signaling takes center stage in B-cell lymphoma. Cancer Res 68:961–964
130. Dierks C et al (2008) Expansion of Bcr-Abl-positive leukemic stem cells is dependent on Hedgehog pathway activation. Cancer Cell 14:238–249
131. Zhao C et al (2009) Hedgehog signalling is essential for maintenance of cancer stem cells in myeloid leukaemia. Nature 458:776–779
132. Read TA et al (2009) Identification of CD15 as a marker for tumor-propagating cells in a mouse model of medulloblastoma. Cancer Cell 15:135–147
133. Hofmann I et al (2009) Hedgehog signaling is dispensable for adult murine hematopoietic stem cell function and hematopoiesis. Cell Stem Cell 4:559–567
134. Gao J et al (2009) Hedgehog signaling is dispensable for adult hematopoietic stem cell function. Cell Stem Cell 4:548–558
135. Siggins SL et al (2009) The Hedgehog receptor Patched1 regulates myeloid and lymphoid progenitors by distinct cell-extrinsic mechanisms. Blood 114:995–1004

136. Merchant A, Joseph G, Wang Q, Brennan S, Matsui W (2010) Gli1 regulates the proliferation and differentiation of HSCs and myeloid progenitors. Blood 115:2391–2396
137. Reya T, Morrison SJ, Clarke MF, Weissman IL (2001) Stem cells, cancer, and cancer stem cells. Nature 414:105–111
138. Sims-Mourtada J et al (2006) Hedgehog: an attribute to tumor regrowth after chemoradiotherapy and a target to improve radiation response. Clin Cancer Res 12:6565–6572
139. Olive KP et al (2009) Inhibition of hedgehog signaling enhances delivery of chemotherapy in a mouse model of pancreatic cancer. Science 324:1457–1461
140. Yoshikawa R et al (2008) Hedgehog signal activation in oesophageal cancer patients undergoing neoadjuvant chemoradiotherapy. Br J Cancer 98:1670–1674
141. Huang S et al (2006) Activation of the hedgehog pathway in human hepatocellular carcinomas. Carcinogenesis 27:1334–1340
142. Lee Y, Kawagoe R, Sasai K, Li Y, Russell HR, Curran T, McKinnon PJ (2007) Loss of suppressor-of-fused function promotes tumorigenesis. Oncogene. 2007 Sep 27;26(44):6442–7. Epub 2007 Apr 23.
143. Kubo M et al (2004) Hedgehog signaling pathway is a new therapeutic target for patients with breast cancer. Cancer Res 64:6071–6074
144. Mancuso M et al (2004) Basal cell carcinoma and its development: insights from radiation-induced tumors in Ptch1-deficient mice. Cancer Res 64:934–941
145. Ellis T et al (2003) Patched 1 conditional null allele in mice. Genesis 36:158–161
146. Adolphe C, Hetherington R, Ellis T, Wainwright B (2006) Patched1 functions as a gatekeeper by promoting cell cycle progression. Cancer Res 66:2081–2088
147. Grachtchouk V et al (2003) The magnitude of hedgehog signaling activity defines skin tumor phenotype. EMBO J 22:2741–2751, doi:10.1093/emboj/cdg271
148. Mao J et al (2006) A novel somatic mouse model to survey tumorigenic potential applied to the Hedgehog pathway. Cancer Res 66:10171–10178
149. Grachtchouk M et al (2000) Basal cell carcinomas in mice overexpressing Gli2 in skin. Nat Genet 24:216–217, doi:10.1038/73417
150. Nilsson M et al (2000) Induction of basal cell carcinomas and trichoepitheliomas in mice overexpressing GLI-1. Proc Natl Acad Sci USA 97:3438–3443
151. Oro AE et al (1997) Basal cell carcinomas in mice overexpressing sonic hedgehog. Science 276:817–821
152. Romer JT et al (2004) Suppression of the Shh pathway using a small molecule inhibitor eliminates medulloblastoma in Ptc1(+/–)p53(–/–) mice. Cancer Cell 6:229–240
153. Nieuwenhuis E et al (2006) Mice with a targeted mutation of patched2 are viable but develop alopecia and epidermal hyperplasia. Mol Cell Biol 26:6609–6622
154. Lee Y et al (2006) Patched2 modulates tumorigenesis in patched1 heterozygous mice. Cancer Res 66:6964–6971
155. Lee Y et al (2007) Loss of suppressor-of-fused function promotes tumorigenesis. Oncogene 26:6442–6447
156. Svard J, Rozell B, Toftgard R, Teglund S (2009) Tumor suppressor gene co-operativity in compound Patched1 and suppressor of fused heterozygous mutant mice. Mol Carcinog 48:408–419
157. Hallahan AR et al (2004) The SmoA1 mouse model reveals that notch signaling is critical for the growth and survival of sonic hedgehog-induced medulloblastomas. Cancer Res 64:7794–7800
158. Schuller U et al (2008) Acquisition of granule neuron precursor identity is a critical determinant of progenitor cell competence to form Shh-induced medulloblastoma. Cancer Cell 14:123–134
159. Yang ZJ et al (2008) Medulloblastoma can be initiated by deletion of Patched in lineage-restricted progenitors or stem cells. Cancer Cell 14:135–145
160. Ward RJ et al (2009) Multipotent CD15+ cancer stem cells in patched-1-deficient mouse medulloblastoma. Cancer Res 69:4682–4690
161. Feldmann G et al (2007) Blockade of hedgehog signaling inhibits pancreatic cancer invasion and metastases: a new paradigm for combination therapy in solid cancers. Cancer Res 67:2187–2196

162. Tian H et al (2009) Hedgehog signaling is restricted to the stromal compartment during pancreatic carcinogenesis. Proc Natl Acad Sci USA 106:4254–4259
163. Teglund S, Toftgard R (2010) Hedgehog beyond medulloblastoma and basal cell carcinoma. Biochim Biophys Acta 1805:181–208
164. Yang L, Xie G, Fan Q, Xie J (2010) Activation of the hedgehog-signaling pathway in human cancer and the clinical implications. Oncogene 29:469–481
165. Chen JK, Taipale J, Cooper MK, Beachy PA (2002) Inhibition of Hedgehog signaling by direct binding of cyclopamine to Smoothened. Genes Dev 16:2743–2748
166. Yauch RL et al (2008) A paracrine requirement for hedgehog signalling in cancer. Nature 455:406–410
167. Sanchez P, Ruiz i Altaba A (2005) In vivo inhibition of endogenous brain tumors through systemic interference of Hedgehog signaling in mice. Mech Dev 122:223–230
168. Berman DM et al (2002) Medulloblastoma growth inhibition by hedgehog pathway blockade. Science 297:1559–1561
169. Tremblay MR et al (2009) Discovery of a potent and orally active hedgehog pathway antagonist (IPI-926). J Med Chem 52:4400–4418
170. Xie J, Garrossian M (2009) Cyclopamine tartrate salt and uses thereof. US Patent WO 2009099625 20090813.
171. Von Hoff DD et al (2009) Inhibition of the hedgehog pathway in advanced basal-cell carcinoma. N Engl J Med 361:1164–1172
172. Hyman JM et al (2009) Small-molecule inhibitors reveal multiple strategies for Hedgehog pathway blockade. Proc Natl Acad Sci USA 106:14132–14137
173. Rudin CM et al (2009) Treatment of medulloblastoma with Hedgehog pathway inhibitor GDC-0449. N Engl J Med 361(12):1173–1178
174. So PL, Fujimoto MA, Epstein EH Jr (2008) Pharmacologic retinoid signaling and physiologic retinoic acid receptor signaling inhibit basal cell carcinoma tumorigenesis. Mol Cancer Ther 7:1275–1284
175. Kim J et al (2010) Itraconazole, a commonly used antifungal that inhibits Hedgehog pathway activity and cancer growth. Cancer Cell 17:388–399
176. Slusarz A et al (2010) Common botanical compounds inhibit the hedgehog signaling pathway in prostate cancer. Cancer Res 70:3382–3390
177. Liu H, Gu D, Xie J (2011) Clinical implications of hedgehog signaling pathway inhibitors. Chin J Cancer 30(1):13–26.

# Chapter 8
# Hedgehog Signaling in Pediatric Brain Tumors

**Joon Won Yoon, Jason Fangusaro, Philip Iannaccone, and David Walterhouse**

**Keywords** Sonic hedgehog • Central nervous system • Cerebellum • Pediatric brain tumors • Medulloblastoma • Hedgehog inhibitors

## Introduction

Sonic hedgehog (SHH) signaling plays roles in patterning and normal development of the mammalian central nervous system (CNS). Precise regulation of the pathway appears to be crucial in the CNS since dysregulation of SHH signaling has been associated with CNS birth defects and brain tumors. In this chapter, we focus on (1) the role of SHH signaling in mammalian CNS development, (2) the role of SHH signaling in pediatric brain tumors, and (3) potential clinical applications of Hedgehog (HH) pathway inhibitors in the treatment of pediatric brain tumors. We use the following conventions in this chapter: upper case = human protein (e.g. SHH, GLI1), lower case = mouse protein (e.g. shh, gli1), upper case italics = human gene (e.g. *SHH*, *GLI1*), and lower case italics = mouse gene (e.g. *shh*, *gli1*). When we are discussing a pathway in a general way without specific reference to gene, protein, or species, we use upper case without italics.

## Role of SHH Signaling in CNS Development

During development, the CNS arises from the neural plate, which is composed of a single layer of cells derived from midline ectoderm. Neuroepithelial cells in the neural plate undergo rapid proliferation and morphologic changes to form the neural tube.

J.W. Yoon (✉)
Department of Pediatrics, Developmental Biology Program, Children's Memorial Research Center, Northwestern University, Feinberg School of Medicine and Children's Memorial Hospital, 2300 Children's Plaza, Box 204, Chicago, IL 60614, USA
e-mail: j-yoon2@northwestern.edu

J. Xie (ed.), *Hedgehog Signaling Activation in Human Cancer and its Clinical Implications*, DOI 10.1007/978-1-4419-8435-7_8, © Springer Science+Business Media, LLC 2011

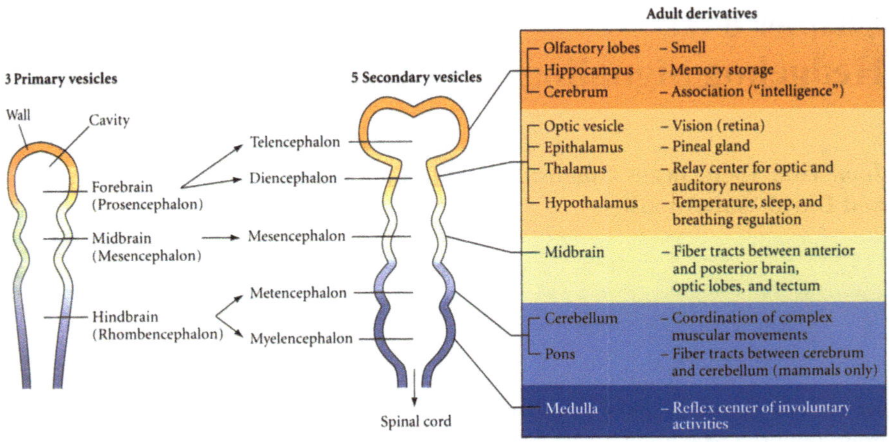


**Fig. 8.1** Early human brain development. The three primary brain vesicles, five secondary brain vesicles, and their adult brain derivatives are shown. Reproduced and adapted from Developmental Biology, eighth edition with permission from Sinauer Associates, Inc.

Further proliferation of cells in the anterior region of the neural tube causes expansion and creates three primordial brain vesicles called the prosencephalic, mesencephalic, and rhombencephalic vesicles. These primary vesicles subsequently develop into secondary vesicles: the telencephalon and diencephalon (forebrain), mesencephalon (midbrain), and metencephalon and myelencephalon (hindbrain) (Fig. 8.1).

At a molecular genetic level, the events described above require complex interactions between key signaling pathways, including the SHH, Wingless (WNT), bone morphogenetic protein (BMP), fibroblast growth factor (FGF), and transforming growth factor beta (TGF-β) pathways and their target genes. Activation of these signaling pathways has been associated with fundamental events during CNS patterning. These events include (1) establishment of polarity within the developing nervous system, (2) rapid expansion of cells in the region of the developing brain, (3) establishment of inter-brain boundaries, and (4) establishment of regional specificity. Remarkably, SHH signal transduction appears to be critically involved in each of these developmental events in a spatial- and temporal-dependent manner.

## *Dorso-Ventral Polarity and SHH Signaling*

During early mammalian development, CNS patterning requires the establishment of axes in the neural tube. shh signaling contributes to establishing the dorso-ventral axis as the neural tube fuses at embryonic (E) day 8.5 in the mouse. shh is secreted by the notochord which lies immediately ventral to the neural tube and is also expressed by the ventral floor plate in the developing neural tube [1, 2]. Expression of *shh* in the ventral region of the developing neural tube establishes a gradient of shh within the neural tube, highest ventrally and lowest dorsally (Fig. 8.2).

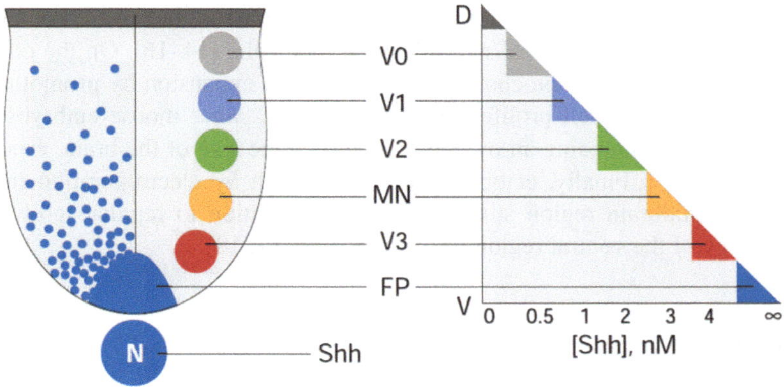


**Fig. 8.2** A shh gradient regulates transcription factors that establish ventral specification in the developing CNS. The shh gradient (shown on the *right*) induces ventral floor plate and specifies five ventral cell types (shown on the *left*). *D* dorsal, *FP* floor plate, *MN* motor neuron, *N* notochord, *V* ventral, *V0–V3* ventral inter neurons 0–3. Reproduced and adapted from EMBO reports 4(8):761–765 (2003) with permission from the Nature Publishing Group

This gradient differentially regulates the expression of transcription factors that specify polarity and ultimately cell fates in the developing CNS, including *pax*6, the homeobox gene *nkx*2.2, and the floor plate marker *hnf3*-β [3, 4]. The importance of shh to early CNS development has been demonstrated in $shh^{-/-}$ mice, in which the notochord degenerates and the ventral floor plate and motor neurons fail to form. The ability of shh to induce differentiation of ventral cell types in the nervous system has been demonstrated by aberrantly expressing *shh* in the dorsal CNS. Aberrant expression of *shh* in the dorsal CNS activates dorsal expression of *hnf3-β* and causes aberrant dorso-ventral patterning [5–7].

Although shh induces differentiation of ventral neural precursor cells [6, 8, 9], further differentiation into motor neurons, interneurons, glial cells (oligodendrocytes), and other cell types in the CNS appears to require complex and incompletely understood interactions between shh signaling and the wnt, bmp, fgf, and tgf-β signaling pathways [7]. For example, shh regulates expression of *fgf8* receptors [10–13], and fgf8 together with shh induce dopaminergic neurons in the ventral region [10].

## *Rapid Expansion of Cells in the Region of the Developing Brain*

Early mammalian brain development is characterized morphologically by rapid growth and expansion of the neural tube, which results in the formation of the brain vesicles (Fig. 8.1). The three primary vesicles have formed by week 4 of human development. This morphologic change results from both increased proliferation and reduced apoptosis of neuroepithelial cells. Several experimental approaches have been used to show roles for shh in regulating both proliferation and survival of cells that contribute to brain development. First, studies placing a transplanted notochord, the source of shh, near the neural tube demonstrate increased proliferation, differentiation, and survival of nearby neural tube cells [14–16]. On the contrary, surgical removal of the notochord disrupts midbrain expansion by promoting cell death and inhibiting cell proliferation [17]. Second, *shh*$^{-/-}$ mouse embryos show multiple defects, including an overall reduction in the size of the brain, especially the forebrain [5]. Finally, ectopic expression of shh by electroporation into the developing midbrain region stimulates cell proliferation to regulate growth and morphology of the ventral region of the midbrain [17, 18].

## Brain Boundaries

Studies in vertebrate embryos suggest that shh signaling specifically regulates genes at the midbrain–hindbrain boundary. Evidence suggests that shh functions to maintain this distinct boundary once it has formed rather than establishing the boundary [19]. Indeed, ectopic expression of *shh* by microinjection into one blastomere at the 2–4 cell stage expands the domain of expression of the shh target gene *sal* at the midbrain–hindbrain boundary [20, 21]. Disruption of shh signaling at the boundary by a mutant form of patched (ptc1) that cannot bind shh causes the midbrain–hindbrain boundary to become broader and indistinct, with midbrain and hindbrain cells inter-mixing across the expanded border [22]. Dorso-ventral cell fates are also affected in this model. As seen in other regions and periods during CNS development, blocking endogenous shh activity in the midbrain transforms cell fates from ventral to dorsal and correlates with the movement of dorsal cells into the ventral midbrain [19].

## Regional Specification of the Developing Brain

*shh* is expressed along the entire anterior–posterior axis of the developing neural tube. It is believed that the establishment of regional specificity along the anterior–posterior axis of the developing CNS is achieved by differential expression of *shh* together with other key signaling pathways, such as the *wnt*, *bmp*, *fgf*, and *tgf-β* pathways in a regional specific manner. Roles for shh in regional specification is reviewed in the following sections.

### Forebrain

shh signaling appears to regulate the size, ventral cell fate specification, and ventral patterning of the developing telencephalon. Targeted loss of shh in *shh*$^{-/-}$ mice results in multiple morphologic defects in the forebrain, including a reduction in size, fused telencephalic vesicles, and fused optic vesicles [5]. On the contrary, ectopic expression of *shh* by retroviral injection in early mouse embryos (E9.0) enhances proliferation and causes a substantial expansion in the size of the telencephalon [23].

If neural explants from the telencephalic region are incubated with *shh*-expressing cells, the neural plate in the prospective forebrain region differentiates into motor neurons which are normally observed in the ventral CNS [8]. shh treatment at E10.5 also represses expression of dorsal telencephalic markers such as *emx1* and *tbr-1* [24].

Ventralization of the telencephalon is also tightly regulated at the level of the gli family transcription factors. For example, shh inhibits expression of the repressor form of *gli3* (*gli3-R*) in the ventral telencephalon, presumably to promote active shh signaling and ventralization as well as to inhibit dorsalization [25]. Indeed, "extra-toes" mice carry a mutation in *gli3*, and E12.5 mutant embryos lack expression of dorsal marker *bmp* genes in the telencephalon, even though shh expression is unaltered [25]. Also, ventral marker genes, such as *dlx*2 and *gsh*2 are expressed in the dorsal telencephalon in "extra-toes" mice [23].

## Midbrain and Hindbrain

shh signaling appears to regulate cell proliferation, apoptosis, and cell fate specification in the developing midbrain and hindbrain. Loss of *shh* expression causes decreased cell proliferation and increased apoptosis in the midbrain region. Similar results are obtained when the shh signal is reduced by surgical separation of the notochord from the floor plate, injection of cyclopamine into the lumen of the midbrain, or in the setting of $shh^{-/-}$ mice [17, 26]. These cellular changes collectively alter early expansion of the brain, causing a reduction in the size of the midbrain and ultimate collapse of the brain vesicles. Of interest, while the growth of the developing midbrain in E8.5 $shh^{-/-}$ mice is significantly reduced, the sizes of the diencephalon and hindbrain are unaffected [26].

shh signaling also specifies dopaminergic neuron cell fate in the developing midbrain [27]. Recent evidence using cultured midbrain suggests that higher level shh signaling inhibits cell proliferation and dopaminergic neuron specification, pointing out that shh signaling functions in a concentration-dependent manner to establish cellular and morphologic phenotypes [28]. In addition, in the hindbrain, shh signaling defines the ventral region and promotes hindbrain growth.

shh signaling in the midbrain and hindbrain regions is mediated through gli family transcription factors [29, 30]. Ectopic expression of *gli1* in the midbrain and hindbrain regions activates the ventral markers *ptc1* and *hnf3-β* [31]. Conditional *gli2* knockout mouse embryos at E9.0 and E11.5 demonstrate that the activator form of gli2 (gli2-A) promotes growth of the ventral midbrain and hindbrain regions, whereas the repressor form of gli3 (gli3-R) is continuously required for the overall growth of the dorsal midbrain and hindbrain, presumably by inhibiting shh signaling in these regions [27–29]. Detailed analysis in the developing hindbrain reveals that *gli*2 –/– mouse embryos show a more severe ventral defect in the hindbrain than in the spinal cord, since gli3 can compensate for the loss of gli2 in the spinal cord but not in the hindbrain [32].

Interactions between the shh signaling pathway and the bmp pathway appear to specify the ventral region of the hindbrain. Aberrant expression of *bmp-7* in the floor plate region inhibits *shh* expression and interrupts dorso-ventral patterning of the hindbrain, suggesting that bmp-7 regulates shh signaling in this domain [33].

## Cerebellum

The cerebellum originates from the metencephalon. It is the largest part of the hindbrain and is connected to the other parts of the brain through projection fibers. Through these fibers, the cerebellum receives input from sensory systems and integrates the signals to coordinate and accurately time movement. The cerebellum is composed of many different types of neurons, including Purkinje neurons, granule neurons, Bergmann glia, astrocytes, interneurons, and neurons of the deep nuclei. Cerebellar development has been reviewed previously in detail [34–37].

shh signaling plays an integral role in the developmental biology of the Purkinje neurons, Bergmann glia, and granule neurons in the cerebellum (Fig. 8.3).

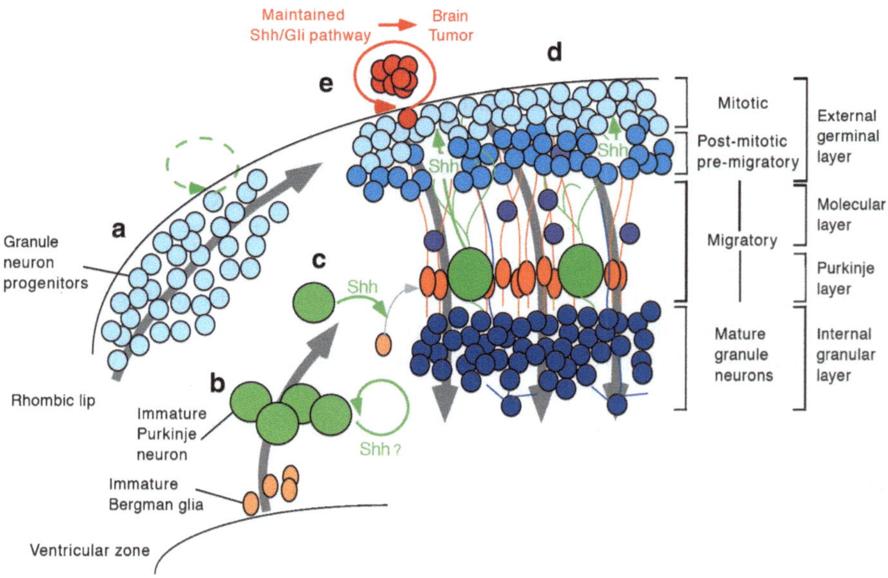


**Fig. 8.3** The role of shh in cerebellar development. (**a**) Granule neuronal precursors (CGPs in text; *light blue*) migrate tangentially from the rhombic lip toward the EGL. During migration, the shh pathway may be transiently active in an autocrine manner (*dashed green arrow*). (**b**) Purkinje neurons (*green*) and Bergmann glia (*pale orange*) migrate from the ventricular zone toward the Purkinje layer. Purkinje neurons may initially use the shh pathway in an autocrine manner (*green arrow*). (**c**) shh from Purkinje neurons (*green arrows*) induces Bergmann glia maturation (*bold orange*). (**d**) In the later stage EGL, granule neuronal precursors (CGPs; *light blue*) proliferate in the outer zone and mature glia send extensions (*orange lines*) toward the inner EGL. Post-mitotic granule cells (*bold blue*) then migrate (*purple cells*) on glial fibers to form the internal granular layer (*dark blue cells*). (**e**) Constitutive shh signaling in EGL cells or failure to induce their differentiation may contribute to the development of medulloblastoma (*red arrows and cells*). Reproduced and adapted from Development 126, 3089–3100 (1999) with permission from the Company of Biologists, Ltd

During early embryonic development, Purkinje cells are derived from progenitor cells located in the ventricular zone of the neural tube, and granule neurons arise from the thickened alar plate of the embryonic rhombencephalon called the rhombic lip. Both cell types migrate into the region of the developing cerebellum, where cerebellar granule precursors (CGPs), also called granule cell precursors, granule neuron progenitors, or granule cell neuron precursors, form the external granular (or germinal) layer (EGL). Once the cells arrive in the EGL, dramatic growth of the neonatal mouse cerebellum ensues, increasing over 1,000-fold in volume. This period of growth is driven by the rapid proliferation of CGPs [38]. As a result of this rapid proliferation, granule neurons become the most abundant neurons in the cerebellum. In fact, more than 50% of the neurons in the entire mouse brain are comprised of granule neurons [39]. Amplified CGPs located in the EGL eventually exit mitosis, differentiate, and migrate internally to form the internal granular layer [39]. Bergmann glia interact with post-mitotic CGPs during their migration.

shh signaling regulates the proliferation and differentiation of CGPs and induces maturation of Bergmann glial cells. In situ hybridization studies for *shh*, *ptc1*, and *gli1* demonstrate that the shh ligand is produced by the Purkinje cells, and the *ptc1* receptor and *gli1* transcription factor are expressed in CGPs in the EGL. This expression pattern suggests paracrine signaling from Purkinje cells to CGPs in the developing cerebellum. Indeed, blocking shh activity with neutralizing anti-shh antibody disrupts CGP proliferation in the EGL [39–41]. Also, CGPs treated with shh in vitro remain undifferentiated while untreated CGPs undergo spontaneous differentiation, suggesting a role for shh in preventing their differentiation [39].

shh drives CGP proliferation by activating $G_1$-cyclins and N-myc [42, 43]. *atoh1* (also called *math1*) is a basic helix-loop-helix transcription factor, which is highly expressed in CGPs [44] that directly activates expression of the shh mediator *gli2*, thereby significantly promoting shh signaling [45]. Conditional knockout of *atoh1* in the post-natal mouse cerebellum reduces the size of the EGL, since CGPs cannot respond to the shh signal. In addition, *gli2* expression is significantly inhibited even in the presence of constitutively activated shh signaling in *atoh1* null conditional mutants, supporting the concept that atoh1 is a critical regulator of *gli2* and therefore shh signaling in the cerebellum [45].

Interactions of shh with several other proteins and signaling pathways are required for CGPs to exit from the proliferative cycles and begin differentiation. The extracellular matrix protein vitronectin is continuously expressed in the developing cerebellum very close to the *shh*-expressing cell population. Physical interaction of vitronectin with shh inhibits shh-mediated proliferation of CGPs and promotes their differentiation [46]. bmp-2 antagonizes shh-mediated CGP proliferation through smad5 [47] and through tieg-1, which inhibits N-myc [48]. The BTB/POZ domain-containing protein REN also antagonizes shh by negatively regulating gli1 and gli2 activity in CGPs, thereby promoting growth arrest, enhancing differentiation, and activating apoptosis of CGPs [49]. Finally, the fgf signaling pathway suppresses shh-induced proliferation of CGPs by down-regulating expression of *gli1*, *N-myc*, and *cyclin D1* [39, 50] and promotes differentiation of CGPs in the presence shh, suggesting an inhibitory role on shh during CGP differentiation [50].

It remains uncertain whether shh plays a direct role in the migration of CGPs. shh increases migration of granule cell explants [40], whereas blocking shh activity with neutralizing anti-shh antibody inhibits the migration of the cells. However, two complementary models, which prevent shh signaling by Purkinje cells show significantly compromised expansion of the CGPs and post-mitotic granule cells, but migration is not affected [38].

## Aberrant Activation of SHH Signaling in Pediatric Brain Tumors

Collectively, CNS tumors represent the most common solid tumors among children and are a leading cause of pediatric cancer-related morbidity and mortality [51–53]. Pediatric brain tumors are a heterogeneous group of malignancies that differ in scope, behavior, and biology compared to adult CNS tumors. The majority of adult brain tumors are high-grade gliomas, meningiomas, and metastases from extra-CNS solid tumors [54–56]. Metastases to the brain and meningiomas are rarely seen in pediatrics, and high-grade gliomas only represent 10–15% of all childhood brain tumors [53]. On the contrary, astrocytomas and medulloblastomas are the most common CNS tumors in children, accounting for approximately 60 and 20% of pediatric brain tumors, respectively [54–56].

Despite multimodal therapies for childhood brain tumors, including surgical resection, cytotoxic chemotherapy, and radiation therapy, there remains a significant group of patients who succumb to their disease. In addition, many children who survive sustain significant late effects related to their original tumor and therapies, including neurocognitive deficits, endocrine dysfunction, ototoxicity, and the development of secondary malignancies [57–60]. A significant amount of research is underway evaluating the molecular, biologic, and cytogenetic characteristics of pediatric CNS tumors. The hope is that future targeted therapies tailored to the specific aberrant molecular pathways within a tumor will not only improve survival, but also may help to minimize some of the late effects. The childhood brain tumor that has advanced the furthest along this research trajectory is medulloblastoma, in large part based on research directed at the SHH signal transduction pathway.

### *Genetic Alteration of Components of the SHH Pathway in Pediatric Brain Tumors*

Based on the fundamental roles that the SHH pathway plays in cell proliferation and cell fate specification during CNS development, it is not surprising that constitutive activation of the pathway is associated with brain tumors. Constitutive pathway activation has been described in association with continuous somatic expression of the SHH ligand in a variety of cancers outside the CNS. In pediatric brain tumors,

**Table 8.1** Dysregulation of SHH signaling in childhood brain tumors

| Tumor | Gene | Type of abnormality | References |
|---|---|---|---|
| Medulloblastoma | *PTCH1* | Loss of function mutation | [68, 96, 112, 142] |
| | *SMO* | Activating mutation | [141] |
| | *SuFu* | Loss of function mutation | [101, 143] |
| | *GLI2* | Amplification | [102] |
| Ependymoma | *IHH* | Overexpression | [121] |
| | *GLI2* | Overexpression | [120, 121] |
| | *GLI1* | Overexpression | [120] |
| Pilocytic astrocytoma | *PTCH1* | Overexpression | [124] |
| | *GLI1* | Overexpression | [124] |
| Craniopharyngioma | *PTCH1* | Loss of function mutation | [119] |

constitutive pathway activation is more typically independent of HH ligands and is the result of mutations in downstream components of the HH pathway. These genetic alterations may be inherited constitutional mutations associated with cancer predisposition syndromes, such as basal cell nevus syndrome, or may be somatic. The most common genetic alterations in HH pathway components in childhood brain tumors are summarized in Table 8.1.

## *Medulloblastoma: Clinical Aspects*

Medulloblastoma is the most common malignant brain tumor in childhood [54, 61]. It is a highly malignant embryonal tumor that is believed to arise from CGPs in the cerebellum. It is considered a central primitive neuroectodermal tumor (PNET) based on the histologic appearance of the cells mimicking embryonic neuroectoderm. The name "medulloblastoma" implies that the primary tumor is located within the cerebellum. Central PNET can occur in other locations within the CNS, including the supratentorium, brainstem, and spine; however, in these locations the tumor is not referred to as medulloblastoma and only represents approximately 2–3% of all pediatric brain tumors [62, 63]. These non-cerebellar central PNETs are thought to be biologically distinct from medulloblastoma based upon genetic and biologic studies as well as worse clinical outcomes compared to medulloblastoma [62, 64]. The cell of origin for these CNS non-cerebellar PNETs is not yet known.

Medulloblastoma is more commonly seen in males than females. The peak age at diagnosis is typically between 5 and 7 years old, however, it can be seen from birth to young adulthood [54, 65]. In fact, age at the time of diagnosis is one of the few important clinical prognostic factors known in medulloblastoma. The etiology of the majority of medulloblastomas is unknown, however, there are a few rare genetic disorders that predispose some patients to medulloblastoma, including basal cell nevus syndrome, Li–Fraumeni syndrome, ataxia telangiectasia, and Turcot's syndrome [66–69].

Children with medulloblastoma typically present with a relative short history of symptoms related to obstructive hydrocephalus, including early morning headaches, emesis, and papilledema. They can also present with signs of cerebellar dysfunction, such as truncal ataxia and unsteady gait [54]. Finally, some patients may present with symptoms related to metastatic foci of disease in other parts of the brain and spine, such as seizures and signs of spinal cord compression.

Medulloblastoma has an inherent tendency to spread throughout the CNS. Therefore, staging at diagnosis is essential and includes a complete brain and spine MRI as well as evaluation of lumbar cerebro-spinal fluid (CSF) cytology to evaluate for metastasis. Typically, the spine MRI and lumbar CSF collection is performed 10–14 days postoperatively in an effort to avoid false positives secondary to surgical debris. The modified Chang staging system is used to stage these patients at diagnosis, where M0 patients have no signs of metastasis, M1 patients only have tumor cells on lumbar cytology, M2 patients have macroscopic spread of tumor to distant parts of the brain, M3 patients have macroscopic metastases to the spine, and M4 patients have spread outside the CNS, which is exceedingly rare in the modern era [70].

Despite a significant understanding of medulloblastoma biology, biologic characteristics have not yet been incorporated into up-front treatment strategies and prognostication. Currently, the major prognostic factors utilized to stratify patients with medulloblastoma include age at diagnosis, extent of tumor resection, absence or presence of CNS dissemination/metastases, and tumor histology. Disease characteristics that render a patient at high risk include residual disease greater than 1.5 $cm^2$ after primary surgery, metastasis to distant parts of the brain or spine, and anaplastic histology [61]. Patients with one or more of these characteristics are typically treated with an intensified regimen that includes both craniospinal irradiation and chemotherapy. In addition, patients less than 3 years old at the time of diagnosis are also considered at high risk due to their worse clinical outcomes and are treated with unique therapy approaches. These strategies often delay or avoid irradiation, since this group of patients is highly susceptible to the deleterious effects of radiation therapy.

Patients who are greater than 3 years of age with high risk disease are currently treated with a combination of full dose craniospinal irradiation (36 Gy to the neuraxis and boost to the posterior fossa up to 54 Gy) and chemotherapy. These patients have a 5-year progression-free survival ranging from 40 to 70% [54, 61, 71, 72]. Those patients who are less than 3 years old at diagnosis are often treated with a combination of high-dose chemotherapy followed by autologous hematopoietic cell rescue with or without adjuvant radiation therapy. These patients have 5-year progression-free survival ranging from 30 to 60% in published prospective series, and many of these patients have avoided radiation therapy completely [73–75]. Superior survival has been published using the German HIT protocol for this group of young patients. This protocol uses a chemotherapy alone regimen that includes intrathecal methotrexate. However, over half of the patients reported in this series had desmoplastic histology which is believed to confer a better prognosis in younger patients [75]. Also, there is concern that the intrathecal methotrexate contributed to neurocognitive sequelae seen in these patients. For these reasons, this approach has not been universally adopted.

The current approaches to patients with standard risk of medulloblastoma, that is, age greater than 3 years at diagnosis, less than 1.5 $cm^2$ residual disease postoperatively, no signs of metastasis (as seen on MRI of the brain and spine and lumbar fluid cytology), and a classic histology, include a dose reduction of craniospinal irradiation (23.4 Gy to the neuraxis and a boost to the posterior fossa up to 54 Gy) and adjuvant chemotherapy. This strategy maintains good outcomes, and preliminary data suggest it may reduce neurocognitive sequelae [76]. These patients have a 5-year progression-free survival of approximately 85% [71].

Desmoplastic medulloblastoma is a less common histologic variant of medulloblastoma, most commonly seen in patients with basal cell nevus syndrome. The association between desmoplastic histology and basal cell nevus syndrome suggests that abnormalities in SHH signaling contribute to the development of this form of medulloblastoma. This histologic variant accounts for approximately 10–20% of sporadic medulloblastomas as well [77]. Histologically, desmoplastic medulloblastomas are characterized by pale nodular areas surrounded by densely packed cells and a significant reticulin network between these areas. The nodular areas are made up of more mature tumor cells with fewer mitoses and more abundant cytoplasm. The densely packed cells, surrounding the nodules appear more typical of classic medulloblastoma [54, 75, 78]. Interestingly, this subtype shows superior survival compared to other subtypes in patients with or without basal cell nevus syndrome [75, 77, 79].

Despite all of the aforementioned strategies, approximately 30% of patients with medulloblastoma will relapse, and unfortunately most of these patients will succumb to their disease [80, 81]. If they have not yet received radiation therapy, as is the case in some very young children, a small percentage may be salvaged using radiation therapy. Unfortunately, based on the international experience, most patients who have already received craniospinal irradiation do not appear to be curable once they recur; however, there are some data to suggest that the use of high-dose chemotherapy with autologous hematopoietic cell rescue may be of value in a select group of patients [81, 82]. Ongoing phase I and phase II trials are attempting to utilize molecularly targeted agents in an effort to improve survival at the time of recurrence.

## *Medulloblastoma: Biologic Aspects*

Studies continue to more completely understand the molecular biology and cytogenetics of medulloblastoma. The most common cytogenetic abnormality in medulloblastoma is isochromosome 17q, which is present in up to 40% of cases [83, 84]. Numerous reports have also identified gains of chromosomes 4, 6, 7, 8, and 18 as well as losses of chromosomes 1, 2, 6, 8, 9, 10, 11, and 16 [84–86]. Approximately 5% of medulloblastomas also harbor a high level of amplification of the *N-MYC* oncogene [84].

Many groups have suggested using these biologic and molecular aberrations as a means of risk stratification. Pfister et al. developed a five-tier system based on screening 80 medulloblastoma samples by array-based comparative genomic hybridization and an independent validation of 260 samples via fluorescence in situ hybridization. This hierarchical medulloblastoma molecular staging system from worst to best outcome includes (1) *c-MYC/N-MYC* amplification, (2) 6q gain, (3) 17q gain, (4) 6q and 17q balanced translocation, and (5) 6q deletion [87]. These data show quite convincingly that molecular and cytogenetic abnormalities are powerful tools for prognostication, and one day may be more useful than the traditional risk categorization based solely on clinical characteristics.

Another group evaluated gene-expression profiles of 46 human medulloblastoma samples. Unsupervised analysis divided the samples into five distinct groups (A–E) enriched for specific genetic alterations that were later confirmed by gene sequence analyses and fluorescence in situ hybridization [88]. Some of the specific abnormalities include WNT pathway mutations and chromosome 6 deletions in subgroup B and SHH pathway mutations in subgroup D [88]. This type of analysis and separation of tumors by genetic alterations may help stratify patients for the most appropriate targeted therapies.

To date, three main molecular signaling pathways have been implicated in medulloblastoma development, including the WNT pathway, the Notch pathway, and the SHH pathway. Better understanding of the WNT pathway and medulloblastoma development has come from a rare disorder called Turcot's syndrome. Turcot's syndrome, also known as glioma-colonic polyposis syndrome, is a genetic disorder characterized by colonic polyposis and an increased risk of developing colon cancer and malignant neuroepithelial CNS tumors. Most commonly, these patients develop glioblastoma multiforme and medulloblastoma [69, 89, 90]. One of the main mutations identified in this syndrome is a defect in the adenomatous polyposis coli (APC) gene, a tumor suppressor, which is a component of the WNT pathway and helps to coordinate the proliferation and ultimate fate of neural progenitor cells. Differing phenotypes may result from mutations at unique regions within the APC gene. Activation of the WNT pathway inhibits phosphorylation of beta-catenin, allowing its translocation into the nucleus [91]. This in turn increases the expression of a variety of genes that ultimately lead to cell proliferation, inhibition of apoptosis, and differentiation within the CNS. In addition, mutations of beta-catenin and other WNT pathway genes have been described in approximately 10–20% of sporadic medulloblastomas [84, 91]. The accumulation of beta-catenin within the nucleus, suggesting WNT pathway activation, appears to predict a favorable outcome in medulloblastoma [84, 92].

The Notch signaling pathway is vital to a variety of developmental processes, including hematopoiesis, somitogenesis, vasculogenesis, and neurogenesis [93]. This pathway has been implicated in the development of a variety of malignancies, including medulloblastoma. Notch signaling is activated by four transmembrane receptors, including Notch 1–4. Notch 1 is thought to be essential to the normal development of the cerebellum, whereas Notch 2 is implicated in medulloblastoma development [54, 93]. Once the receptors bind their extracellular ligands, proteolytic cleavage leads to the release of the intracellular domain of the receptors into the intracellular compartment and eventual translocation into the nucleus. A variety of downstream targets are then activated, such as *cyclin D1* and apoptosis associated genes [54, 94]. If unregulated, this activation is thought to drive a variety of processes, including neural "stem" cell maintenance, gliogenesis, and oncogenesis [93].

## Medulloblastoma: Dysregulation of SHH Signaling

Evidence linking the SHH signaling pathway and medulloblastoma originated from recognizing that patients with basal cell nevus syndrome are at increased risk for medulloblastoma. Basal cell nevus syndrome is a rare autosomal dominant genetic disorder affecting 1 in 60,000 individuals [77]. It is characterized by skeletal anomalies (frontoparietal bossing, rib and vertebrae abnormalities, and dural calcifications), large body habitus, soft tissue fibromas, radiation sensitivity (increased risk of developing radiation-induced tumors such as meningioma, ependymoma, fibrosarcoma, and basal cell carcinoma), and a high incidence of basal cell carcinoma, rhabdomyosarcoma, and medulloblastoma [67, 77, 95]. Approximately 3–5% of patients with basal cell nevus syndrome will develop medulloblastoma, typically the desmoplastic variant.

Basal cell nevus syndrome is caused by inherited inactivating germ-line mutations in one patched (*PTCH1*) allele [91, 95]. During normal development of CGPs in the cerebellum, PTCH1 maintains the SHH pathway in an inactive state except when it binds SHH. On the contrary, mutant forms of PTCH1 maintain the pathway in a constitutively activated state, even in the absence of ligand. Constitutive SHH pathway activation appears to account for the birth defects and cancer predisposition in patients with basal cell nevus syndrome. Cancers presumably develop in the setting of basal cell nevus syndrome if a somatic inactivating mutation occurs in the remaining wild type *PTCH1* allele in limited cell types, including CGPs.

Loss-of-function somatic mutations in *PTCH1* have been subsequently described in approximately 10–15% of sporadic medulloblastomas [68, 96]. It is now believed that SHH pathway activation occurs in 20–30% of all medulloblastomas, largely accounted for by inactivating *PTCH1* mutations. A variety of loss of function mutations in *PTCH1* have been described in the setting of medulloblastoma, including frame shift mutations, small deletions, duplication insertions, and splice site mutations. Interestingly, the sporadic medulloblastomas that exhibit abnormal signaling via the SHH are not all desmoplastic. Constitutive activation of SHH signaling in cerebellar CGPs is believed to contribute directly to the genesis of human medulloblastoma, based on the fact that $ptc1^{+/-}$ mice develop medulloblastoma [97].

Mutations in components of the SHH signaling pathway that are downstream of *PTCH1* have been more rarely described. Although activating mutations in *SMO* have been widely described in basal cell carcinomas, they appear to be rare in medulloblastoma and only have been reported in recent years [98, 99]. Suppressor of Fused (SuFu) normally binds to GLI1 and inhibits GLI1-mediated transcriptional activation by exporting GLI1 from the nucleus to the cytoplasm [100]. Several somatic mutations in *SuFu* have been described in medulloblastomas, including frame shift and exon skipping mutations [101]. The mutant forms of the SuFu protein lack the carboxy terminal domain and are unable to bind GLI1. Therefore, the SHH pathway remains in an active state since mutant SuFu cannot sequester GLI1 in the cytoplasm. A single case of medulloblastoma with *GLI2* gene amplification has been reported in a patient with the Li–Fraumeni familial cancer syndrome [102]. The significance and the biological role of *GLI2* amplification have not been studied in medulloblastoma.

## *Other Pathways Affecting SHH Signaling in Medulloblastoma*

Although mutations in the SHH signaling genes *PTCH1*, *SMO*, and *SuFu* are believed to directly contribute to the genesis of medulloblastoma, such mutations are observed in only a subset of the tumors, suggesting that there are other mechanisms and gene pathways that can cause or play an important role in the biology of medulloblastoma either independent of SHH signaling or by dysregulating SHH signaling. Indeed, noncanonical activation of GLI family transcription factors, which mediate HH signaling, has been described in the setting of cancer. For example, TGF-β activates the expression of *GLI1* and *GLI2* through SMAD3, independent of SHH signaling in human pancreatic cancer cells [103]. *GLI1* also appears to be activated independent of the canonical HH pathway by the EWS–FLI11 oncoprotein in Ewing sarcoma [104]. A number of studies using mouse models and human specimens demonstrate interactions between SHH signaling and other pathways and genes, both in CGPs and in medulloblastoma. Recent progress in this field is summarized in Table 8.2.

**Table 8.2** Interactions between SHH signaling and other pathways in medulloblastoma

| Gene | Effect on pathway | References |
|---|---|---|
| *Genes that enhance SHH signaling in medulloblastoma* | | |
| *atoh1* | Increases *gli2* expression | [45] |
| *c-myc* | Enhances shh tumorigenicity | [107] |
| *yap1* | Increases gli2 expression | [108] |
| *Genes that inhibit SHH signaling in medulloblastoma* | | |
| *bmp-2,4* | Degrades atoh1 | [105] |
| *bFGF* | Decreases *gli1*, *N-myc*, and *cyclin D1* expression | [50] |
| *pacap* | Inhibits gli1 activity | [109] |
| *p53* | Inhibits gli1 activity | [114] |
| *REN* | Decreases *gli1* expression, nuclear localization of gli1, and gli2 activity | [49] |
| *Targets of SHH signaling in medulloblastoma* | | |
| *igf2* | Increases expression | [116] |
| *insm1* | Increases expression | [118] |
| *irs1* | Stabilizes protein | [117] |
| *nhih* | Increases expression | [118] |

Some proteins and pathways appear to enhance HH pathway activity in CGPs and medulloblastoma. *ATOH1* is highly expressed in CGPs and in a subset of human medulloblastomas. Conditional deletion of *atoh1* in mice downregulates *gli2* expression. In fact, atoh1 directly activates *gli2* by binding a *gli2* intron, therefore, significantly promoting the activity of the shh signaling pathway [45]. The role of atoh1 in the genesis of medulloblastoma has been tested using a mouse medulloblastoma model, carrying an activating mutation in *smo*. In this background, loss of expression of *atoh1*, using an *atoh1* null conditional mutant, significantly inhibits tumor formation, suggesting that atoh1 and activation of shh signaling are required for medulloblastoma formation [45]. bmp-2 and bmp-4 down-regulate expression of *gli1* in medulloblastoma by degrading the atoh1 protein [105].

*MYC* family genes are amplified and overexpressed in 5–10% of medulloblastomas [106]. Overexpression of *c-myc* alone does not appear to cause medulloblastomas in mice. However, *c-myc* greatly enhances the tumorigenicity of shh signaling in CGPs, suggesting cooperation of c-myc with hh signaling in shh-mediated medulloblastoma formation [107]. The molecular mechanism of this cooperation is unknown.

Finally, the transcriptional coactivator yes-associated protein (*yap1*) that is a key factor in Hippo signaling pathway is amplified in some medulloblastomas [108]. It activates transcription of target genes by interacting with a tea domain family transcription factor (tead). The yap1–tead1 complex drives *gli2* transcription by directly interacting with the CATTC consensus sequence in the *gli2* promoter and thus promotes constitutive shh pathway activation in medulloblastoma.

On the contrary, other proteins and pathways appear to inhibit SHH signaling in CGPs and in medulloblastoma. bFGF dramatically downregulates the expression of *gli1*, *N-myc*, and *cyclin D1*, and thereby suppresses shh-induced proliferation of CGPs [39, 50]. The fgf pathway also inhibits *gli1l* expression and proliferation of medulloblastoma cells derived from *ptc1*$^{+/-}$ mice, suggesting an inhibitory role in the genesis of shh-induced medulloblastoma [50].

Pituitary adenylyl cyclase-activating peptide (pacap) activates PKA and functions as a powerful inhibitor of medulloblastoma formation. In fact, double heterozygote *ptc1*$^{+/-}$ *pacap*$^{+/-}$ mice demonstrate a 2.5-fold increase in medulloblastoma incidence [109]. pacap inhibits gli1 in medulloblastoma cells by activating PKA. The tumor suppressor *REN* is deleted in 39% of sporadic human medulloblastomas and inhibits the growth and the tumorigenicity of medulloblastomas [49, 110]. REN inhibits *gli1* expression and gli1 activity in medulloblastoma by blocking Dyrk1-mediated nuclear localization of gli1. REN also impairs gli2-dependent gene transcription.

Finally, inherited germ-line *p53* mutations are associated with the development of medulloblastoma in some patients with Li–Fraumeni syndrome [111]. Normally, only a small subset of *ptc1*$^{+/-}$ mice develop medulloblastoma [112]. However, *ptc1* $^{+/-}$ *p53*$^{-/-}$ mice develop medulloblastoma significantly more frequently (>95%) and at an earlier age [113]. This finding strongly suggests that loss of p53 and constitutive activation of the shh signaling pathway interact functionally. The mechanism of this interaction in medulloblastoma is not known. However, recent studies suggest a feedback loop between p53 and gli1. p53 inhibits the transcriptional activity, nuclear localization, and level of expression of gli1, while gli1 inhibits the activity p53 [114]. Thus, loss of p53 may enhance GLI1 activity and thereby medulloblastoma formation. In addition, increased expression of *p53* has been observed following the transfection of *GLI1* into rat kidney epithelial cells (RK3E cells) and in the subset of medulloblastomas with HH pathway activation, suggesting that GLI1 may regulate *p53* expression [115].

shh signaling appears to regulate several genes in the "insulin regulatory pathway" in both CGPs and medulloblastoma. Insulin-like growth factor 2 (igf2) promotes cell proliferation in developing embryos. Normally, *igf2* is expressed in the meninges and at lower level in CGPs. However, *igf2* is highly expressed in medulloblastomas that develop in *ptc1*$^{+/-}$ mice. In addition, *igf2* expression in CGPs is directly activated by shh in vitro and is inhibited by cyclopamine treatment in medulloblastoma cell lines [116]. Loss of *igf2* expression decreases medulloblastoma formation in *ptc1*$^{+/-}$ mice, suggesting a role as a vital downstream target of the shh signaling pathway in medulloblastoma. Another insulin-related gene that is necessary for proliferation of CGPs and aberrantly activated in medulloblastoma is the insulin receptor substrate 1 (*irs1*) [117]. shh signaling stabilizes the irs1 protein by inhibiting the mTOR pathway that directs the degradation of irs1. Neural basic helix-loop-helix 1 (*nhlh1*) and insulinoma-associated 1/IA1 (*insm1*) are also highly expressed in rapidly expanding CGPs and medulloblastomas [118]. Both *nhih1* and *insm1* are activated transcriptionally by shh signaling in cultured CGPs, and activation of *nhih1* is directly mediated by gli1. Understanding the interactions between genes/proteins in the HH signaling pathway with those of other pathways that modulate HH signaling, as well as identifying vital effects of HH signaling in CGPs and medulloblastomas, will be important when developing targeted therapy and making informed decisions about which agents to test in combination.

## *Dysregulation of SHH Components in Other Pediatric Brain Tumors*

Limited information is available about genetic alterations or aberrant activation of the SHH signaling pathway in other pediatric brain tumors. Craniopharyngiomas, which arise from the embryonic remnants of Rathke's pouch and account for 5.6–6.2% of all pediatric brain tumors [52], have been reported in members of a family with basal cell nevus syndrome [119]. Analysis of the *PTCH1* gene in this family shows an insertion mutation, causing a frame shift. The craniopharyngiomas from these patients have loss of heterozygosity in the *PTCH1* region, suggesting potential involvement of SHH signaling in this tumor [119]. It is of interest that the shh, bmp, and fgf pathways appear to be involved in establishing dorso-ventral polarity in Rathke's pouch during development.

A study conducted to identify an ependymoma-specific gene signature identified overexpression of *GLI1* and *GLI2*, suggesting a potential role for HH signaling in this type of tumor [120]. More recently, gene-expression analysis using 34 ependymoma samples demonstrated that a subset of HH pathway components were highly expressed, including *GLI2* and Indian Hedgehog (*IHH*) [121]. During development, ependymal cells are derived from *nkx6.1*-expressing ventral neuroepithelial cells, which are regulated by shh [122]. A mechanism for HH pathway dysregulation in ependymomas has not been reported.

Pilocytic astrocytoma is a very heterogeneous tumor that is the most frequently occurring brain tumor during childhood [52]. It typically arises in the cerebellum (40–70%) and generally is benign with an excellent survival rate [123]. The level of *PTCH1* mRNA is elevated in approximately 45% of pilocytic astrocytomas and its level is inversely correlated with the age of patient [124]. Higher expression of *PTCH1* and *GLI1* is associated with younger age at diagnosis and more rapid tumor growth, suggesting a role for the pathway in regulating proliferation. It will be important to expand the analyses of HH signaling in these tumors and to see if any genetic alterations are associated with the activation of the SHH pathway in any of these non-medulloblastoma tumors.

## Potential Clinical Applications of HH Pathway Inhibitors in the Treatment of Patients with Pediatric Brain Tumors

Specifically targeting the SHH pathway as cancer therapy becomes possible as our understanding of the pathway improves. An observation reported in 1962 provided the first evidence that blocking the SHH pathway is feasible. It was noted that when pregnant ewes ingested the *Veratrum californicum* plant during their first trimester, they bore lambs with congenital cyclopean-type malformations [125]. Later, a steroidal alkaloid, called cyclopamine, was isolated from this same plant, which induced midline deformities in lamb fetuses [126]. It is now known that cyclopamine directly binds to the heptahelical bundle of SMO, likely changing the protein's conformation and thereby inactivating the SMO protein. In vitro, cyclopamine has been shown to inhibit SHH-dependent expression of *GLI1*, *GLI2*, and *PTCH1* and cause medulloblastoma cell cycle arrest [127]. In murine medulloblastoma tumor allograft and xenograft models, cyclopamine induces rapid tumor cell death [128, 129]. Unfortunately, due to its pharmacokinetic and side effect profile cyclopamine is not ideal for clinical use in humans [130]. Cyclopamine has poor solubility, acid sensitivity, weak potency and is a known teratogen [131]. More recently, a variety of naturally occurring and synthetic small molecule antagonists have been discovered [127, 128, 132].

Most of the small molecule antagonists to the SHH pathway also target SMO [132]. Romer and Curran have evaluated a small molecule inhibitor that binds and inhibits SMO called HhAntag-691 (Genentech). This compound is a benzimidazole derivative, which readily enters the brain of mice [128]. When *ptc1*$^{+/-}$ *p53*$^{-/-}$ mice with medulloblastomas are treated with HhAntag-691, there is dose-dependent down-regulation of several genes overexpressed in medulloblastomas, including *gli1*, *ptc2*, and *atoh1* [128]. Importantly, treatment of these mice improves tumor-free survival with minimal noted toxicities [128, 133].

Rubin and de Sauvage conducted cell-based screens for novel compounds that block SHH-activated gene transcription [134]. A variety of agents were discovered that target the pathway. One in particular, HhAntag, was initially very promising as it had oral bioavailability and showed potent antitumor activity in both murine *ptc1*-mutated medulloblastoma and human xenograft models [133]. Unfortunately, when evaluated in humans, it was determined that this drug had low potency and unpredictable pharmacokinetics. Therefore, newer drugs, such as GDC-0449 (Genentech) have been developed to improve the pharmacokinetics and potency of HhAntag. GDC-0449 also blocks the SHH pathway by binding and inhibiting SMO. Preclinical studies evaluating absorption, distribution, metabolism, and tumor responses have been promising. The compound's characteristics include the following: low plasma clearance, a volume of distribution estimated to be approximately equal to total body water, high protein binding, and oral bioavailability ranging from 13 to 53% in different species [135]. It is currently undergoing evaluation in phase I and phase II studies in adults with a variety of cancers, including basal cell carcinoma, stomach cancer, pancreatic cancer, breast cancer, lung cancer, ovarian cancer, glioblastoma multiforme, and medulloblastoma. There is also an ongoing pediatric brain tumor consortium (PBTC) phase I trial evaluating the use of GDC-0449 in children and adults with recurrent medulloblastoma [136].

There will be many challenges and questions to address while introducing inhibitors of HH signaling into medulloblastoma therapy. Activation of HH signaling is believed vital in tumor initiation and maintenance for some medulloblastomas, often but not exclusively of the desmoplastic histologic subtype. We would expect HH inhibitors to be most effective in this subset of tumors. Therefore, a reproducible and clinically useful method to identify active HH signaling must be established in medulloblastoma. Immunohistochemistry for components of the HH signal transduction pathway, such as PTCH1 and GLI1 may be promising. On the contrary, using gene-expression profiles to identify subsets of patients is still difficult in real time across centers.

Patients with desmoplastic histology tend to be young and have a more favorable prognosis. Moving newer agents into therapy for this subset of patients may be challenging; but may ultimately be of considerable value in improving outcomes and especially limiting toxicity. Typically, phase I and phase II studies are conducted in the setting of recurrence, as is the case with the ongoing Phase I PBTC HH inhibitor study. The youngest patients have been initially excluded based on the concern of a role for HH signaling in post-natal bone growth and development. Indeed, chondrocyte and osteoblast development require IHH, and there have been significant defects noted in long bone development in IHH knockout mouse models [137]. Mutant mice have reduced chondrocyte proliferation, failure of osteoblast development in endochondral bones, and premature closure of growth plates [137]. Interestingly, in humans, IHH mutations have recently been associated with a disorder known as acrocapitofemoraldysplasia [138]. This is a rare autosomal recessive growth disorder characterized by short stature, short limbs, brachydactyly, large head, narrow thorax, and pectus deformities [138, 139]. Osteopontin has also been identified as a target of GLI1 [140].

Efficacy of HH inhibitors may differ at the time of diagnosis versus at the time recurrence. Using a HH inhibitor at the time of diagnosis may optimize chances of observing a therapeutic benefit, since cells have not yet been exposed to agents that may select for resistant clones, however, treating at the time of recurrence may more effectively allow identification of agents that treat drug-resistant clones and more effectively build upon current therapeutic approaches. The optimal approach for testing HH inhibitors remains uncertain.

As with other agents, HH inhibitors may be most beneficial when either paired with traditional chemotherapy and radiotherapy or with other biologically active agents. Identifying other pathways that enhance or inhibit SHH signaling in medulloblastoma may help to inform decisions concerning drug combinations. For example, HH signaling in medulloblastoma activates the "insulin regulatory pathway." Inhibitors of this pathway are currently undergoing testing in a wide range of tumors. Targeting the upstream HH pathway and downstream "insulin regulatory pathway" may enhance efficacy. Identification of critical downstream targets will be essential [115].

A growing body of literature suggests a potential pitfall to the use of a HH inhibitor that targets SMO. There is now evidence that SMO's targets, GLI1, GLI2, and/or GLI3, may be activated in ways other than through the canonical HH pathway. Activation of GLI family transcription factors may then bypass the effect of the SMO inhibitor. A more complete understanding of ways to activate GLI family transcription factors in medulloblastoma and in other cancers is needed. In addition, an amino acid substitution in SMO was recently reported in human medulloblastoma, conferring resistance to GDC-0449, rather than disrupting the pathway which suggests another possible mechanism of resistance to this agent [141].

## Summary

Childhood brain tumors are significantly different from their adult counterparts, since the latent period is very short, growth is fast, and the cell populations causing the tumors arise from the embryonic cells. The causes of childhood brain tumors remain incompletely understood. However, significant progress in the genetics and biology of childhood brain tumors has been made in the past 15 years. In particular, important genes and signaling pathways involved in the development of childhood brain tumors have been identified [106]. We describe recent important discoveries of the role of SHH signaling pathway in brain development and tumorigenesis. SHH signaling appears to play fundamental roles in regulating proliferation and differentiation during development of a variety of cell types in the CNS. A role for SHH signaling in the normal development of CGPs in the cerebellum appears to reflect a role that the pathway plays in the genesis of some medulloblastomas. For this reason, drugs targeting major components of the SHH signaling pathway and interacting genes may prove to be a valuable alternative or adjunctive approach for the treatment of some children with medulloblastoma and potentially children with other brain tumors.

## References

1. Roelink H et al (1994) Floor plate and motor neuron induction by vhh-1, a vertebrate homolog of hedgehog expressed by the notochord. Cell 76:761–775
2. Teillet MA, Lapointe F, Le Douarin NM (1998) The relationships between notochord and floor plate in vertebrate development revisited. Proc Natl Acad Sci USA 95:11733–11738
3. Jacob J, Briscoe J (2003) Gli proteins and the control of spinal-cord patterning. EMBO Rep 4:761–765
4. Marti E, Takada R, Bumcrot DA, Sasaki H, McMahon AP (1995) Distribution of Sonic hedgehog peptides in the developing chick and mouse embryo. Development 121:2537–2547
5. Chiang C et al (1996) Cyclopia and defective axial patterning in mice lacking Sonic hedgehog gene function. Nature 383:407–413
6. Echelard Y et al (1993) Sonic hedgehog, a member of a family of putative signaling molecules, is implicated in the regulation of CNS polarity. Cell 75:1417–1430
7. Hebert JM, Fishell G (2008) The genetics of early telencephalon patterning: some assembly required. Nat Rev Neurosci 9:678–685
8. Ericson J et al (1995) Sonic hedgehog induces the differentiation of ventral forebrain neurons: a common signal for ventral patterning within the neural tube. Cell 81:747–756
9. Patten I, Placzek M (2000) The role of Sonic hedgehog in neural tube patterning. Cell Mol Life Sci 57:1695–1708
10. Aoto K, Nishimura T, Eto K, Motoyama J (2002) Mouse GLI3 regulates Fgf8 expression and apoptosis in the developing neural tube, face, and limb bud. Dev Biol 251:320–332
11. Ericson J, Morton S, Kawakami A, Roelink H, Jessell TM (1996) Two critical periods of Sonic Hedgehog signaling required for the specification of motor neuron identity. Cell 87:661–673
12. Litingtung Y, Chiang C (2000) Specification of ventral neuron types is mediated by an antagonistic interaction between Shh and Gli3. Nat Neurosci 3:979–985
13. Pringle NP et al (1996) Determination of neuroepithelial cell fate: induction of the oligodendrocyte lineage by ventral midline cells and sonic hedgehog. Dev Biol 177:30–42
14. van Straaten HW, Thors F, Wiertz-Hoessels L, Hekking J, Drukker J (1985) Effect of a notochordal implant on the early morphogenesis of the neural tube and neuroblasts: histometrical and histological results. Dev Biol 110:247–254
15. van Straaten HW, Hekking JW, Beursgens JP, Terwindt-Rouwenhorst E, Drukker J (1989) Effect of the notochord on proliferation and differentiation in the neural tube of the chick embryo. Development 107:793–803
16. Charrier JB, Lapointe F, Le Douarin NM, Teillet MA (2001) Anti-apoptotic role of Sonic hedgehog protein at the early stages of nervous system organogenesis. Development 128:4011–4020
17. Britto J, Tannahill D, Keynes R (2002) A critical role for sonic hedgehog signaling in the early expansion of the developing brain. Nat Neurosci 5:103–110
18. Agarwala S, Sanders TA, Ragsdale CW (2001) Sonic hedgehog control of size and shape in midbrain pattern formation. Science 291:2147–2150
19. Fogel JL, Chiang C, Huang X, Agarwala S (2008) Ventral specification and perturbed boundary formation in the mouse midbrain in the absence of Hedgehog signaling. Dev Dyn 237:1359–1372

20. Koster R, Stick R, Loosli F, Wittbrodt J (1997) Medaka spalt acts as a target gene of hedgehog signaling. Development 124:3147–3156
21. Strahle U, Blader P, Henrique D, Ingham PW (1993) Axial, a zebrafish gene expressed along the developing body axis, shows altered expression in cyclops mutant embryos. Genes Dev 7:1436–1446
22. Bayly RD, Ngo M, Aglyamova GV, Agarwala S (2007) Regulation of ventral midbrain patterning by Hedgehog signaling. Development 134:2115–2124
23. Rallu M et al (2002) Dorsoventral patterning is established in the telencephalon of mutants lacking both Gli3 and Hedgehog signaling. Development 129:4963–4974
24. Kohtz JD, Baker DP, Corte G, Fishell G (1998) Regionalization within the mammalian telencephalon is mediated by changes in responsiveness to Sonic Hedgehog. Development 125:5079–5089
25. Tole S, Ragsdale CW, Grove EA (2000) Dorsoventral patterning of the telencephalon is disrupted in the mouse mutant extra-toes(J). Dev Biol 217:254–265
26. Ishibashi M, McMahon AP (2002) A sonic hedgehog-dependent signaling relay regulates growth of diencephalic and mesencephalic primordia in the early mouse embryo. Development 129:4807–4819
27. Hynes M et al (1995) Induction of midbrain dopaminergic neurons by Sonic hedgehog. Neuron 15:35–44
28. Joksimovic M et al (2009) Spatiotemporally separable Shh domains in the midbrain define distinct dopaminergic progenitor pools. Proc Natl Acad Sci USA 106:19185–19190
29. Blaess S, Corrales JD, Joyner AL (2006) Sonic hedgehog regulates Gli activator and repressor functions with spatial and temporal precision in the mid/hindbrain region. Development 133:1799–1809
30. Blaess S, Stephen D, Joyner AL (2008) Gli3 coordinates three-dimensional patterning and growth of the tectum and cerebellum by integrating Shh and Fgf8 signaling. Development 135:2093–2103
31. Hynes M et al (1997) Control of cell pattern in the neural tube by the zinc finger transcription factor and oncogene Gli-1. Neuron 19:15–26
32. Lebel M, Mo R, Shimamura K, Hui CC (2007) Gli2 and Gli3 play distinct roles in the dorsoventral patterning of the mouse hindbrain. Dev Biol 302:345–355
33. Arkell R, Beddington RS (1997) BMP-7 influences pattern and growth of the developing hindbrain of mouse embryos. Development 124:1–12
34. Stecca B, Ruiz i Altaba A (2005) Brain as a paradigm of organ growth: hedgehog–Gli signaling in neural stem cells and brain tumors. J Neurobiol 64:476–490
35. Sotelo C (2004) Cellular and genetic regulation of the development of the cerebellar system. Prog Neurobiol 72:295–339
36. Vaillant C, Monard D (2009) SHH pathway and cerebellar development. Cerebellum 8:291–301
37. Traiffort E, Angot E, Ruat M (2010) Sonic Hedgehog signaling in the mammalian brain. J Neurochem 113:576–590
38. Lewis PM, Gritli-Linde A, Smeyne R, Kottmann A, McMahon AP (2004) Sonic hedgehog signaling is required for expansion of granule neuron precursors and patterning of the mouse cerebellum. Dev Biol 270:393–410
39. Wechsler-Reya RJ, Scott MP (1999) Control of neuronal precursor proliferation in the cerebellum by Sonic Hedgehog. Neuron 22:103–114
40. Dahmane N, Ruiz i Altaba A (1999) Sonic hedgehog regulates the growth and patterning of the cerebellum. Development 126:3089–3100
41. Wallace VA (1999) Purkinje-cell-derived Sonic hedgehog regulates granule neuron precursor cell proliferation in the developing mouse cerebellum. Curr Biol 9:445–448
42. Kenney AM, Rowitch DH (2000) Sonic hedgehog promotes G(1) cyclin expression and sustained cell cycle progression in mammalian neuronal precursors. Mol Cell Biol 20:9055–9067
43. Kenney AM, Cole MD, Rowitch DH (2003) Nmyc upregulation by sonic hedgehog signaling promotes proliferation in developing cerebellar granule neuron precursors. Development 130:15–28

44. Jarman AP, Grau Y, Jan LY, Jan YN (1993) Atonal is a proneural gene that directs chordotonal organ formation in the Drosophila peripheral nervous system. Cell 73:1307–1321
45. Flora A, Klisch TJ, Schuster G, Zoghbi HY, Zoghbi HY (2009) Deletion of Atoh1 disrupts Sonic Hedgehog signaling in the developing cerebellum and prevents medulloblastoma. Science 326:1424–1427
46. Pons S, Trejo JL, Martinez-Morales JR, Marti E (2001) Vitronectin regulates Sonic hedgehog activity during cerebellum development through CREB phosphorylation. Development 128:1481–1492
47. Rios I, Alvarez-Rodriguez R, Marti E, Pons S (2004) Bmp2 antagonizes sonic hedgehog-mediated proliferation of cerebellar granule neurones through Smad5 signalling. Development 131:3159–3168
48. Alvarez-Rodriguez R, Barzi M, Berenguer J, Pons S (2007) Bone morphogenetic protein 2 opposes Shh-mediated proliferation in cerebellar granule cells through a TIEG-1-based regulation of Nmyc. J Biol Chem 282:37170–37180
49. Argenti B et al (2005) Hedgehog antagonist REN(KCTD11) regulates proliferation and apoptosis of developing granule cell progenitors. J Neurosci 25:8338–8346
50. Fogarty MP, Emmenegger BA, Grasfeder LL, Oliver TG, Wechsler-Reya RJ (2007) Fibroblast growth factor blocks Sonic hedgehog signaling in neuronal precursors and tumor cells. Proc Natl Acad Sci USA 104:2973–2978
51. Miller B (ed) ((1993) SEER cancer statistics review. National Institutes of Health, Bethesda, MD
52. Rickert CH, Paulus W (2001) Epidemiology of central nervous system tumors in childhood and adolescence based on the new WHO classification. Childs Nerv Syst 17:503–511
53. Fangusaro J (2009) Pediatric high-grade gliomas and diffuse intrinsic pontine gliomas. J Child Neurol 24:1409–1417
54. Dhall G (2009) Medulloblastoma. J Child Neurol 24:1418–1430
55. Ohgaki H (2009) Epidemiology of brain tumors. Methods Mol Biol 472:323–342
56. Surawicz TS et al (1999) Descriptive epidemiology of primary brain and CNS tumors: results from the Central Brain Tumor Registry of the United States, 1990–1994. Neuro Oncol 1:14–25
57. Benesch M et al (2006) Late sequela after treatment of childhood low-grade gliomas: a retrospective analysis of 69 long-term survivors treated between 1983 and 2003. J Neurooncol 78:199–205
58. Broniscer A et al (2004) Second neoplasms in pediatric patients with primary central nervous system tumors: the St. Jude Children's Research Hospital experience. Cancer 100:2246–2252
59. Kretschmar CS, Warren MP, Lavally BL, Dyer S, Tarbell NJ (1990) Ototoxicity of preradiation cisplatin for children with central nervous system tumors. J Clin Oncol 8:1191–1198
60. Shaw S (2009) Endocrine late effects in survivors of pediatric brain tumors. J Pediatr Oncol Nurs 26:295–302
61. Packer RJ, Rood BR, MacDonald TJ (2003) Medulloblastoma: present concepts of stratification into risk groups. Pediatr Neurosurg 39:60–67
62. Fangusaro J et al (2008) Intensive chemotherapy followed by consolidative myeloablative chemotherapy with autologous hematopoietic cell rescue (AuHCR) in young children with newly diagnosed supratentorial primitive neuroectodermal tumors (sPNETs): report of the Head Start I and II experience. Pediatr Blood Cancer 50:312–318
63. Fangusaro JR et al (2008) Brainstem primitive neuroectodermal tumors (bstPNET): results of treatment with intensive induction chemotherapy followed by consolidative chemotherapy with autologous hematopoietic cell rescue. Pediatr Blood Cancer 50:715–717
64. Li MH, Bouffet E, Hawkins CE, Squire JA, Huang A (2005) Molecular genetics of supratentorial primitive neuroectodermal tumors and pineoblastoma. Neurosurg Focus 19:E3
65. Brandes AA et al (2009) Adult neuroectodermal tumors of posterior fossa (medulloblastoma) and of supratentorial sites (stPNET). Crit Rev Oncol Hematol 71:165–179
66. Guran S, Tunca Y, Imirzalioglu N (1999) Hereditary TP53 codon 292 and somatic P16INK4A codon 94 mutations in a Li–Fraumeni syndrome family. Cancer Genet Cytogenet 113:145–151

67. Mathur MN, Thompson JF, O'Brien CJ, Davidson TI, McCarthy WH (1993) Naevoid basal cell carcinoma (Gorlin's) syndrome. Aust N Z J Surg 63:413–415
68. Pietsch T et al (1997) Medulloblastomas of the desmoplastic variant carry mutations of the human homologue of Drosophila patched. Cancer Res 57:2085–2088
69. Sarin S, Bernath A (2008) Turcot syndrome (glioma polyposis): a case report. South Med J 101:1273–1274
70. Chang CH, Housepian EM, Herbert C Jr (1969) An operative staging system and a megavoltage radiotherapeutic technic for cerebellar medulloblastomas. Radiology 93:1351–1359
71. Packer RJ et al (1999) Treatment of children with medulloblastomas with reduced-dose craniospinal radiation therapy and adjuvant chemotherapy: a Children's Cancer Group Study. J Clin Oncol 17:2127–2136
72. Packer RJ et al (1994) Outcome for children with medulloblastoma treated with radiation and cisplatin, CCNU, and vincristine chemotherapy. J Neurosurg 81:690–698
73. Chi SN et al (2004) Feasibility and response to induction chemotherapy intensified with high-dose methotrexate for young children with newly diagnosed high-risk disseminated medulloblastoma. J Clin Oncol 22:4881–4887
74. Dhall G et al (2008) Outcome of children less than three years old at diagnosis with non-metastatic medulloblastoma treated with chemotherapy on the "Head Start" I and II protocols. Pediatr Blood Cancer 50:1169–1175
75. Rutkowski S et al (2005) Treatment of early childhood medulloblastoma by postoperative chemotherapy alone. N Engl J Med 352:978–986
76. Ris MD, Packer R, Goldwein J, Jones-Wallace D, Boyett JM (2001) Intellectual outcome after reduced-dose radiation therapy plus adjuvant chemotherapy for medulloblastoma: a Children's Cancer Group study. J Clin Oncol 19:3470–3476
77. Amlashi SF, Riffaud L, Brassier G, Morandi X (2003) Nevoid basal cell carcinoma syndrome: relation with desmoplastic medulloblastoma in infancy. A population-based study and review of the literature. Cancer 98:618–624
78. Levy RA, Blaivas M, Muraszko K, Robertson PL (1997) Desmoplastic medulloblastoma: MR findings. AJNR Am J Neuroradiol 18:1364–1366
79. Rutkowski S, Van Hoff K, Emser A, Garre M, Walker D, Grundy R, Dhall G, Finlay J, Grill J (2008) Prognostic factors and survival of young children with medulloblastoma: an international meta-analysis. Nuerooncology 10:437
80. Beutler D et al (2005) Three-year recurrence-free survival in a patient with recurrent medulloblastoma after resection, high-dose chemotherapy, and intrathecal Yttrium-90-labeled DOTA0-D-Phe1-Tyr3-octreotide radiopeptide brachytherapy. Cancer 103:869–873
81. Butturini AM et al (2009) High-dose chemotherapy and autologous hematopoietic progenitor cell rescue in children with recurrent medulloblastoma and supratentorial primitive neuroectodermal tumors: the impact of prior radiotherapy on outcome. Cancer 115:2956–2963
82. Dunkel IJ et al (1998) High-dose carboplatin, thiotepa, and etoposide with autologous stem-cell rescue for patients with recurrent medulloblastoma. Children's Cancer Group. J Clin Oncol 16:222–228
83. Mendrzyk F et al (2006) Isochromosome breakpoints on 17p in medulloblastoma are flanked by different classes of DNA sequence repeats. Genes Chromosomes Cancer 45:401–410
84. Pfister S, Hartmann C, Korshunov A (2009) Histology and molecular pathology of pediatric brain tumors. J Child Neurol 24:1375–1386
85. Rossi A, Caracciolo V, Russo G, Reiss K, Giordano A (2008) Medulloblastoma: from molecular pathology to therapy. Clin Cancer Res 14:971–976
86. Rossi MR et al (2006) Array CGH analysis of pediatric medulloblastomas. Genes Chromosomes Cancer 45:290–303
87. Pfister S et al (2009) Outcome prediction in pediatric medulloblastoma based on DNA copy-number aberrations of chromosomes 6q and 17q and the MYC and MYCN loci. J Clin Oncol 27:1627–1636
88. Thompson MC et al (2006) Genomics identifies medulloblastoma subgroups that are enriched for specific genetic alterations. J Clin Oncol 24:1924–1931

89. McLaughlin MR, Gollin SM, Lese CM, Albright AL (1998) Medulloblastoma and glioblastoma multiforme in a patient with Turcot syndrome: a case report. Surg Neurol 49:295–301
90. Qualman SJ, Bowen J, Erdman SH (2003) Molecular basis of the brain tumor-polyposis (Turcot) syndrome. Pediatr Dev Pathol 6:574–576
91. Guessous F, Li Y, Abounader R (2008) Signaling pathways in medulloblastoma. J Cell Physiol 217:577–583
92. Ellison DW et al (2005) Beta-Catenin status predicts a favorable outcome in childhood medulloblastoma: the United Kingdom Children's Cancer Study Group Brain Tumour Committee. J Clin Oncol 23:7951–7957
93. Lasky JL, Wu H (2005) Notch signaling, brain development, and human disease. Pediatr Res 57:104R–109R
94. Sjolund J, Manetopoulos C, Stockhausen MT, Axelson H (2005) The Notch pathway in cancer: differentiation gone awry. Eur J Cancer 41:2620–2629
95. Crean SJ, Cunningham SJ (1996) Gorlin's syndrome: main features and recent advances. Br J Hosp Med 56:392–397
96. Raffel C et al (1997) Sporadic medulloblastomas contain PTCH mutations. Cancer Res 57:842–845
97. Kim JY et al (2003) Medulloblastoma tumorigenesis diverges from cerebellar granule cell differentiation in patched heterozygous mice. Dev Biol 263:50–66
98. Xie J et al (1998) Activating Smoothened mutations in sporadic basal-cell carcinoma. Nature 391:90–92
99. Reifenberger J et al (1998) Missense mutations in SMOH in sporadic basal cell carcinomas of the skin and primitive neuroectodermal tumors of the central nervous system. Cancer Res 58:1798–1803
100. Ding Q et al (1999) Mouse suppressor of fused is a negative regulator of sonic hedgehog signaling and alters the subcellular distribution of Gli1. Curr Biol 9:1119–1122
101. Taylor MD et al (2002) Mutations in SUFU predispose to medulloblastoma. Nat Genet 31:306–310
102. Rieber J et al (2009) Novel oncogene amplifications in tumors from a family with Li–Fraumeni syndrome. Genes Chromosomes Cancer 48:558–568
103. Dennler S et al (2007) Induction of sonic hedgehog mediators by transforming growth factor-beta: Smad3-dependent activation of Gli2 and Gli1 expression in vitro and in vivo. Cancer Res 67:6981–6986
104. Beauchamp E et al (2009) GLI1 is a direct transcriptional target of EWS–FLI1 oncoprotein. J Biol Chem 284:9074–9082
105. Zhao H, Ayrault O, Zindy F, Kim JH, Roussel MF (2008) Post-transcriptional down-regulation of Atoh1/Math1 by bone morphogenic proteins suppresses medulloblastoma development. Genes Dev 22:722–727
106. Dubuc AM et al (2010) The genetics of pediatric brain tumors. Curr Neurol Neurosci Rep 10:215–223
107. Rao G, Pedone CA, Coffin CM, Holland EC, Fults DW (2003) c-Myc enhances sonic hedgehog-induced medulloblastoma formation from nestin-expressing neural progenitors in mice. Neoplasia 5:198–204
108. Fernandez LA et al (2009) YAP1 is amplified and up-regulated in hedgehog-associated medulloblastomas and mediates Sonic hedgehog-driven neural precursor proliferation. Genes Dev 23:2729–2741
109. Lelievre V et al (2008) Disruption of the PACAP gene promotes medulloblastoma in ptc1 mutant mice. Dev Biol 313:359–370
110. Di Marcotullio L et al (2004) REN(KCTD11) is a suppressor of Hedgehog signaling and is deleted in human medulloblastoma. Proc Natl Acad Sci USA 101:10833–10838
111. Malkin D et al (1990) Germ line p53 mutations in a familial syndrome of breast cancer, sarcomas, and other neoplasms. Science 250:1233–1238
112. Goodrich LV, Milenkovic L, Higgins KM, Scott MP (1997) Altered neural cell fates and medulloblastoma in mouse patched mutants. Science 277:1109–1113

113. Wetmore C, Eberhart DE, Curran T (2001) Loss of p53 but not ARF accelerates medulloblastoma in mice heterozygous for patched. Cancer Res 61:513–516
114. Stecca B, Ruiz i Altaba A (2009) A GLI1-p53 inhibitory loop controls neural stem cell and tumour cell numbers. EMBO J 28:663–676
115. Yoon JW, Gilbertson R, Iannaccone S, Iannaccone P, Walterhouse D (2009) Defining a role for Sonic hedgehog pathway activation in desmoplastic medulloblastoma by identifying GLI1 target genes. Int J Cancer 124:109–119
116. Corcoran RB, Bachar Raveh T, Barakat MT, Lee EY, Scott MP (2008) Insulin-like growth factor 2 is required for progression to advanced medulloblastoma in patched1 heterozygous mice. Cancer Res 68:8788–8795
117. Parathath SR, Mainwaring LA, Fernandez LA, Campbell DO, Kenney AM (2008) Insulin receptor substrate 1 is an effector of sonic hedgehog mitogenic signaling in cerebellar neural precursors. Development 135:3291–3300
118. De Smaele E et al (2008) An integrated approach identifies Nhlh1 and Insm1 as Sonic Hedgehog-regulated genes in developing cerebellum and medulloblastoma. Neoplasia 10:89–98
119. Musani V et al (2006) Mutation in exon 7 of PTCH deregulates SHH/PTCH/SMO signaling: possible linkage to WNT. Int J Mol Med 17:755–759
120. Modena P et al (2006) Identification of tumor-specific molecular signatures in intracranial ependymoma and association with clinical characteristics. J Clin Oncol 24:5223–5233
121. Palm T et al (2009) Expression profiling of ependymomas unravels localization and tumor grade-specific tumorigenesis. Cancer 115:3955–3968
122. Fu W, Baker NE (2003) Deciphering synergistic and redundant roles of Hedgehog. Decapentaplegic and Delta that drive the wave of differentiation in Drosophila eye development. Development 130:5229–5239
123. Burkhard C et al (2003) A population-based study of the incidence and survival rates in patients with pilocytic astrocytoma. J Neurosurg 98:1170–1174
124. Rush SZ, Abel TW, Valadez JG, Pearson M, Cooper MK (2010) Activation of the Hedgehog pathway in pilocytic astrocytomas. Neuro Oncol 12(8):790–798
125. Binns W, James LF, Shupe JL, Thacker EJ (1962) Cyclopian-type malformation in lambs. Arch Environ Health 5:106–108
126. Chen JK, Taipale J, Cooper MK, Beachy PA (2002) Inhibition of Hedgehog signaling by direct binding of cyclopamine to Smoothened. Genes Dev 16:2743–2748
127. Stanton BZ, Peng LF (2010) Small-molecule modulators of the Sonic Hedgehog signaling pathway. Mol Biosyst 6:44–54
128. Romer J, Curran T (2005) Targeting medulloblastoma: small-molecule inhibitors of the Sonic Hedgehog pathway as potential cancer therapeutics. Cancer Res 65:4975–4978
129. Berman DM et al (2002) Medulloblastoma growth inhibition by hedgehog pathway blockade. Science 297:1559–1561
130. Kumar SK et al (2008) Targeted inhibition of hedgehog signaling by cyclopamine prodrugs for advanced prostate cancer. Bioorg Med Chem 16:2764–2768
131. Kiselyov AS, Tkachenko SE, Balakin KV, Ivachtchenko AV (2007) Small-molecule modulators of Hh and Wnt signaling pathways. Expert Opin Ther Targets 11:1087–1101
132. Tremblay MR et al (2009) Discovery of a potent and orally active hedgehog pathway antagonist (IPI-926). J Med Chem 52:4400–4418
133. Romer JT et al (2004) Suppression of the Shh pathway using a small molecule inhibitor eliminates medulloblastoma in Ptc1(+/−)p53(−/−) mice. Cancer Cell 6:229–240
134. Rubin LL, de Sauvage FJ (2006) Targeting the Hedgehog pathway in cancer. Nat Rev Drug Discov 5:1026–1033
135. Wong H et al (2009) Preclinical assessment of the absorption, distribution, metabolism and excretion of GDC-0449 (2-chloro-*N*-(4-chloro-3-(pyridin-2-yl)phenyl)-4-(methylsulfonyl) benzamide), an orally bioavailable systemic Hedgehog signalling pathway inhibitor. Xenobiotica 39:850–861
136. Gajjar AJ, Stewart C, Ellison DW, Curran T, Phillips P, Goldman G, Packer R, Kun LE, Boyett JM, Gilbertson RJ (2010) A phase I pharmacokinetic trial of sonic hedgehog (SHH)

antagonist GDC-0449 in pediatric patients with recurrent or refractory medulloblastoma: a Pediatric Brain Tumor Consortium study (PBTC 25). J Clin Oncol 28 (Suppl. 18): CRA9501
137. St-Jacques B, Hammerschmidt M, McMahon AP (1999) Indian hedgehog signaling regulates proliferation and differentiation of chondrocytes and is essential for bone formation. Genes Dev 13:2072–2086
138. Hellemans J et al (2003) Homozygous mutations in IHH cause acrocapitofemoral dysplasia, an autosomal recessive disorder with cone-shaped epiphyses in hands and hips. Am J Hum Genet 72:1040–1046
139. Mortier GR, Kramer PP, Giedion A, Beemer FA (2003) Acrocapitofemoral dysplasia: an autosomal recessive skeletal dysplasia with cone shaped epiphyses in the hands and hips. J Med Genet 40:201–207
140. Yoon JW et al (2002) Gene expression profiling leads to identification of GLI1-binding elements in target genes and a role for multiple downstream pathways in GLI1-induced cell transformation. J Biol Chem 277:5548–5555
141. Yauch RL et al (2009) Smoothened mutation confers resistance to a Hedgehog pathway inhibitor in medulloblastoma. Science 326:572–574
142. Vorechovsky I et al (1997) Somatic mutations in the human homologue of Drosophila patched in primitive neuroectodermal tumours. Oncogene 15:361–366
143. Lee Y et al (2007) Loss of suppressor-of-fused function promotes tumorigenesis. Oncogene 26:6442–6447

# Chapter 9
# Hedgehog Signaling in Mammary Gland Development and Breast Cancer

**Adriana P. Visbal and Michael T. Lewis**

## Introduction

The *Hedgehog* (*Hh*) gene was discovered in *Drosophila* as a gene specifying segment polarity [1]. Since that initial discovery, the Hedgehog signaling network has been shown to regulate development of most structures and organs in both invertebrate and vertebrate species [2]. For example, in mammals, the network is known to pattern the limb and the dorso-ventral axis of the neural tube to stimulate proliferation of neural precursor cells and to regulate hair follicle regeneration [3]. Like Wnt and Notch signaling, Hedgehog signaling has also been implicated in the maintenance and self-renewal of several adult stem/progenitor cell types, including those of the mammary gland [4–6].

## Mammalian Hedgehog Signaling

Although autocrine signaling can occur, mammalian Hedgehog signaling generally takes place between two cells, a signaling cell and a receiving cell [7–9].

In the absence of a Hedgehog ligand, the PTCH receptor catalytically inhibits Smoothened (SMO), the primary signal transducer. Inhibition of SMO allows the phosphorylation and cleavage of two members of the GLI-family of transcription factors (GLI2 and GLI3). This cleavage is promoted by a complex of kinases, and the scaffold protein Suppressor of fused, which phosphorylate the GLI proteins and

M.T. Lewis (✉)
The Lester and Sue Smith Breast Center, Baylor College of Medicine, One Baylor Plaza, BCM600, Room N1210, Houston, TX 77030, USA
and
Departments of Molecular and Cellular Biology and Radiology,
Baylor College of Medicine, One Baylor Plaza, BCM600, Room N1210, Houston, TX 77030, USA
e-mail: mtlewis@bcm.edu

J. Xie (ed.), *Hedgehog Signaling Activation in Human Cancer and its Clinical Implications*,
DOI 10.1007/978-1-4419-8435-7_9, 

target them for processing. The cleaved, repressor forms of GLI can then translocate into the nucleus and block transcription of target genes.

Whereas in *Drosophila* there is only a single gene encoding one Hedgehog ligand, mammals have three genes encoding three distinct proteins: *Sonic Hedgehog* (*Shh*), *Desert Hedgehog* (*Dhh*), and *Indian Hedgehog* (*Ihh*). Hedgehog proteins are translated and autocatalytically cleaved to an N-terminal signaling fragment in the signaling cell, and are secreted as palmitoylated and cholesterol-modified ligands via function of the Dispatched protein. Once the ligand is secreted by the signaling cell, it binds the Patched (PTCH) family of 12-pass trans-membrane receptors (PTCH1 or PTCH2) on the receiving cell. Ligand binding to PTCH can be facilitated by Commodo, Brother of Commodo, and Growth Arrest Specific-1, or it can be inhibited by Hedgehog Interacting Protein by sequestration of the ligand [10–13].

Upon ligand binding to a PTCH receptor, inhibition of SMO by PTCH is released. Once SMO is allowed to function, it inhibits the phosphorylation and cleavage of the GLI family transcription factors to yield full-length transcriptional activators. Of the three GLI proteins (GLI1, GLI2, and GLI3), GLI1 acts exclusively as a transcriptional activator, as it lacks the proteolytic cleavage site. GLI2 and GLI3 can function either as activators or repressors, depending on whether or not they are cleaved. In vivo data suggests that GLI2 functions primarily as a transcriptional activator, while GLI3 is found primarily in its repressor form [14–16].

The ability to activate Hedgehog signaling in mammalian cells has been associated with the presence of a primary cilium on the receiving cell. As such, proteins involved in cilium formation and intraflagellar transport (IFT) play vital roles in Hedgehog signaling. Mouse IFT proteins have been shown to be required for GLI activator and GLI repressor functions, and mice lacking these proteins exhibit Hedgehog loss-of-function phenotypes [17, 18]. *Drosophila* has homologs to IFT components as well, yet they do not appear to be required for Hedgehog signaling since flies lacking these components do not exhibit Hedgehog loss-of-function phenotypes [19–21]. Localization of PTCH1 and SMO to cilia is mutually exclusive. Ligand-bound PTCH1 is internalized upon binding, and only then can SMO translocate to the primary cilium to activate GLI proteins [17, 22, 23]. The role of primary cilia in Hedgehog signaling is reviewed in detail by Wong et al. [24].

## Non-canonical Hedgehog Network Functions

In addition to canonical GLI-mediated transcriptional responses induced by Hedgehog ligands, several non-canonical functions of Hedgehog network genes have been identified.

Among the most versatile Hedgehog network components is the PTCH1 protein. In addition to preventing signaling in the absence of ligand, and transducing the Hedgehog signal in their presence, PTCH1 has a number of additional functions. First, the PTCH1 receptor can act to sequester the Hedgehog ligand and restrict its range of diffusion such that otherwise signaling-competent cells may not come into contact with ligand [25, 26]. Second, the PTCH1 receptor has been demonstrated to act as a dependence receptor in cell types that require the presence of ligand-bound PTCH1 in order to survive. In these cells, unbound PTCH1 receptor initiates apoptotic cell death [27–29]. Third, the PTCH1 was shown to act as a "gatekeeper" of the cell cycle by controlling the transition from G2 to M phase. PTCH1 performs this function by binding phosphorylated cyclin B1 and sequestering it in the cytoplasm. Upon SHH ligand addition, this interaction is disrupted allowing cyclin B1 to localize to the nucleus followed by cell cycle progression [30]. Functional significance of this interaction was supported in a skin-specific PTCH1 loss-of-function model, in which PTCH1 loss led to basal cell carcinoma (BCC) lesions accompanied by nuclear accumulation of cyclin D1 and cyclin B1 [31]. Fourth and finally, recent data using mammary epithelial cells suggest that PTCH1 can activate ERK1/2 independent of SMO-mediated signaling in mammary epithelium via one or more SH3-domain-mediated protein–protein interactions, including potential interactions with Grb2, c-src, and p85β (PIK3R2) [32].

SMO may also have multiple functions. SMO is a seven-pass trans-membrane protein which is a nonredundant component and serves as an essential, rate-limiting mediator of canonical signal transduction. SMO shares structural and evolutionary homology with G protein-coupled receptors (GPCRs) such as rhodopsin and β-adrenergic receptor, and is most closely related to the Frizzled family of proteins that mediate Wnt signaling. While the Wnt-Frizzled signaling network is known to transduce signal via heterotrimeric G proteins [33], evidence to support a role for heterotrimeric G-proteins in activated Hedgehog signaling has only recently emerged [34–36].

The first solid evidence that SMO-coupled heterotrimeric G protein activation can occur was an in vitro study by Riobo et al. which demonstrated that mammalian SMO can couple with selected $G_{\alpha}$ subunits primarily of the $G_{\alpha i}$ family upon co-transfection [37]. This study indicated that mammalian SMO can couple with $G_{i1}$, $G_{i2}$, $G_{i3}$, $G_{0}$, $G_{z}$, and $G_{15}$ in Sf9 (insect) cells. On the contrary, SMO could not couple with $G_{s}$, $G_{q}$, $G_{qzIc}$, $G_{qG66D}$, $G_{qG66Dx5}$, $G_{12}$, $G_{13}$, or $G_{16}$, although a different study found that $G_{12}$ and $G_{13}$ mediated the SHH–SMO response in neuroblastoma cells [38]. While G protein activation could be partially or completely blocked by cyclopamine, a relatively specific inhibitor of SMO, activation of $G_{z}$ was largely insensitive. Four of the $G_{\alpha}$ subunits shown to couple with SMO ($G_{i1}$, $G_{i2}$, $G_{i3}$, and $G_{0}$) are irreversibly inhibited by pertussis toxin (PTX)-mediated ADP-ribosylation.

Of potential importance for the interpretation of Hedgehog network function, G protein coupling by SMO could be genetically separated from GLI activation using a truncated SMO protein [39]. Thus, it may be possible that these two functions can be uncoupled in vivo under the right conditions. Consistent with the possibility of uncoupled function, SHH was found to mediate axon guidance in a SMO-dependent, but GLI-independent, manner via activation of SRC family kinases [40]. Additionally, Hedgehog ligands were found to activate a pro-angiogenic response in endothelial cells in a GLI-independent manner. These GLI-independent effects could be suppressed by PTX treatment, suggesting an independent GPCR function for SMO is critical for at least some aspects of SMO activity [41].

There have been other indications that SMO might function as a GPCR. For example, SMO function can be enhanced by catalytically active G protein-coupled receptor kinase 2 (GRK2), but not catalytically inactive GRK2. GRK2 activity promoted association of SMO with β-arrestin-2 [42, 43]. GPRK2 was also found to participate in Hedgehog signaling in *Drosophila*, further suggesting that SMO function as a GPCR might be evolutionarily conserved [44]. Subsequently, *Drosophila* SMO was found to interact with $G\alpha_i$ in vivo, and this interaction is apparently essential for Hedgehog signal transduction [45].

With respect to the GLI transcription factors, the GLI proteins were long believed to act solely as the transcriptional activators of the Hedgehog network. However, ligand- and SMO-independent function for the GLI family of transcription factors have also been suggested, with emerging evidence indicating that these transcription factors are regulated by other signaling networks including that of TGF-β, as well as by activation of AKT- and RAS-mediated signaling [46–49].

## Hedgehog Network Control of Mammary Gland Development and Breast Cancer

### *Mammary Gland Development*

In both mouse and human, mammary gland development begins during embryonic life with growth of a rudimentary ductal tree. After this initial hormone-independent growth, the rudimentary ducts remain relatively growth-quiescent from birth until puberty. With the onset of ovarian hormone secretion during puberty, the rudimentary ducts begin to grow and elongate as secondary and tertiary ducts, led by rapid and

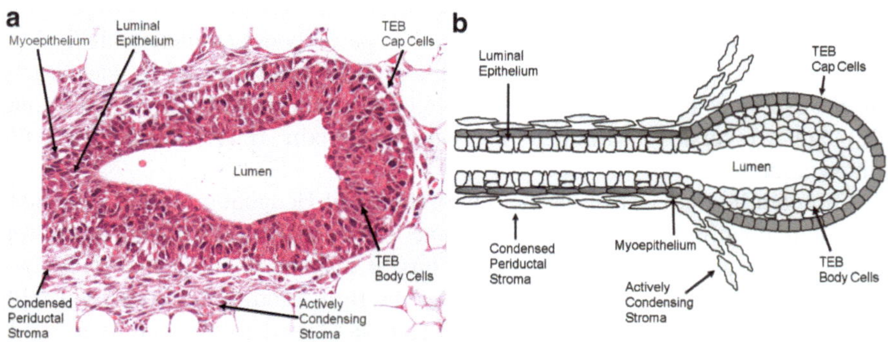


**Fig. 9.1** The terminal end bud (TEB). (**a**) Histological preparation showing the histoarchitecture of a typical TEB. Distinct cap cell and body cell layers are identified, as well as areas of actively condensing and condensed periductal stroma. (**b**) Schematic diagram of a TEB and mature subtending duct showing a layer of heterogeneous luminal epithelial cells surrounded by a layer of myoepithelial cells and a condensed mammary stroma

invasive growth of the terminal end bud (TEB) (Fig. 9.1a, b). The TEB is a bulb-shaped structure consisting of multiple layers of rapidly dividing, immature epithelial cells. The outer cell layer is composed of cap cells, which differentiate into the duct-associated myoepithelial cells as the duct elongates. The inner layers are composed of body cells that give rise to and/or differentiate into ductal and alveolar progenitors as well as the more differentiated luminal ductal epithelial cell types. Upon reaching the edges of the available mammary fat pad, TEB structures regress, leaving an arborized tree of differentiated ducts that are open to the surface of the skin via the nipple.

While strikingly similar, mouse and human mammary gland development in the virgin are not identical. In humans, a relatively high level of alveolar development also occurs due to the presence of a luteal phase of the menstrual cycle. On the contrary, most commonly used mouse strains lack a luteal phase to their estrus cycle, and thus show a relative lack of alveolar development in virgin animals. As a consequence, the relative proportion of ductal vs. alveolar cell types in the mouse vs. human mammary gland is quite different in the nonpregnant female.

Regardless of the degree of alveolar development in virgins, neither the human nor mouse mammary gland can be considered fully differentiated until pregnancy and lactation. With pregnancy, alveolar development culminates, followed by the production and secretion of milk during lactation [50]. Upon weaning of young, the mammary gland involutes. Involution is characterized by extensive apoptosis and epithelial remodeling that yields a mammary ductal tree that is similar to, but not identical, with that of the adult virgin [51, 52].

## *Hedgehog Network Gene Expression and Function in Mammary Gland Development*

Most major insights into Hedgehog network function in breast development and breast cancer come from gene expression analyses in situ, and the study of genetically engineered mice.

In mice, the Hedgehog ligands are expressed in mammary epithelium at several stages of development (Fig. 9.2a, b). *Ihh* and *Dhh* are detectable in the pubertal gland by in situ hybridization and *Ihh* mRNA levels are upregulated during pregnancy and lactation, suggesting a role in alveolar differentiation or function [53, 54]. Expression of *Ihh* mRNA becomes undetectable in early involution, but returns to roughly pre-pregnant levels after gland remodeling [55]. Despite the developmentally regulated changes in gene expression as a function of pregnancy, embryonic tissue transplantation of individual knockouts of the *Ihh* and *Shh* genes showed no overt developmental phenotype [56, 57]. The *Dhh* knockout has not been analyzed for a mammary phenotype, but homozygous females are able to feed their pups successfully, suggesting no major impairments in mammary gland function [58]. Hedgehog ligands can compensate for one another functionally, thus, the lack of phenotype in single gene knockouts might be explained by functional redundancy.

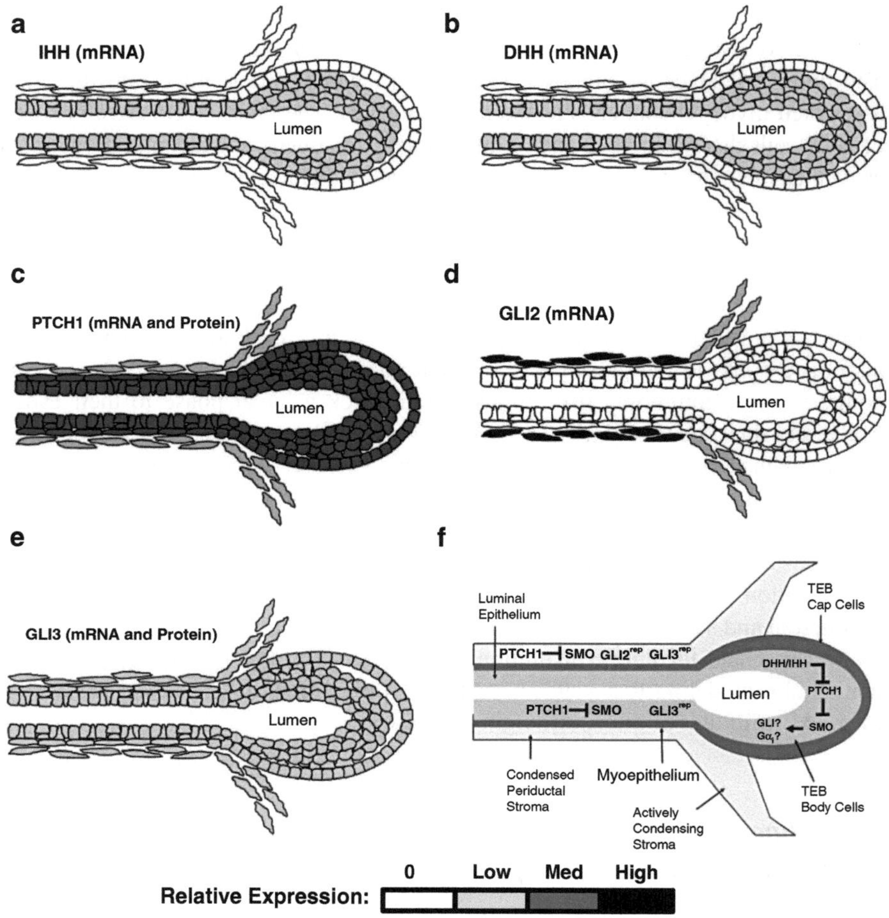


**Fig. 9.2** Schematic representation of Hedgehog network gene expression patterns and associated working model for Hedgehog network function in ductal development. A scale of relative expression is shown below the figure. (**a**) Low-level expression of *Ihh* mRNA by in situ hybridization. (**b**) Low-level expression of *Dhh* mRNA by in situ hybridization. (**c**) Robust expression of *Ptch1* mRNA and protein in both epithelium and stroma. (**d**) Stroma-limited expression of *Gli2* mRNA. (**e**) Widespread low-level expression of *Gli3* mRNA and protein in both epithelium and stroma. (**f**) Schematic diagram of a TEB and subtending duct overlayed with a working model for Hedgehog network gene function during ductal development based on gene expression and phenotypic analysis of genetically engineered mice. Expression and function of SMO is postulated, but not yet supported by in vivo experimentation

*Ptch1* mRNA and protein are expressed in both epithelial and stromal compartments of the virgin mouse mammary gland (Fig. 9.2c). Coordinate with the increase in *Ihh* mRNA levels during pregnancy and lactation, expression of *Ptch1* mRNA also appears elevated during pregnancy and lactation [53, 55]. Further, coordinate with *Ihh* expression in involution, *Ptch1* message becomes undetectable in early involution but becomes detectable again after gland remodeling.

Loss-of-function studies using several different *Ptch1* mutants have demonstrated a role for *Ptch1* for normal patterning and elongation of the mammary ductal tree [53, 59, 60]. Heterozygous loss of *Ptch1* (*Ptch1*$^{Tm1Mps}$ allele) led to mild TEB defects, and abnormal proliferation and accumulation of luminal epithelial cells within the ducts. The accumulated cells were lost during pregnancy, but returned during involution, thus demonstrating conditional haploinsufficiency for *Ptch1* in the nonpregnant animal [53]. Homozygous *Ptch1*$^{mes}$/*Ptch1*$^{mes}$ (a hypomorphic allele) mice showed failure of ductal elongation in a high percentage of glands examined, as well as ductal dysplasia in "escape" ducts that did form. Finally, complete loss of Ptch1 in the epithelium using a *Cre*-dependent conditional null allele led to dilated and dysmorphic ducts, but did not compromise ductal elongation or cause accumulation of cells within the ducts as did *Ptch1* heterozygosity. Subsequent transplantation studies demonstrated that *Ptch1* was required in both mammary epithelium and stroma for appropriate mammary gland development [53, 60]. Additionally, *Ptch1* appears to be required in the pituitary since failure of ductal elongation in homozygous *Ptch1*$^{mes}$/*Ptch1*$^{mes}$ mice could be rescued by a pituitary isograft [60].

Loss of Ptch1 function in heterozygous *Ptch1*$^{Tm1Mps/+}$ mice has been associated with forfeiture of quiescence, increased proliferation, and expansion of a progenitor cell pool via differential regulation of the *TP63* promoter [59]. However, because these studies were conducted in intact mice rather than via epithelial transplantation, it is still somewhat unclear whether this effect is due to local or systemic functions, or some combination thereof. Nevertheless, the *Ptch1*$^{Tm1Mps/+}$ studies complement work using a constitutively activated Smo transgene driven by the Mouse Mammary Tumor Virus promoter (*MMTV-SmoM2*) selectively in mammary epithelium. Glands in virgin *MMTV-SmoM2* mice showed increased proliferation and ductal dysplasias, but did not show accumulation of cells within the duct, thereby demonstrating that *Ptch1* loss was not functionally equivalent to activation of SMO strictly in the epithelium [61]. However, consistent with the *Ptch1*$^{Tm1Mps/+}$ data, *MMTV-SmoM2* activation led to an approximately twofold *decrease* in regenerative stem cell frequency, but an approximately twofold *increase* in a progenitor cell pool capable of anchorage-independent growth as mammospheres.

These data remain to be reconciled with data using human mammary cells in which activated Hedgehog signaling was associated with self-renewal of normal mammary epithelial stem cells [62, 63]. Treatment of normal cells with SHH ligand increased the mammosphere-forming potential and self-renewal of multipotent epithelial cells capable of giving rise to mixed luminal and myoepithelial colonies. Treatment with cyclopamine, as expected, decreased the frequency of these cells. In addition, genetic knockdown of *Gli1* and *Gli2* resulted in reduced anchorage-independent growth in mammosphere cultures, while overexpression of *Gli2* in human cells led to hyperplasia upon transplantation into immunocompromised hosts. One plausible explanation for the apparent discrepancy between the mouse and human data is the observation that alveolar progenitor cells are capable of giving rise to both luminal and a distinct population of alveolus-associated myoepithelial cells [64, 65]. Given that the relative frequency of alveolar cell types is far greater in human vs. mouse (as mentioned above), and that activated Hedgehog signaling may play a role in alveolar differentiation [55], it is possible that ligand treatment acts to promote self-renewal of a bipotent alveolar progenitor cell capable of anchorage-independent growth, rather than the most primitive regenerative stem cell.

*Ptch2*, although detectable by gene expression analysis in mature virgin mice, has not been studied in mammary gland development. *Ptch2* knockout mice are viable, and no defects in lactation were reported. As with Hedgehog ligands, functional compensation by *Ptch1* is possible [66].

SMO protein is not detectable by immunolocalization using currently available antibodies in either the mouse or human mammary gland, but mRNA can be detected in the mouse [61]. Whether *Smo* is required in mammary gland development, and thus whether canonical (or non-canonical) signaling is required during mammary gland development is still an open question. The requirement for *Smo* is currently being tested in our laboratory using a conditional, *Cre*-dependent disruption allele.

Expression and functional analyses of the *Gli* family of transcription factor genes in mammary epithelium have yielded somewhat ambiguous results. Using a *Gli1-lacZ* knock-in reporter line (*Gli1*$^{lzki}$), lacZ activity was only detected in lymphatic vessels in both the embryonic and adult mammary gland, but not in mammary ducts or alveoli [67]. Consistent with lack of expression in mammary epithelium, homozygous *Gli1*$^{lzki}$ loss-of-function had no phenotype as a single gene mutation. On the contrary, targeted overexpression of Gli1 in mammary epithelium via the MMTV promoter (MMTV-Gli1) led to impaired lobuloalveolar development and lactation defects in transgenic female mice [68].

*Gli2* mRNA was detected by in situ hybridization and localized exclusively to the periductal stroma during virgin development (Fig. 9.2d) [55]. However, expression was both epithelial and stromal during pregnancy and lactation, and levels appeared to increase coordinately with those of *Ihh* and *Ptch1* mRNAs consistent with activation of the canonical signaling network during these phases of development [55]. Transplantation analysis of whole embryonic mammary glands derived from a targeted disruption mutant for *Gli2* revealed that *Gli2* function is required for normal mammary duct development. However, transplantation of epithelial fragments from homozygous mutant embryos into cleared fat pads of immune compromised mice failed to recapitulate the dysplastic ductal phenotypes observed in whole gland transplantation, indicating that *Gli2* functions primarily in the mammary stroma to affect mammary epithelial cell behavior [55]. Because ligand levels appear to be low during virgin phases of development, GLI2 protein is predicted to be in its repressor state (Fig. 9.2d, f).

In a different study [67], both *Gli2* and *Gli3* were found to be expressed in stromal cells in the virgin (Fig. 9.2d, e), and in myoepithelial cells after pregnancy. *Gli3* was also expressed in luminal epithelial cells, and was shown to be essential in the somites for proper formation of the embryonic mammary buds from surface ectoderm [67, 69]. There was no requirement for either *Gli1* or *Gli2* in embryonic mammary gland development. Together, these data suggested that the Hedgehog network must remain inactive for appropriate embryonic and pubertal ductal development of the mammary gland [67]. However, as stated above, the phenotypic effect of the complete loss of Hedgehog signaling by disruption of Smo has not yet been evaluated fully.

## *Role of Primary Cilia in Mammary Gland Hedgehog Signaling*

The existence of cilia in mammary epithelium had not been carefully examined until recent work by McDermott et al. [70]. This study examined the distribution of cilia in the mouse mammary gland and found that epithelial, myoepithelial, and stromal cells all contained cilia. Cilia distribution in luminal epithelial cells was highest during pubertal development (~17%) and less abundant as the animal matured, leveling out at ~4% in both the virgin and pregnant adult animal. Disruption of cilia led to a reduction in branching morphogenesis in the mammary ductal tree. Decreased branching was accompanied by an increase in Wnt signaling, but a decrease in canonical Hedgehog signaling as defined by reduced *Gli1* mRNA expression [70].

## *The Hedgehog Network in Breast Cancer*

The first evidence for a role of the Hedgehog network in cancer came from the study of patients with Gorlin's syndrome. Individuals with this disease carry inherited loss-of-function mutations in the *Ptch1* gene, and are strongly predisposed to BCC [71, 72]. Since then, activated Hedgehog signaling has been implicated in medulloblastoma, glioblastoma, rhabdomyosarcoma, and melanoma as well as in cancers of the breast, pancreas, lung, prostate, gastrointestinal system, and hematopoietic system, among others. The role of the Hedgehog network in these malignancies is reviewed in other chapters of this book (see also [73]).

A potential role for activated Hedgehog signaling in breast cancer was postulated almost immediately upon identification of mutations in *Ptch1* associated with Gorlin's syndrome and BCC. However, Gorlin's syndrome patients do not show increased risk of breast cancer. Thus, significant evidence supporting such a role has been lacking (reviewed in [74, 75]).

Analysis of mutations in breast cancers has thus far shown little evidence that mutation of Hedgehog signaling genes are common [76], with most studies failing to identify mutations [77, 78]. More recent genomic sequencing efforts identified three missense mutations in the *Gli1* gene in 11 breast cancer samples examined. The functional significance of these mutations is not known [79]. At the genomic level, array comparative genomic hybridization analyses indicate that genomic loss at the *Ptch1* locus was the fourth most commonly detected change among the tumor suppressor genes identified in the study, occurring in 19% of human breast cancers and 33% of breast cancer cell lines [80]. Amplification of the *Gli1* gene has also been demonstrated [81].

Data related to expression of Hedgehog network genes in human breast cancer are currently ambiguous, most likely due to use of a variety of unvalidated, or poorly validated, commercially available immunoreagents. The discrepancies among the various expression studies recounted below remain to be resolved.

Initial immunohistochemical staining studies suggested that Hedgehog signaling is activated in a majority of human invasive breast cancers (IBC) based on ectopic expression of PTCH1 and GLI1 [82], which were not detected in normal tissue. Another small study using patient-matched samples showed SHH, PTCH1, and GLI1 expression [83] in both normal and cancer tissues, with about half of the samples showing increased expression of at least one protein in cancerous epithelium relative to normal. A recent publication [84] analyzed 21 normal breast samples and 121 invasive ductal carcinomas by immunohistochemistry for expression of IHH, PTCH1, SMO, GLI1, GLI2, and GLI3. For all six proteins, expression was higher in invasive ductal carcinoma relative to normal breast epithelia, with several correlations between expression of individual genes with clinical biomarkers and behaviors. Expression of GLI1 protein has been associated with the emergence of estrogen receptor negative (ER–) breast cancer, as well as with increased survival of ER– breast cancer cells. Further, high GLI1 expression has been associated with poor clinical outcome [85, 86].

Some aspects of these studies conflict with other published analyses. Using a relatively large panel of normal, ductal carcinomas in situ (DCIS), and IBC samples [61], SMO protein expression was undetectable in normal breast, but was ectopically expressed in ~70% of DCIS and ~30% of IBC [61]. Specificity of the SMO antibody was controlled by detection of MMTV-SmoM2 transgene expression in the mouse mammary gland, whereas wild type animals showed no detectable expression. Increased SMO protein expression in human breast cancer was consistent with Q-PCR results of Mukherjee et al. [83], which showed increased mRNA expression in ~40% of their samples.

Contrary to SMO, PTCH1 protein was detectable at moderate levels throughout the epithelium, and in isolated stromal cells of the normal breast. This pattern was observed with two independent antibodies, and was consistent with expression patterns in the mouse mammary gland assayed by both immunolocalization and in situ hybridization [53, 61]. Consistent with the observation of genomic loss at the *Ptch1* locus, PTCH1 protein expression was decreased or absent in ~50% of DCIS and IBC. Reduced PTCH1 expression in this study was in remarkable agreement with an independent study by Wolf et al. [87] in which epigenetic silencing of the *Ptch1* gene, as well as reduced protein expression in clinical specimens, was demonstrated in ~50% of the samples examined.

If, in fact, Hedgehog signaling is ectopically activated in human breast cancer, activated signaling may influence breast cancer development in a number of ways in addition to those mentioned above. Perhaps most important from both a biological and clinical perspective, Hedgehog signaling has been implicated in the regulation of self-renewal of the CD44$^+$;CD24$^-$ population of breast cancer stem cells (tumor-initiating cells) [62, 63]. One recent study suggests that activated signaling (via expression of activated SMO) can compromise the tumor suppressor function of TP53 (p53) by promoting MDM2-mediated TP53 degradation [88]. With respect to metastasis, osteolytic behavior of MDA-MB-231 cells was dependent on *Gli2* [89].

Despite provocative indications of an importance of activated Hedgehog signaling in human breast cancer, and the development of ductal hyperplasia and dysplasia in several genetically engineered mouse models [53, 55, 60, 61], long-term tumor formation studies in either *Ptch1* heterozygotes or MMTV:*SmoM2* transgenic mice did not indicate increased frequency of mammary tumors [61, 90]. As the sole exception, conditional overexpression of *Gli1* was recently shown to result in tumor development [91] with multiple histopathologies and expression of basal cell type markers. Together, these data suggest that inappropriate activation, or de-repression, of Hedgehog network target genes may have important consequences for tumor growth and behavior, but that activated signaling may not be the primary driver of tumorigenesis as it is in cancers such as BCC and medulloblastoma.

**Table 9.1** Mammary gland phenotypes in Hedgehog pathway mouse models

| | Mouse model | Effect on pathway | Mammary gland phenotype | Epithelial/stromal role | References |
|---|---|---|---|---|---|
| Ligands | *Shh*$^{-/-}$ (*Rescue Transplants*) | Decreased activity | No overt phenotype | – | [56, 57] |
| | *Ihh*$^{-/-}$ (*Rescue Transplants*) | Decreased activity | No overt phenotype | – | [56] |
| Receptors | *Ptch1*$^{Tm1Mps/+}$ | Increased activity | Aberrant TEBs, ductal dysplasia. No embryonic phenotype | Both | [53, 59] |
| | *Ptch1*$^{Mes}$ | ? | Ductal dysplasia | Both | [60] |
| | *MMTV-Cre*; *Ptch1*$^{fl/fl}$ | Increased activity | Dilated ducts | Epithelial | [60] |
| Transducer | *MMTV-SmoM2* | Increased activity | Increased branching and budding | Epithelial | [61] |
| Effectors | *MMTV-Gli1* | Increased activity | Impaired lobuloalveolar development | Epithelial | [91] |
| | *Gli1*$^{lzkl/lzkl}$ | Decreased activity | No overt phenotype | – | [67] |
| | *Gli3*$^{ext/+}$ | Increased activity | Aberrant mammary placode formation | Both | [67] |
| | *Gli2*$^{lzki/lzki}$ | Decreased activity | No overt phenotype | – | [67] |
| | *Gli2*$^{-/-}$ (*Rescue Transplants*) | Decreased activity | Aberrant ductal morphogenesis | Stromal | [55] |

## Hedgehog Signaling as a Therapeutic Target in Breast Cancer

A number of Hedgehog signaling antagonists have been identified and characterized (Table 9.2) (exhaustively reviewed in [92] and references therein). Antagonists include a group of plant-derived steroidal alkaloids (e.g. cyclopamine and jervine) first identified as potent teratogens in sheep, rodents, and other vertebrates [93–95]. Compounds in this chemical class directly bind SMO to inhibit downstream signaling [96–98]. In addition to these naturally occurring antagonists, several other Hedgehog signaling agonists and antagonists have been identified or synthesized that target either the SHH ligand, SMO, or the downstream GLI transcription factors (Table 9.2). Most recently, a screen of FDA-approved compounds identified the antifungal agent itraconazole as a potent inhibitor of SMO, but in a manner distinct from that of cyclopamine and related compounds [99].

In preclinical studies, cyclopamine and CUR0199691 have been used in vitro to treat a limited panel of breast cancer cell lines [82, 83, 100]. Results are generally consistent across studies, with cyclopamine doses of 10 μM or higher leading to significant inhibition of cell growth via both reduction of proliferation and induction of apoptosis. However, the specificity of these compounds at the doses required for inhibition is in question due to the observation that these two compounds showed activity against two different subsets of cell lines. Further, activity did not correlate with detectable expression of *Smo* mRNA, nor did activity correlate with the ability of cell lines to respond to treatment with recombinant dual lipid-modified SHH ligand [100].

While in vivo tests of Hedgehog signaling inhibitors have yet to be reported for the treatment of breast cancers, preclinical studies in mice using SMO inhibitors have shown promising effects in prevention of metastases from pancreatic cancer [101], as well as inhibition of tumor growth in medulloblastoma [102].

Several of the Hedgehog signaling modulators listed in Table 9.2 are entering clinical trials (http://www.clinicaltrials.gov). GCD-0449 has completed a phase I clinical trial in patients with locally advanced or metastatic BCC with measurable responses in 29 of 33 patients (including two complete responses), with no dose-limiting toxicities [103, 104]. This compound is now in phase II trials in patients with a variety of cancers including advanced or metastatic BCC, pancreatic cancer, gastric cancers, colorectal, and ovarian cancers. In advanced breast cancer, GCD-0449 is being investigated in combination with a gamma secretase inhibitor (RO4929097) to block Notch signaling based on data indicating that both Hedgehog and Notch networks regulate mammary stem cell self-renewal.

The Pfizer SMO inhibitor PF-04449913 has entered a phase I clinical trial in patients with hematological malignancies including CML for use either alone or in combination with Dasatinib (a c-src inhibitor). The BMS compound BMS-833923 (XL139) has entered a phase I clinical trial for patients with BCC. The infinity compound IPI-926 has entered a phase 1 study in patients with advanced and/or

**Table 9.2** Selected small molecule modulators of Hedgehog signaling

| Activity | Name | Target | Company | Usage | A.K.A |
|---|---|---|---|---|---|
| Agonists | *Purmorphamine* | SMO | | Basic/preclinical | |
| | *SAG* | SMO | | Basic/preclinical | |
| Antagonists | *Cyclopamine* | SMO | Natural product | Basic/preclinical | |
| | *Jervine* | SMO | Natural product | Basic/preclinical | |
| | *SANT1* | SMO | | Basic/preclinical | |
| | *SANT2* | SMO | | Basic/preclinical | |
| | *SANT19* | SMO | | Basic/preclinical | |
| | *SANT74* | SMO | | Basic/preclinical | |
| | *SANT75* | SMO | | Basic/preclinical | |
| | *Cur61414* | SMO | Curis | Basic/preclinical | |
| | *Cur0199691* | SMO | Curis | Basic/preclinical | HhAntag 691 |
| | *GDC-0449* | SMO | Curis/Genentech/Roche | Clinical | RG3616 |
| | *IPI926* | SMO | Infinity Pharmaceuticals | Clinical | |
| | *LDE225* | SMO | Novartis | Clinical (planned) | |
| | *XL-139* | SMO | Bristol-Meyers Squibb | Clinical | BMS-833923 |
| | *PF-04449913* | SMO | Pfizer | Clinical (planned) | |
| | *Itraconazole* | SMO | Janssen Pharmaceutica Products | Antifungal/preclinical | Sporanox |
| | *Unknown* | SMO | Amgen | | |
| | *Unknown* | SMO | Eli Lilly | | |
| | *Unknown* | SMO | Takeda | | |
| | *GANT58* | GLI | | Basic/preclinical | |
| | *GANT61* | GLI | | Basic/preclinical | |
| | *Physalin F* | GLI | | Basic/preclinical | |
| | *Physalin B* | GLI | | Basic/preclinical | |

(continued)

**Table 9.2** (continued)

| Activity | Name | Target | Company | Usage | A.K.A |
|---|---|---|---|---|---|
| | *NMDA298-1* | GLI | | Basic/preclinical | |
| | *JK-184* | GLI | | Basic/preclinical | |
| | *Robotnikinin* | SHH | | Basic/preclinical | |
| | *HPI-1* | Unknown | | | |
| | *HPI-2* | Unknown | | | |
| | *HPI-3* | Unknown | | | |
| | *HPI-4* | Unknown | | | |

metastatic solid tumor malignancies. Similarly, the Novartis compound LDE225 is also in phase I trials in patients with advanced solid tumors as well as BCC and medulloblastoma. Since advanced solid tumors will include breast cancers, results from these phase I and phase II trials should be informative with respect to potential efficacy against advanced breast cancers.

## References

1. Nusslein-Volhard C, Wieschaus E (1980) Mutations affecting segment number and polarity in Drosophila. Nature 287:795–801
2. McMahon AP, Ingham PW, Tabin CJ (2003) Developmental roles and clinical significance of Hedgehog signaling. Curr Top Dev Biol 53:1–114
3. Jia J, Jiang J (2006) Decoding the Hedgehog signal in animal development. Cell Mol Life Sci 63:1249–1265
4. Duncan AW et al (2005) Integration of Notch and Wnt signaling in hematopoietic stem cell maintenance. Nat Immunol 6:314–322
5. Dontu G et al (2004) Role of Notch signaling in cell-fate determination of human mammary stem/progenitor cells. Breast Cancer Res 6:R605–R615
6. Beachy PA, Karhadkar SS, Berman DM (2004) Tissue repair and stem cell renewal in carcinogenesis. Nature 432:324–331
7. Handrigan GR, Richman JM (2009) Autocrine and paracrine Shh signaling are necessary for tooth morphogenesis, but not tooth replacement in snakes and lizards (Squamata). Dev Biol 337:171–186. doi:S0012-1606(09)01279-2 [pii] 10.1016/j.ydbio.2009.10.020
8. Singh RR et al (2010) Hedgehog signaling pathway is activated in diffuse large B-cell lymphoma and contributes to tumor cell survival and proliferation. Leukemia 24:1025–1036. doi:leu201035 [pii] 10.1038/leu.2010.35
9. Yang L et al (2008) Sonic Hedgehog is an autocrine viability factor for myofibroblastic hepatic stellate cells. J Hepatol 48:98–106. doi:S0168-8278(07)00528-4 [pii] 10.1016/j.jhep.2007.07.032
10. Allen BL, Tenzen T, McMahon AP (2007) The Hedgehog-binding proteins Gas1 and Cdo cooperate to positively regulate Shh signaling during mouse development. Genes Dev 21:1244–1257. doi:21/10/1244 [pii] 10.1101/gad.1543607
11. Martinelli DC, Fan CM (2007) Gas1 extends the range of Hedgehog action by facilitating its signaling. Genes Dev 21:1231–1243. doi:21/10/1231 [pii] 10.1101/gad.1546307
12. Tenzen T et al (2006) The cell surface membrane proteins Cdo and Boc are components and targets of the Hedgehog signaling pathway and feedback network in mice. Dev Cell 10:647–656. doi:S1534-5807(06)00165-1 [pii] 10.1016/j.devcel.2006.04.004
13. Yao S, Lum L, Beachy P (2006) The ihog cell-surface proteins bind Hedgehog and mediate pathway activation. Cell 125:343–357. doi:S0092-8674(06)00378-3 [pii] 10.1016/j.cell.2006.02.040
14. Hooper JE, Scott MP (2005) Communicating with Hedgehogs. Nat Rev Mol Cell Biol 6:306–317
15. Varjosalo M, Taipale J (2007) Hedgehog signaling. J Cell Sci 120:3–6
16. Riobo NA, Manning DR (2007) Pathways of signal transduction employed by vertebrate Hedgehogs. Biochem J 403:369–379
17. Corbit KC et al (2005) Vertebrate Smoothened functions at the primary cilium. Nature 437:1018–1021
18. Huangfu D, Anderson KV (2005) Cilia and Hedgehog responsiveness in the mouse. Proc Natl Acad Sci USA 102:11325–11330

19. Ray K et al (1999) Kinesin-II is required for axonal transport of choline acetyltransferase in Drosophila. J Cell Biol 147:507–518
20. Han YG, Kwok BH, Kernan MJ (2003) Intraflagellar transport is required in Drosophila to differentiate sensory cilia but not sperm. Curr Biol 13:1679–1686
21. Avidor-Reiss T et al (2004) Decoding cilia function: defining specialized genes required for compartmentalized cilia biogenesis. Cell 117:527–539
22. Ocbina PJ, Anderson KV (2008) Intraflagellar transport, cilia, and mammalian Hedgehog signaling: analysis in mouse embryonic fibroblasts. Dev Dyn 237:2030–2038. doi:10.1002/dvdy.21551
23. Rohatgi R, Scott MP (2007) Patching the gaps in Hedgehog signalling. Nat Cell Biol 9:1005–1009
24. Wong SY, Reiter JF (2008) The primary cilium at the crossroads of mammalian Hedgehog signaling. Curr Top Dev Biol 85:225–260. doi:S0070-2153(08)00809-0 [pii] 10.1016/S0070-2153(08)00809-0
25. Johnson RL, Milenkovic L, Scott MP (2000) In vivo functions of the patched protein: requirement of the C terminus for target gene inactivation but not Hedgehog sequestration. Mol Cell 6:467–478
26. Lewis MT (2001) Hedgehog signaling in mouse mammary gland development and neoplasia. J Mammary Gland Biol Neoplasia 6:53–66
27. Guerrero I, Ruiz i Altaba A (2003) Development. Longing for ligand: Hedgehog, patched, and cell death. Science 301:774–776
28. Lewis MT, Veltmaat JM (2004) Next stop, the twilight zone: Hedgehog network regulation of mammary gland development. J Mammary Gland Biol Neoplasia 9:165–181
29. Thibert C et al (2003) Inhibition of neuroepithelial patched-induced apoptosis by sonic Hedgehog. Science 301:843–846
30. Barnes EA, Kong M, Ollendorff V, Donoghue DJ (2001) Patched1 interacts with cyclin B1 to regulate cell cycle progression. EMBO J 20:2214–2223
31. Adolphe C, Hetherington R, Ellis T, Wainwright B (2006) Patched1 functions as a gatekeeper by promoting cell cycle progression. Cancer Res 66:2081–2088
32. Chang H, Li Q, Moraes RC, Lewis MT, Hamel PA (2010) Activation of Erk by sonic Hedgehog independent of canonical Hedgehog signalling. Int J Biochem Cell Biol 42:1462–1471. doi:S1357-2725(10)00167-6 [pii] 10.1016/j.biocel.2010.04.016
33. Malbon CC (2005) G proteins in development. Nat Rev Mol Cell Biol 6:689–701
34. Hammerschmidt M, McMahon AP (1998) The effect of pertussis toxin on zebrafish development: a possible role for inhibitory G-proteins in Hedgehog signaling. Dev Biol 194:166–171
35. Fremion F et al (1999) The heterotrimeric protein Go is required for the formation of heart epithelium in Drosophila. J Cell Biol 145:1063–1076
36. DeCamp DL, Thompson TM, de Sauvage FJ, Lerner MR (2000) Smoothened activates Galphai-mediated signaling in frog melanophores. J Biol Chem 275:26322–26327
37. Riobo NA, Saucy B, Dilizio C, Manning DR (2006) Activation of heterotrimeric G proteins by Smoothened. Proc Natl Acad Sci USA 103:12607–12612
38. Kasai K et al (2004) The G12 family of heterotrimeric G proteins and Rho GTPase mediate Sonic Hedgehog signalling. Genes Cells 9:49–58
39. Riobo NA, Lu K, Emerson CP Jr (2006) Hedgehog signal transduction: signal integration and cross talk in development and cancer. Cell Cycle 5:1612–1615. doi:3130 [pii]
40. Yam PT, Langlois SD, Morin S, Charron F (2009) Sonic Hedgehog guides axons through a noncanonical Src-family-kinase-dependent signaling pathway. Neuron 62:349–362
41. Chinchilla P, Xiao L, Kazanietz MG, Natalia AN, Riobo GM (2010) Hedgehog proteins activate pro-angiogenic responses in endothelial cells through non-canonical signaling pathways. Cell Cycle 9:570–579. doi:10591 [pii]
42. Chen W et al (2004) Activity-dependent internalization of smoothened mediated by beta-arrestin 2 and GRK2. Science 306:2257–2260
43. Meloni AR et al (2006) Smoothened signal transduction is promoted by G protein-coupled receptor kinase 2. Mol Cell Biol 26:7550–7560

44. Molnar C, Holguin H, Mayor F Jr, Ruiz-Gomez A, de Celis JF (2007) The G protein-coupled receptor regulatory kinase GPRK2 participates in Hedgehog signaling in Drosophila. Proc Natl Acad Sci USA 104:7963–7968
45. Ogden SK et al (2008) G protein Galphai functions immediately downstream of Smoothened in Hedgehog signalling. Nature 456:967–970
46. Nolan-Stevaux O et al (2009) GLI1 is regulated through Smoothened-independent mechanisms in neoplastic pancreatic ducts and mediates PDAC cell survival and transformation. Genes Dev 23:24–36
47. Katoh Y, Katoh M (2009) Integrative genomic analyses on GLI1: positive regulation of GLI1 by Hedgehog-GLI, TGFbeta-Smads, and RTK-PI3K-AKT signals, and negative regulation of GLI1 by Notch-CSL-HES/HEY, and GPCR-Gs-PKA signals. Int J Oncol 35:187–192
48. Das S et al (2009) The Hedgehog pathway transcription factor, GLI1 promotes malignant behavior of cancer cells by upregulating osteopontin. J Biol Chem 284:22888–22897
49. Wang C, Ruther U, Wang B (2007) The Shh-independent activator function of the full-length Gli3 protein and its role in vertebrate limb digit patterning. Dev Biol 305:460–469
50. Russo J, Russo IH (2004) Development of the human breast. Maturitas 49:2–15. doi:10.1016/j.maturitas.2004.04.011 S0378-5122(04)00204-X [pii]
51. Master SR et al (2002) Functional microarray analysis of mammary organogenesis reveals a developmental role in adaptive thermogenesis. Mol Endocrinol 16:1185–1203
52. Watson CJ, Khaled WT (2008) Mammary development in the embryo and adult: a journey of morphogenesis and commitment. Development 135:995–1003
53. Lewis MT et al (1999) Defects in mouse mammary gland development caused by conditional haploinsufficiency of Patched-1. Development 126:5181–5193
54. Kouros-Mehr H, Werb Z (2006) Candidate regulators of mammary branching morphogenesis identified by genome-wide transcript analysis. Dev Dyn 235:3404–3412
55. Lewis MT et al (2001) The Gli2 transcription factor is required for normal mouse mammary gland development. Dev Biol 238:133–144. doi:10.1006/dbio.2001.0410 S0012160601904105 [pii]
56. Gallego MI, Beachy PA, Hennighausen L, Robinson GW (2002) Differential requirements for shh in mammary tissue and hair follicle morphogenesis. Dev Biol 249:131–139
57. Michno K, Boras-Granic K, Mill P, Hui CC, Hamel PA (2003) Shh expression is required for embryonic hair follicle but not mammary gland development. Dev Biol 264:153–165
58. Bitgood MJ, Shen L, McMahon AP (1996) Sertoli cell signaling by Desert Hedgehog regulates the male germline. Curr Biol 6:298–304. doi:S0960-9822(02)00480-3 [pii]
59. Li N et al (2008) Reciprocal intraepithelial interactions between TP63 and Hedgehog signaling regulate quiescence and activation of progenitor elaboration by mammary stem cells. Stem Cells 26:1253–1264. doi:2007-0691 [pii] 10.1634/stemcells.2007-0691
60. Moraes RC et al (2009) Ptch1 is required locally for mammary gland morphogenesis and systemically for ductal elongation. Development 136:1423–1432. doi:dev.023994 [pii] 10.1242/dev.023994
61. Moraes RC et al (2007) Constitutive activation of smoothened (SMO) in mammary glands of transgenic mice leads to increased proliferation, altered differentiation and ductal dysplasia. Development 134:1231–1242
62. Liu S et al (2006) Hedgehog signaling and Bmi-1 regulate self-renewal of normal and malignant human mammary stem cells. Cancer Res 66:6063–6071
63. Tanaka H et al (2009) The Hedgehog signaling pathway plays an essential role in maintaining the CD44+CD24–/low subpopulation and the side population of breast cancer cells. Anticancer Res 29:2147–2157. doi:29/6/2147 [pii]
64. Boulanger CA, Wagner KU, Smith GH (2005) Parity-induced mouse mammary epithelial cells are pluripotent, self-renewing and sensitive to TGF-beta1 expression. Oncogene 24:552–560
65. Matulka LA, Triplett AA, Wagner KU (2007) Parity-induced mammary epithelial cells are multipotent and express cell surface markers associated with stem cells. Dev Biol 303:29–44. doi:S0012-1606(06)01435-7 [pii] 10.1016/j.ydbio.2006.12.017

66. Nieuwenhuis E et al (2006) Mice with a targeted mutation of patched2 are viable but develop alopecia and epidermal hyperplasia. Mol Cell Biol 26:6609–6622. doi:26/17/6609 [pii] 10.1128/MCB.00295-06
67. Hatsell SJ, Cowin P (2006) Gli3-mediated repression of Hedgehog targets is required for normal mammary development. Development 133:3661–3670
68. Fiaschi M, Rozell B, Bergstrom A, Toftgard R, Kleman MI (2007) Targeted expression of GLI1 in the mammary gland disrupts pregnancy-induced maturation and causes lactation failure. J Biol Chem 282:36090–36101. doi:M704280200 [pii] 10.1074/jbc.M704280200
69. Veltmaat JM et al (2006) Gli3-mediated somitic Fgf10 expression gradients are required for the induction and patterning of mammary epithelium along the embryonic axes. Development 133:2325–2335
70. McDermott KM, Liu BY, Tlsty TD, Pazour GJ (2010) Primary cilia regulate branching morphogenesis during mammary gland development. Curr Biol 20:731–737. doi:S0960-9822(10)00236-8 [pii] 10.1016/j.cub.2010.02.048
71. Concordet JP et al (1996) Spatial regulation of a zebrafish patched homologue reflects the roles of sonic Hedgehog and protein kinase A in neural tube and somite patterning. Development 122:2835–2846
72. Levanat S et al (1996) A two-hit model for developmental defects in Gorlin syndrome. Nat Genet 12:85–87. doi:10.1038/ng0196-85
73. Teglund S, Toftgard R (2010) Hedgehog beyond medulloblastoma and basal cell carcinoma. Biochim Biophys Acta 1805:181–208. doi:S0304-419X(10)00006-5 [pii] 10.1016/j.bbcan.2010.01.003
74. O'Toole SA, Swarbrick A, Sutherland RL (2009) The Hedgehog signalling pathway as a therapeutic target in early breast cancer development. Expert Opin Ther Targets 13:1095–1103
75. Kasper M, Jaks V, Fiaschi M, Toftgard R (2009) Hedgehog signalling in breast cancer. Carcinogenesis 30:903–911
76. Xie J et al (1997) Mutations of the PATCHED gene in several types of sporadic extracutaneous tumors. Cancer Res 57:2369–2372
77. Wicking C et al (1998) No evidence for the H133Y mutation in SONIC HEDGEHOG in a collection of common tumour types. Oncogene 16:1091–1093. doi:10.1038/sj.onc.1201644
78. Vorechovsky I, Benediktsson KP, Toftgard R (1999) The patched/Hedgehog/smoothened signalling pathway in human breast cancer: no evidence for H133Y SHH PTCH and SMO mutations. Eur J Cancer 35:711–713
79. Sjoblom T et al (2006) The consensus coding sequences of human breast and colorectal cancers. Science 314:268–274
80. Naylor TL et al (2005) High resolution genomic analysis of sporadic breast cancer using array-based comparative genomic hybridization. Breast Cancer Res 7:R1186–R1198
81. Nessling M et al (2005) Candidate genes in breast cancer revealed by microarray-based comparative genomic hybridization of archived tissue. Cancer Res 65:439–447. doi:65/2/439 [pii]
82. Kubo M et al (2004) Hedgehog signaling pathway is a new therapeutic target for patients with breast cancer. Cancer Res 64:6071–6074
83. Mukherjee S et al (2006) Hedgehog signaling and response to cyclopamine differ in epithelial and stromal cells in benign breast and breast cancer. Cancer Biol Ther 5:674–683
84. Xuan Y, Lin Z (2009) Expression of Indian Hedgehog signaling molecules in breast cancer. J Cancer Res Clin Oncol 135:235–240
85. Xu L et al (2009) Gli1 promotes cell survival and is predictive of a poor outcome in ERalpha-negative breast cancer. Breast Cancer Res Treat 123:59–71. doi:10.1007/s10549-009-0617-5
86. ten Haaf A et al (2009) Expression of the glioma-associated oncogene homolog (GLI) 1 in human breast cancer is associated with unfavourable overall survival. BMC Cancer 9:298. doi:1471-2407-9-298 [pii] 10.1186/1471-2407-9-298
87. Wolf I et al (2007) Unmasking of epigenetically silenced genes reveals DNA promoter methylation and reduced expression of PTCH in breast cancer. Breast Cancer Res Treat 105:139–155. doi:10.1007/s10549-006-9440-4

88. Abe Y et al (2008) Hedgehog signaling overrides p53-mediated tumor suppression by activating Mdm2. Proc Natl Acad Sci USA 105:4838–4843. doi:0712216105 [pii] 10.1073/pnas.0712216105
89. Sterling JA et al (2006) The Hedgehog signaling molecule Gli2 induces parathyroid hormone-related peptide expression and osteolysis in metastatic human breast cancer cells. Cancer Res 66:7548–7553. doi:66/15/7548 [pii] 10.1158/0008-5472.CAN-06-0452
90. Goodrich LV, Milenkovic L, Higgins KM, Scott MP (1997) Altered neural cell fates and medulloblastoma in mouse patched mutants. Science 277:1109–1113
91. Fiaschi M, Rozell B, Bergstrom A, Toftgard R (2009) Development of mammary tumors by conditional expression of GLI1. Cancer Res 69:4810–4817. doi:0008-5472.CAN-08-3938 [pii] 10.1158/0008-5472.CAN-08-3938
92. Peukert S, Miller-Moslin K (2010) Small-molecule inhibitors of the Hedgehog signaling pathway as cancer therapeutics. ChemMedChem 5:500–512. doi:10.1002/cmdc.201000011
93. Keeler RF (1970) Teratogenic compounds of *Veratrum californicum* (Durand) X. Cyclopia in rabbits produced by cyclopamine. Teratology 3:175–180
94. Keeler RF (1975) Teratogenic effects of cyclopamine and jervine in rats, mice and hamsters. Proc Soc Exp Biol Med 149:302–306
95. Dunn MK, Mercola M, Moore DD (1995) Cyclopamine, a steroidal alkaloid, disrupts development of cranial neural crest cells in Xenopus. Dev Dyn 202:255–270
96. Incardona JP, Gaffield W, Kapur RP, Roelink H (1998) The teratogenic Veratrum alkaloid cyclopamine inhibits sonic Hedgehog signal transduction. Development 125:3553–3562
97. Cooper MK, Porter JA, Young KE, Beachy PA (1998) Teratogen-mediated inhibition of target tissue response to Shh signaling. Science 280:1603–1607
98. Chen JK, Taipale J, Cooper MK, Beachy PA (2002) Inhibition of Hedgehog signaling by direct binding of cyclopamine to Smoothened. Genes Dev 16:2743–2748
99. Kim J et al (2010) Itraconazole, a commonly used antifungal that inhibits Hedgehog pathway activity and cancer growth. Cancer Cell 17:388–399. doi:S1535-6108(10)00070-X [pii] 10.1016/j.ccr.2010.02.027
100. Zhang X et al (2009) Cyclopamine inhibition of human breast cancer cell growth independent of Smoothened (Smo). Breast Cancer Res Treat 115:505–521. doi:10.1007/s10549-008-0093-3
101. Feldmann G et al (2008) An orally bioavailable small-molecule inhibitor of Hedgehog signaling inhibits tumor initiation and metastasis in pancreatic cancer. Mol Cancer Ther 7:2725–2735
102. Sasai K et al (2007) Medulloblastomas derived from Cxcr6 mutant mice respond to treatment with a smoothened inhibitor. Cancer Res 67:3871–3877. doi:0008-5472.CAN-07-0493 [pii] 10.1158/0008-5472.CAN-07-0493
103. Von Hoff DD et al (2009) Inhibition of the Hedgehog pathway in advanced basal-cell carcinoma. N Engl J Med 361:1164–1172. doi:NEJMoa0905360 [pii] 10.1056/NEJMoa0905360
104. Rudin CM et al (2009) Treatment of medulloblastoma with Hedgehog pathway inhibitor GDC-0449. N Engl J Med 361:1173–1178. doi:NEJMoa0902903 [pii] 10.1056/NEJMoa0902903

# Chapter 10
# Hedgehog Signaling and Cancer Treatment Resistance

**Yu-Jen Chen, Hui-Fen Liao, and Clifford Chao**

## Introduction

The hedgehog (HH) signaling is critical for growth and differentiation during embryonic development and is required for the maintenance of somatic stem cells [1]. In adult cells, HH signaling has been implicated in the maintenance of homeostasis of stem or progenitor cells in a number of epithelial tissues, including intestinal epithelia [2]. HH signaling also contributes to physiologic processes of epithelial repair and regeneration after injury [3]. However, aberrant activation of HH signaling in tumors from a wide range of tissues may allow escape from regulatory mechanisms that cause the return to quiescence that normally follows regeneration [4, 5]. Activation of HH signaling by binding of secreted HH ligands (Sonic, Indian, and Desert) to the membrane receptor Patched (PTCH) results in the nuclear translocation of the Gli family and initiation of HH-related gene expression [1, 6], including genes controlling the cell cycle, cell adhesion, signal transduction, angiogenesis, and apoptosis [7]. Several studies have shown that unregulated progenitor cell proliferation induced by abnormal Sonic hedgehog (SHH) signaling has a role in carcinogenesis [5, 8, 9]. For example, small cell lung carcinoma (SCLC), one of the identified malignancies with HH overexpression, could block the growth by the treatment of smoothened (Smo) inhibitor cyclopamine or a monoclonal antibody blocking SHH [10].

Y.-J. Chen (✉)
School of Medicine, National Yang Ming University, Taipei, Taiwan
and
Department of Radiation Oncology, Mackay Memorial Hospital, Taipei, Taiwan
and
Institute of Pharmacology, Taipei Medical University, Taipei, Taiwan
e-mail: chenmdphd@yahoo.com

J. Xie (ed.), *Hedgehog Signaling Activation in Human Cancer and its Clinical Implications*,
DOI 10.1007/978-1-4419-8435-7_10, © Springer Science+Business Media, LLC 2011

## Cancer Treatment Resistance

Cancer treatment, including radiotherapy (RT) and chemotherapy are considered effective for many types of cancers for clinical benefit and improvement in survival. Chemoradiotherapy (CRT) has been recommended as a standard treatment strategy for solid cancers, such as those arising from head and neck [11], cervix uteri [12], lung [13], esophagus [14], pancreas [15], stomach [16], and rectum [17]. Although genotoxic agents are strongly linked to tumorigenesis, the cytotoxic effect of DNA damage is also a critical facet of cancer therapy. In fact, the majority of human tumors treated with genotoxic agents possibly induced secondary primary malignancies and facilitated therapy-resistant forms [18, 19]. The self-renewal of cancer stem cells (CSCs), DNA repair, drug trafficking system, and other factors expressed in cancer cells are considered to prevent the injury by the therapy. For example, radiation therapy enhanced HSP90 chaperone function, causing radio-resistant lung cancer cells [20]. CRT, concurrent combination of RT and chemotherapy, in cervical cancer has been evaluated that improved treatment outcomes and/or maximize efficiency in comparison with RT alone [21]. Articles reported that the causes for chemotherapy failure in cancer treatment reside in multiple levels, such as poor vascularization, hypoxia, intratumoral high interstitial fluid pressure, and phenotypic resistance to drug-induced toxicity through upregulated xenobiotic metabolism or DNA repair mechanisms and silencing of apoptotic pathways [22, 23]. Factors that have been demonstrated with cancer treatment resistance are listed as follows:

1. *Growth factors* (*GFs*). In the cervical cancer patients, tumors with higher GF were more sensitive to radiation than those with low GF [24]. For instance, radiation-activated epidermal growth factor receptor (EGFR) signaling in cancers, such as non-small cell lung cancer (NSCLC), leading to radioresistance by inducing cell proliferation and enhanced DNA repair [20].
2. *Protumorigenic signaling cascade*. Some tumors cause rapid proliferation phenomenon, accelerated proliferation, by stimulation of irradiation during the course of RT. Accelerated proliferation increases dose required to control tumor cells and is an important cause of acquiring tumor radioresistance [21]. The phosphatidylinositol-3-kinase (PI3K)–Akt pathway involved in several human cancers is frequently upregulated [25], which may cause a tumorigenic phenotype with increased cell proliferation, metastasis, and angiogenesis. Akt inhibitors may significantly reduce viability of certain CSCs [26] and sensitize them to chemotherapeutic agents [27]. Inhibition of the Akt pathway further causes delayed repair of ionizing radiation (IR)-induced DNA double-strand breaks (DSB) formation and radiosensitization [28], indicating that the activation of Akt signaling may underlie at least some cases of radiation resistance [29].
3. *Hypoxia*. Research on an in vivo solid tumor demonstrated that contained a certain proportion of hypoxic fractions [30]. The existence of hypoxic cells is well recognized as one of the major factors causing radiation resistance which possibly results in local failure after RT [31].

4. *P53 and the factors relating apoptosis*. Apoptosis is an active mode of cell death which occurs in response to DNA damage by ionizing radiation, ultraviolet irradiation, and certain chemotherapeutic agents [32]. Mutations in proto-oncogenes or tumor suppressors, like Ras and p53, alters apoptosis signaling and changes the tumor microenvironment which traits of tumor cell resistance to therapy [33, 34] and subsequent tumor recurrence [35]. A large number of experimental studies have shown that apoptosis induced by irradiation is a determining factor of radiosensitivity [36].
5. *Molecular targeting agents (EGFR, COX2, and Mn-SOD)*. Molecular targeting agents, such as EGFR, COX2, and Mn-SOD, may be possible to efficiently increase radiosensitivity of cancer cells when given with RT and also eradicate subclinical metastases by themselves [21].

## The Role of Hedgehog Signaling in Cancer Treatment Resistance

HH ligands (Sonic, Desert, and Indian) bind to and antagonize the cell surface receptor PTCH, relieving the PTCH-mediated suppression of the transmembrane protein smoothened (Smo). Smo then initiates an intracellular signaling cascade that leads to the activation and nuclear translocation of the Gli family (Gli-1, 2, and 3). Gli family mediates transcription of genes controlling proliferation, differentiation, and survival of cells [1, 6]. Aberrations in hedgehog signaling have been found in cancers [37], resulting in overexpression of HH signaling pathway and an increase in endogenous production of HH ligands [4, 5]. Therefore, suppression of HH signaling might be a valid therapeutic option for overcoming drug resistance and for increasing the success of chemotherapy. Cui et al. analyzed 60 glioma samples, indicating that overexpression of Gli-1 is correlated with glioma recurrence after chemotherapy including VCR, VP16, CDDP, and ACNU [8]. Blockade of HH pathway enhanced cytotoxicity of chemotherapeutic agents in glioma cells through downregulating the expressions of MDR1, MRP1, MVP, MGMT, Bcl-2, and Survivin genes [38].

A growing body of evidence indicates that HH signaling plays an important role in regulating cancer treatment resistance. For example, IR-induced DSBs activate the PI3K-related kinases ATM and ATR, which regulate apoptosis, cell cycle progression, and DNA repair [39]. Research on basal cell nevus syndrome (BCNS; also known as Gorlin syndrome) patients and *Ptc1*$^{+/-}$ mice have shown a defect in the IR-induced activation of the ATR-Chk1 checkpoint signaling pathway (a pathway that serves as a barrier to the development of tumors), resulting in dramatically increases the incidence of tumors in *Ptcl*$^{+/-}$ mice [40]. Likewise, transient expression of Gli-1 disrupts Chk1 activation in human cells, suggesting that SHH signaling attenuates the activation of a genotoxin-triggered ATR–Chk1 checkpoint signal transduction pathway, and inappropriate SHH pathway activation promotes tumorigenesis by disabling a key signaling pathway that helps maintain genomic stability and inhibits tumorigenesis [40].

## Induction of Tumor Regrowth and Cancer Stem Cells

Stem cells and CSCs share some features, including signaling pathways to regulate self-renewal and differentiation [41]. Similar to normal stem cells, CSC are thought to be relatively quiescent, to be resistant to drugs and toxins, and to possess the DNA repair capacity [42]. For radiation sensitivity in cancer cells, the vast majority of experimental and clinical studies support four determinant phenomena in radiobiology: repair of DNA damage, redistribution of cells in the cell cycle, repopulation, and reoxygenation of hypoxic tumor areas [43]. The effectiveness of each radiation fraction decreases with increasing repopulation of tumor cells, suggesting repopulation by an RT-resistant progeny [44]. The mechanisms that underlie accelerated repopulation are poorly understood, but may involve the proliferation of previously quiescent treatment-resistant clonogenic cells or, CSC [45].

Signaling pathways, such as the Bmi-1, Notch, Wnt, and SHH pathways [46, 47] that support the dysregulated self-renewal and proliferation of CSC may be targets for regulating tumor regrowth and improving treatment outcomes [48, 49]. The investigations of CSC signaling activation during tumor repopulation suggest that the SHH pathway is an important target to regulate proliferation of surviving clonogens after concurrent chemoradiotherapy (CCRT) [50]. A significant upregulation of SHH and Gli-1 expression was observed in the majority of residual tumors after chemoradiotherapy, suggesting that HH signaling may contribute to cancer resistance [4]. Treatment of cancer cell lines with the HH-inhibitory compound cyclopamine results in downregulation of the proliferation marker Ki67 and reduced proliferation rates [5, 50], indicating that HH pathway activation may be essential for tumor growth and maintenance. Smo knockout studies in chronic myeloid leukemia (CML) CSCs (Bcr-Abl-driven $Lin^-/Sca1^+/c\text{-}Kit^+$) cells reduced their ability to form tumors in irradiated mice [51]. Smo antagonists inhibit the growth of these CML CSCs in vitro and prolong survival in vivo, importantly also in cells with resistance to the currently used Bcr-Abl inhibitors imatinib or nilotinib, suggesting that combination therapy might help to prevent relapses in this chronic disease [51]. Sims-Mourtada et al. indicated that the SHH signaling pathway was extensively activated in both chemoradiotherapy-resistant residual esophageal carcinoma specimens and animal tumor xenografts, causing the promotion of tumor repopulation and contribute to chemoradiation resistance through upregulation of the G1-cyclin-Rb axis [4].

## Anti-apoptosis and Cell Cycle Regulation

Research on SHH pathway has demonstrated that SHH contributes to the survival of cells by opposing the execution of intrinsic and extrinsic apoptotic cascades [52, 53]. Research in lymphocytes demonstrated that this effect of SHH signaling may go through the prevention of Fas-induced apoptosis [54]. Many malignant cells remain resistant to tumor necrosis factor-related apoptosis-inducing ligand (TRAIL) induced apoptosis, and blockade of HH using pharmacological and genetic tools sensitized the cells to TRAIL cytotoxicity [55]. Small interfering RNA (siRNA)-targeted knockdown of Gli-3, but not Gli-1 or Gli-2, restored receptors death receptor 4 (DR4) expression and TRAIL sensitivity, suggesting that modulation of HH–Gli pathway might be a new therapeutic approach for TRAIL-resistant neoplasms [55]. In B-cell chronic lymphocytic leukemia (B-CLL), selective downregulation of Gli-1 by antisense oligodeoxynucleotides results in decreased Bcl-2 expression and increased apoptosis, suggesting that Gli-1 may regulate Bcl-2 and, thereby, modulate cell survival [56]. For tumor cells, inhibition of SHH signaling has been shown to induce apoptosis in tumors through the activation of both intrinsic and extrinsic apoptosis cascades [57], such as SHH signaling increased Bcl-2 expression in BCCs [58].

Additionally, the sensitivity of cells to the cytotoxic effects of radiation is cell cycle dependent, with the S-phase being more resistant and the G1-S boundary and G2/M phase being more sensitive [4]. Sims-Mourtada et al. demonstrated that the treatment of SEG-1 cells with ionizing radiation alone led to a slight but not significant reduction in the radiation-resistant S-phase fraction. Treatment with HH inhibitors alone led to a significant reduction in the S-phase fraction, and the combination of radiation and HH inhibitors caused a greater reduction in the S-phase fraction compared with untreated cells [4]. In a mouse xerograft model, SHH–Gli-1 signaling pathway was shown a high association with the increase in proliferation and repopulation of esophageal cancer observed after CCRT [4].

## Repair of DNA Damage

DNA damage includes endogenous (such as oxidative metabolites) and external exposures (such as environmental pollution), causing single-strand breaks (SSB) and DSBs that may limit survival and the regenerative potential of cells [29]. Repair of DNA DSB can occur via nonhomologous end joining (NHEJ) or homologous recombination (HR). HR is required for a sister chromatid present in the S/G2 phase of replicating cells to provide an error-free template for DNA repair [59] while NHEJ is an error-prone repair mechanism that enzymatically modifies the two ends of a DNA break so that they are compatible for direct ligation [60]. UV light is known to induce DNA repair in irradiated cells through the upregulation of damaged DNA-binding (DDB) proteins, DDB1 (127 kDa) and DDB2 (48 kDa), which mediate a key process in nucleotide excision repair after UV damage [61, 62]. EGFR is also involved in DNA synthesis and DNA repair through its interactions with DNA proliferating cell nuclear antigen (PCNA) [63] and DNA-dependent protein kinase (DNA-PK), which is required for DNA repair [64]. Other factors, such as ionizing radiation, heat, $H_2O_2$, and cisplatin treatment, induce Ku70/80 and phosphatase I translocation to the nucleus and increase DNA-PK activity for initiation of DNA repair [64–66].

Drug resistance can, in some cases, be attributed to increased DNA repair response but may also result from a variety of other alterations, including decreased apoptotic signaling in response to this form of DNA damage [67, 68]. Articles reported that p53- and DNA mismatch repair (MMR) deficiency are two key genetic changes that have been associated with resistance to cisplatin [68, 69]. Frappart et al. have demonstrated that inactivation of the DNA repair factors, together with p53 loss, led to rapid medulloblastoma formation [60]. Genomic analysis of the tumors showed recurring chromosome 13 alterations via chromosomal loss or translocations involving regions containing *Ptch1*. Sequence analysis of the remaining *Ptch1* allele showed a variety of inactivating mutations in all tumors analyzed, highlighting the critical tumor suppressor function of this hedgehog-signaling regulator and Ptch1 tumor suppressor activity [60]. Moreover, mutations of multiple genes involved in the SHH pathway (including *PTCH1*, *SUFU*, *Smo*) or the Wingless (*WNT*) pathway (such as *AXIN1* or β-*CATENIN*) have also been found in sporadic human medulloblastomas, the most common malignant pediatric brain tumor, highlighting the importance of these pathways for preventing cancer [70]. Couvé-Privat et al. (2002) demonstrated that the presence of relatively high levels of ultraviolet-specific mutations in the *Smo* proto-oncogene in BCC from DNA repair-deficient xeroderma pigmentosum patients has confirmed its importance in BCC development [71]. Research on engineered loss of Pten or expression of a constitutively active Akt can synergize with engineered dysregulation of SHH signaling in mouse models to generate medulloblastoma [72]. Both pathways were targeted by somatic changes arising in medulloblastoma with defective HR, which showed abnormalities in Pten and PI3K signaling in combination with biallelic inactivation of *Ptch1* [60]. Shafaee et al. reported that cyclopamine increased the cytotoxic effects of paclitaxel and ionizing radiation in HH expressing pancreatic carcinoma cells [73]. Although potential interactions between DNA repair mechanisms and the HH pathway is suspected, the radiosensitizing mechanism of cyclopamine is still not fully understood [74].

## Stimulation of Multiple Drug Resistant Transporter System

Multidrug resistance (MDR) is a common problem in cancer chemotherapy, resulting from enhanced drug efflux from cancer cells mediated by members of the ATP-binding cassette (ABC) transporter family [75]. Permeability-glycoprotein (P-gp), a product of the multidrug resistance gene 1 (*mdr1*), is one of the best characterized MDR molecules, which highly expressed in solid tumors and, moreover, in CSCs [75]. Recently, research demonstrated that imatinib mesylat (IM), a specific tyrosine kinase inhibitor commonly used in CML, was a substrate of P-gp so that *mdr1* gene overexpression can confer resistance to it [76, 77].

Constitutive activation of the HH pathway has been shown to contribute to the growth and maintenance of various cancers [78]. Previous studies have shown that the HH pathway regulates cell cycle progression and apoptotic resistance; this likely contributes to HH-induced chemoresistance [4, 5]. Sims-Mourtada et al. show that HH signaling regulates the expression of the ABC transporter protein P-gp and breast cancer resistance protein (BCRP), and blockade of SHH activation by cyclopamine or a Gli-1 specific siRNA resulted in decreased expression of these transporters [74, 78]. In addition, simultaneous treatment of SHH ligand and cyclosporine A, a broad inhibitor of ABC transporter function, blocked this decrease of drug uptake in SEG-1 esophageal adenocarcinoma cells for [3] H-labeled Taxol, MTX, and VP-16 [78]. These findings suggest that SHH signaling may promote MDR via increasing drug efflux by ABC transporters [75].

## Development of HH Regulating Therapeutics

It has been shown that topical application of cyclopamine inhibited the growth of human BCC [79]; however, concerns of neurological disturbances may limit the systemic application of this drug. Cyclopamine and other compounds, such as cyclopamine derivatives IPI-926 and Cyc-T, and synthetic compounds GDC-0449, XL-139, LDE-225, SANT1, and Cur-61414, action in binding to and antagonizing Smo. SHH-neutralizing antibodies and Robotnikinin were reported that block the SHH pathway by directly targeting SHH, while arsenic, HPI-1, HPI-2, GANT-56, and GANT-61 were potent Gli inhibitors [80–82]. Several small molecule compounds that prevent HH signaling by binding to and inhibiting Smo are currently under development, including Cur-61414 [83], which has shown promising results in the inhibition of BCC and pancreatic cancer in preclinical models. Vismodegib (GDC-0449, discovered by Genentech Inc. under collaboration with Curis Inc.) [84] is a small, orally administrable molecule with suppression effect on HH signaling by binding to and interfering with Smo. Preclinical studies of vismodegib in mouse models of medulloblastoma and in xenograft models of colorectal and pancreatic cancer, and phase I clinical trials in patients with advanced BCC and MB highlighted an objective response to vismodegib [84, 85]. Because of its low toxicity (with only one grade 4 adverse side effects) and specificity for the HH pathway, vismodegib is currently undergoing phase II clinical trials for the treatment of more solid tumors, and may also be used in combination treatments with conventional chemotherapy [84]. Although systemic inhibitors of HH signaling have been undergoing clinical trials, the discovery may provide potentially a novel therapeutic strategy in tumors because HH signaling blockade may not only impair tumor proliferation, but may increase chemotherapeutic efficacy, and result in improved treatment responses. The therapeutic effects of HH pathway blockade in combination with current CRT regimens are perspective to be investigated in the future. The differential regulation and timing of HH activity in normal and tumor tissue after CRT should also be investigated to optimize the most beneficial therapeutic index.

## Concluding Remark

HH signaling increases the resistance of cancer cells to radiotherapy, chemotherapy, and CRT. Research results demonstrate that HH signaling confer treatment resistance of cancer cells through four aspects, including the induction of tumor regrowth and CSCs, anti-apoptosis and cell cycle regulation, modulation of DNA damage repair, and stimulation of MDR transporter system. Inhibition of HH activity may sensitize tumor cells to radiation and chemotherapeutic drugs to improve the treatment outcome [4].

## References

1. Ingham PW, McMahon AP (2001) Hedgehog signaling in animal development: paradigms and principles. Genes Dev 15:3059–3087
2. Adolphe C et al (2004) An in vivo comparative study of sonic, desert and Indian hedgehog reveals that hedgehog pathway activity regulates epidermal stem cell homeostasis. Development 131:5009–5019
3. Beachy PA, Karhadkar SS, Berman DM (2004) Tissue repair and stem cell renewal in carcinogenesis. Nature 432:324–331
4. Sims-Mourtada J et al (2006) Hedgehog: an attribute to tumor regrowth after chemoradiotherapy and a target to improve radiation response. Clin Cancer Res 12:6565–6572
5. Berman DM et al (2003) Widespread requirement for Hedgehog ligand stimulation in growth of digestive tract tumours. Nature 425:846–851
6. Ruel L, Rodriguez R, Gallet A, Lavenant-Staccini L, Therond PP (2003) Stability and association of Smoothened, Costal2 and Fused with Cubitus interruptus are regulated by Hedgehog. Nat Cell Biol 5:907–913
7. Yoon JW et al (2002) Gene expression profiling leads to identification of GLI1-binding elements in target genes and a role for multiple downstream pathways in GLI1-induced cell transformation. J Biol Chem 277:5548–5555
8. Thayer SP et al (2003) Hedgehog is an early and late mediator of pancreatic cancer tumorigenesis. Nature 425:851–856
9. Watkins DN et al (2003) Hedgehog signalling within airway epithelial progenitors and in small-cell lung cancer. Nature 422:313–317
10. Yilmaz OH et al (2006) PTEN dependence distinguishes haematopoietic stem cells from leukaemia-initiating cells. Nature 441:475–482
11. Chufal KS et al (2010) Exploring new potentials and generating hypothesis for management of locally advanced head neck cancer: analysis of pooled data from two phase II trials. J Cancer Res Ther 6:185–193
12. Durand RE, Aquino-Parsons C (2004) Predicting response to treatment in human cancers of the uterine cervix: sequential biopsies during external beam radiotherapy. Int J Radiat Oncol Biol Phys 58:555–560
13. Stinchcombe TE, Bogart J, Wigle DA, Govindan R (2010) Annual review of advances in lung cancer clinical research: a report for the year 2009. J Thorac Oncol 5:935–939
14. Yi Y et al (2010) Predictors of sensitivity to chemoradiotherapy of esophageal squamous cell carcinoma. Tumour Biol 31:333–340
15. Wilkowski R et al (2009) Chemoradiotherapy with concurrent gemcitabine and cisplatin with or without sequential chemotherapy with gemcitabine/cisplatin vs chemoradiotherapy with concurrent 5-fluorouracil in patients with locally advanced pancreatic cancer – a multi-centre randomised phase II stud. Br J Cancer 101:1853–1859

16. Dikken JL et al (2010) Impact of the extent of surgery and postoperative chemoradiotherapy on recurrence patterns in gastric cancer. J Clin Oncol 28:2430–2436
17. Lin JZ et al (2010) Phase II study of pre-operative radiotherapy with capecitabine and oxaliplatin for rectal cancer and carcinoembryonic antigen as a predictor of pathological tumour response. J Int Med Res 38:645–654
18. Mimeault M, Hauke R, Batra SK (2008) Recent advances on the molecular mechanisms involved in the drug resistance of cancer cells and novel targeting therapies. Clin Pharmacol Ther 83:673–691
19. McClendon AK, Dean JL, Ertel A, Knudsen ES (2010) Differential impact of tumor suppressor pathways on DNA damage response and therapy-induced transformation in a mouse primary cell model. PLoS One 5:e8558
20. Provencio M, Sánchez A, Garrido P, Valcárcel F (2010) New molecular targeted therapies integrated with radiation therapy in lung cancer. Clin Lung Cancer 11:91–97
21. Nakano T, Ohno T, Ishikawa H, Suzuki Y, Takahashi T (2010) Current advancement in radiation therapy for uterine cervical cancer. J Radiat Res 51:1–8
22. Goldie JH, Coldman AJ (1984) The genetic origin of drug resistance in neoplasms: implications for systemic therapy. Cancer Res 44:3643–3653
23. Silva AS, Gatenby RA (2010) A theoretical quantitative model for evolution of cancer chemotherapy resistance. Biol Direct 5:25–42
24. Nakano T, Oka K (1993) Differential values of Ki-67 index and mitotic index of proliferating cell population: an assessment of cell cycle and prognosis in radiation therapy for cervical cancer. Cancer 72:2401–2408
25. Hambardzumyan D, Squatrito M, Carbajal E, Holland EC (2008) Glioma formation, cancer stem cells, and Akt signaling. Stem Cell Rev 4:203–210
26. Eyler CE et al (2008) Brain cancer stem cells display preferential sensitivity to Akt inhibition. Stem Cells 26:3027–3036
27. Westhoff M-A et al (2009) The pyridinylfuranopyrimidine inhibitor, PI-103, chemosensitizes glioblastoma cells for apoptosis by inhibiting DNA repair. Oncogene 28:3586–3596
28. Kao GD, Jiang Z, Fernandes AM, Gupta AK, Maity A (2007) Inhibition of phosphatidylinositol-3-OH kinase/Akt signaling impairs DNA repair in glioblastoma cells following ionizing radiation. J Biol Chem 282:21206–21212
29. Frosina G (2010) The bright and the dark sides of DNA repair in stem cells. J Biomed Biotechnol 2010:845396
30. Urtasun RC et al (1986) Binding of 3H-misonidazole to solid human tumors as a measure of tumor hypoxia. Int J Radiat Oncol Biol Phys 12:1263–1267
31. Nordsmark M, Overgaard M, Overgaard J (1996) Pretreatment oxygenation predicts radiation response in advanced squamous cell carcinoma of the head and neck. Radiother Oncol 41:31–39
32. Kawamura H et al (2009) Ceramide induces myogenic differentiation and apoptosis in Drosophila Schneider cells. J Radiat Res 50:161–169
33. Karnoub AE, Weinberg RA (2008) Ras oncogenes: split personalities. Nat Rev Mol Cell Biol 9:517–531
34. Eke I et al (2010) PINCH1 regulates Akt1 activation and enhances radioresistance by inhibiting PP1$\alpha$. J Clin Invest 120:2516–2527
35. Baumann M, Krause M, Hill R (2008) Exploring the role of cancer stem cells in radioresistance. Nat Rev Cancer 8:545–554
36. Crook T et al (1992) Clonal p53 mutation in primary cervical cancer: association with human papilloma virus-negative tumors. Lancet 339:1070–1073
37. Kubo A, Corley DA (2004) Marked multi-ethnic variation of esophageal and gastric cardia carcinomas within the United States. Am J Gastroenterol 99:582–588
38. Cui D, Xu Q, Wang K, Che X (2010) Gli-1 is a potential target for alleviating multidrug resistance of gliomas. J Neurol Sci 288:156–166
39. Abraham RT (2001) Cell cycle checkpoint signaling through the ATM and ATR kinases. Genes Dev 15:2177–2196

40. Leonard JM, Ye H, Wetmore C, Karnitz LM (2008) Sonic Hedgehog signaling impairs ionizing radiation-induced checkpoint activation and induces genomic instability. J Cell Biol 183:385–391
41. Cho RW, Clarke MF (2008) Recent advances in cancer stem cells. Curr Opin Genet Dev 18:48–53
42. McDonald JW et al (1999) Transplanted embryonic stem cells survive, differentiate and promote recovery in injured rat spinal cord. Nat Med 5:1410–1412
43. Pajonk F, Vlashi E, Mcbride WH (2010) Radiation resistance of cancer stem cells: the 4 R's of radiobiology revisited. Stem Cells 28:639–648
44. Ponta H, Sherman L, Herrlich PA (2003) CD44: from adhesion molecules to signaling regulators. Nat Rev Mol Cell Biol 4:33–45
45. Jin L et al (2006) Targeting of CD44 eradicates human acute myeloid leukemic stem cells. Nat Med 12:1167–1174
46. Li C et al (2007) Identification of pancreatic cancer stem cells. Cancer Res 67:1030–1037
47. O'Brien CA et al (2007) A human colon cancer cell capable of initiating tumour growth in immunodeficient mice. Nature 445:106–110
48. Ricci-Vitiani L et al (2007) Identification and expansion of human colon-cancerinitiating cells. Nature 445:111–115
49. Singh SK et al (2004) Identification of human brain tumor initiating cells. Nature 432:396–401
50. Karhadkar SS et al (2004) Hedgehog signalling in prostate regeneration, neoplasia and metastasis. Nature 431:707–712
51. Zhao C et al (2009) Hedgehog signalling is essential for maintenance of cancer stem cells in myeloid leukaemia. Nature 458:776–779
52. Chen YJ, Wang YF, Chao KSC (2009) Cancer stem cells and sonic hedgehog signaling in head and neck cancer: potential targets for overcoming chemoradiation resistance. J Clin Oncol Soc 25:81–88
53. Jeong J et al (2004) Hedgehog signaling in the neural crest cells regulates the patterning and growth of facial primordia. Genes Dev 18:937–951
54. Sacedon R et al (2005) Sonic hedgehog is produced by follicular dendritic cells and protects germinal center B cells from apoptosis. J Immunol 174:1456–1461
55. Kurita S et al (2010) GLI3-dependent repression of DR4 mediates hedgehog antagonism of TRAIL-induced apoptosis. Oncogene 29(34):4848–4858
56. Hegde GV et al (2008) Hedgehog-induced survival of B-cell chronic lymphocytic leukemia cells in a stromal cell microenvironment: a potential new therapeutic target. Mol Cancer Res 6:1928–1936
57. Athar M et al (2004) Inhibition of smoothened signaling prevents ultraviolet B-induced basal cell carcinomas through regulation of Fas expression and apoptosis. Cancer Res 64:7545–7552
58. Bigelow RL et al (2004) Transcriptional regulation of bcl-2 mediated by the sonic hedgehog signaling pathway through GLI-1. J Biol Chem 279:1197–1205
59. Wyman C, Kanaar R (2006) DNA double-strand break repair: all's well that ends well. Annu Rev Genet 40:363–383
60. Frapparta PO et al (2009) Recurrent genomic alterations characterize medulloblastoma arising from DNA double-strand break repair deficiency. Proc Natl Acad Sci USA 106:1880–1885
61. Keeney S, Chang GJ, Linn S (1993) Characterization of a human DNA damage binding protein implicated in xeroderma pigmentosum E. J Biol Chem 268:21293–21300
62. Sun NK, Sun CL, Lin CH, Pail LM, Chao CC (2010) Damaged DNA-binding protein 2 (DDB2) protects against UV irradiation in human cells and *Drosophila*. J Biomed Sci 17:27–40
63. Wang SC et al (2006) Tyrosine phosphorylation controls PCNA function through protein stability. Nat Cell Biol 8:1359–1368
64. Dittmann K et al (2005) Radiation-induced epidermal growth factor receptor nuclear import is linked to activation of DNA-dependent protein kinase. J Biol Chem 280:31182–31189
65. Hsu SC, Miller SA, Wang Y, Hung MC (2009) Nuclear EGFR is required for cisplatin resistance and DNA repair. Am J Transl Res 1:249–258

66. Rabik CA, Dolan ME (2007) Molecular mechanisms of resistance and toxicity associated with platinating agents. Cancer Treat Rev 33:9–23
67. Stubbert LJ, Smith JM, McKay BC (2010) Decreased transcription-coupled nucleotide excision repair capacity is associated with increased p53- and MLH1-independent apoptosis in response to cisplatin. BMC Cancer 10:207–216
68. Siddik ZH (2003) Cisplatin: mode of cytotoxic action and molecular basis of resistance. Oncogene 22:7265–7279
69. Bhana S, Lloyd DR (2008) The role of p53 in DNA damage-mediated cytotoxicity overrides its ability to regulate nucleotide excision repair in human fibroblasts. Mutagenesis 23:43–50
70. Dellovade T, Romer JT, Curran T, Rubin LL (2006) The hedgehog pathway and neurological disorders. Annu Rev Neurosci 29:539–563
71. Couvé-Privat S et al (2002) Significantly high levels of ultraviolet-specific mutations in the *smoothened* gene in basal cell carcinomas from DNA repair-deficient xeroderma pigmentosum patients. Cancer Res 62:7186–7189
72. Hambardzumyan D et al (2008) PI3K pathway regulates survival of cancer stem cells residing in the perivascular niche following radiation in medulloblastoma in vivo. Genes Dev 22:436–448
73. Shafaee Z, Schmidt H, Du W, Posner M, Weichselbaum R (2006) Cyclopamine increases the cytotoxic effects of paclitaxel and radiation but not cisplatin and gemcitabine in Hedgehog expressing pancreatic cancer cells. Cancer Chemother Pharmacol 58:765–770
74. Chen YJ, Sims-Mourtadal J, Izzo J, Chao KS (2007) Targeting the hedgehog pathway to mitigate treatment resistance. Cell Cycle 6:1826–1830
75. Gottesman MM, Pastan I (1993) Biochemistry of multidrug resistance mediated by the multidrug transporter. Annu Rev Biochem 62:385–427
76. Widmer N, Colombo S, Buclin T, Decosterd LA (2003) Functional consequence of MDR1 expression on imatinib intracellular concentrations. Blood 102:1142
77. Gao L et al (2006) STI571 combined with vincristine greatly suppressed the tumor formation of multidrug-resistant K562 cells in a human-nude mice xenograft model. Chin Med J 119:911–918
78. Sims-Mourtada J, Izzo JG, Ajani J, Chao KS (2007) Sonic hedgehog promotes multiple drug resistance by regulation of drug transport. Oncogene 26:5674–5679
79. Tas S, Avci O (2004) Rapid clearance of psoriatic skin lesions induced by topical cyclopamine. A preliminary proof of concept study. Dermatology 209:126–131
80. Stanton BZ, Peng LF (2010) Small-molecule modulators of the Sonic Hedgehog signaling pathway. Mol Biosyst 6:44–54
81. Kim J, Lee JJ, Kim J, Gardner D, Beachy PA (2010) Arsenic antagonizes the hedgehog pathway by preventing ciliary accumulation and reducing stability of the Gli2 transcriptional effector. Proc Natl Acad Sci USA 107(30):13432–13437
82. Yang L, Xie G, Fan Q, Xie J (2010) Activation of the hedgehog-signaling pathway in human cancer and the clinical implications. Oncogene 29:469–481
83. Williams JA et al (2003) Identification of a small molecule inhibitor of the hedgehog signaling pathway: effects on basal cell carcinomalike lesions. Proc Natl Acad Sci USA 100:4616–4621
84. De Smaele E, Ferretti E, Gulino A (2010) Vismodegib, a small-molecule inhibitor of the hedgehog pathway for the treatment of advanced cancers. Curr Opin Investig Drugs 11:707–718
85. Doggrell SA (2010) The hedgehog pathway inhibitor GDC-0449 shows potential in skin and other cancers. Expert Opin Investig Drugs 19:451–454

# Chapter 11
# Small-Molecule Inhibitors of the Hedgehog Pathway

**Ari J. Firestone and James K. Chen**

## Introduction

The Hedgehog (Hh) pathway is one of several intercellular communication mechanisms used to establish and maintain complex tissues in multicellular organisms [33]. Initiated by a family of secreted, lipid-modified polypeptides – Sonic (Shh), Indian (Ihh), and Desert (Dhh) Hhs in mammals – this signaling pathway regulates tissue patterning during embryogenesis, sustains tissue function in postnatal life, and enables tissue repair and regeneration in response to injury. For example, Hh signaling and the resulting transcription of Hh target genes contributes to dorsal–ventral polarity in the neural tube [65], anterior–posterior digit identity in the limb bud [46], bulge stem cell renewal in the cycling hair follicle [45], and post-hepatectomy regeneration of the liver [56]. Dysregulated Hh pathway activity consequently can result in developmental defects such as holoprosencephaly [6, 66] and polydactyly [87], as well as abnormal hair growth [72] and impaired wound healing [44] in postnatal life.

Consistent with its function in controlling cell proliferation and differentiation, the Hh pathway can also promote the onset and/or progression of several human cancers when it is inappropriately upregulated in children and adults [77]. Autocrine, paracrine, and ligand-independent Hh pathway activation have been linked to cancers of the skin [34], brain [82], lung [84], pancreas [8], prostate [37], and blood [93], to name but a few examples, correlating with the role of Hh target gene expression in the development of these tissues and organs. This mechanistic connection between ontogeny and oncogenesis has generated significant interest in small molecules that can block Hh signaling, including natural products previously eschewed as potent teratogens and synthetic antagonists discovered in high-throughput screens. In the past decade, the development of Hh pathway inhibitors has advanced from cell-based models to preclinical animal studies to human clinical trials, resulting in tumor responses that inspire new hope for cancer patients.

J.K. Chen (✉)
Department of Chemical and Systems Biology, Stanford University School of Medicine,
269 Campus Drive, CCSR 3155, Stanford, CA 94305, USA
e-mail: jameschen@stanford.edu

J. Xie (ed.), *Hedgehog Signaling Activation in Human Cancer and its Clinical Implications*,
DOI 10.1007/978-1-4419-8435-7_11, 

We discuss, in this chapter, current models of Hh signaling, genetic lesions that can lead to uncontrolled pathway activation and tumorigenesis, and the small-molecule blockade of these oncogenic transformations. In particular, we consider the mechanisms by which known Hh pathway inhibitors act, their efficacy in pre-clinical or clinical studies, and potential limitations in their therapeutic use. We conclude with an examination of the challenges associated with therapeutically targeting the Hh pathway and possible opportunities for future research.

## Hedgehog Signaling Mechanisms and Cancer

As with any communication system, Hh signaling can be deconstructed into three basic steps: (1) signal generation, (2) a signal transport, and (3) a signal reception (Fig. 11.1) [33]. Dysregulation of any of these processes can cause aberrant inhibition

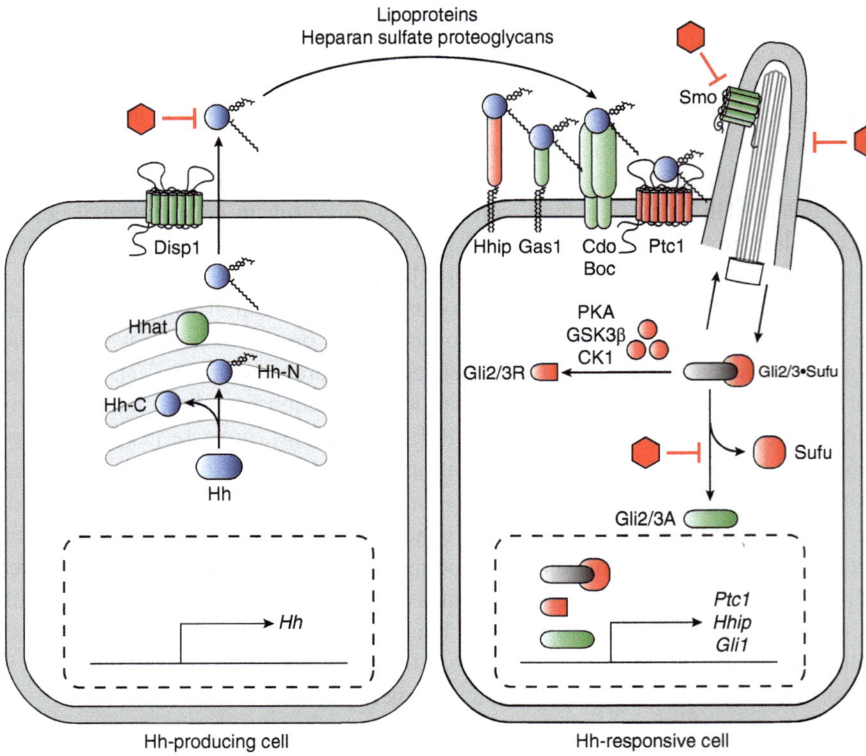


**Fig. 11.1** Pharmacological modulation of Hh signaling mechanisms. Schematic representation of Hh protein biogenesis, secretion, transport, and reception. Hh ligands are depicted in *blue*, positive regulators of Hh signaling in *green*, and negative regulators in *red*. Signaling proteins or processes that can currently be inhibited by small molecules are indicated by the *red hexagons*

or activation of Hh target gene expression, resulting in human disorders and disease. The biochemical reactions and interactions that underlie Hh signal generation and transport appear to be largely conserved across metazoan organisms; Hh reception mechanisms in mammals, however, have evolutionarily diverged from those in their invertebrate counterparts, and we limit our brief discussion here to the former. First, production of the Hh ligand itself is a surprisingly complex process, involving the autoproteolytic cleavage of a polypeptide precursor to generate the signaling protein functionalized with a C-terminal cholesteryl ester [60]. The N -terminus of the Hh ligand is subsequently palmitoylated by Hh acetyltranferase (Hhat) to yield the fully active morphogen [13]. Secretion of the lipid-modified ligand is mediated at least in part by a 12-pass transmembrane protein Dispatched1 (Disp1) [49].

Perhaps because of these dual lipid modifications, movement of the secreted Hh morphogen through tissues relies on active transport mechanisms. Genetic and biochemical studies in *Drosophila* have suggested roles for lipoproteins and heparan sulfate proteoglycans in intercellular Hh transport [59, 79], and similar mechanisms appear to be operative in other organisms. Reception of the Hh signal is then mediated in mammals by the 12-pass transmembrane protein Patched1 (Ptc1) [50], which is structurally related to Disp1 but functionally disparate. Ptc1 binds directly to the Hh morphogens and this interaction can be either facilitated by other Hh-binding proteins at the cell surface, such as growth arrest-specific 1 (Gas1) and the Ig/fibronectin superfamily members Cdo and Boc [1, 78, 89], or competitively inhibited, as is the case with the membrane protein Hh-interacting protein (Hhip) [19]. When unbound by Hh ligands, Ptc1 localizes to the primary cilium, a microtubule-based sensory organelle that protrudes from the plasma membrane [68]. Members of the Gli transcription factor family (Gli2 and Gli3) traffic through the primary cilium [28], and in the absence of Hh signaling, these factors are sequentially phosphorylated by protein kinase A (PKA), glycogen synthase kinase 3 (GSK3), and casein kinase 1 (CK1) and then either proteolytically processed into N-terminal transcriptional repressors (Gli2R and Gli3R) or degraded by the proteosome [57, 83]. These enzymatic reactions may occur within the primary cilium or the centriole/basal body from which it extends, since genetic perturbations that block ciliogenesis can also prevent Gli repressor formation [47].

The binding of Ptc1 by Hh ligands results in the inhibition of this transmembrane protein, its trafficking out of the cilium, and activation of a G protein-coupled receptor (GPCR)-like protein called Smoothened (Smo) [67]. Active Smo accumulates within the primary cilium [20, 67], and through a poorly understood process it abrogates Gli2 and Gli3 proteolysis and converts the full-length factors into transcriptional activators (Gli2A and Gli3A). These events also release the Gli proteins from the repressive effects of Suppressor of Fused (Sufu) [30], a negative regulator of the Hh pathway that binds directly to full-length Gli proteins and restrains their activity through several mechanisms [17, 25, 41, 74]. Gli2A is the primary activator of Hh target gene expression under normal physiological conditions, and several regulators of cell proliferation and differentiation are transcribed in response to Hh ligands. Hh target genes also include pathway components, thereby creating negative (Ptc1 and Hhip) and positive (Gli1, a constitutively active Gli homolog) feedback loops.

Consistent with the mechanistic complexity of Hh signaling, there are several ways by which genetic lesions can promote uncontrolled Hh pathway activation and tumorigenesis. The role for Hh target gene expression and cancer was first recognized when it was discovered that mutations in the *Ptc1* locus are the primary cause of nevoid basal cell carcinoma syndrome (NBCCS), also known as Gorlin's syndrome [34]. As implied by its name, individuals afflicted with NBCCS are markedly predisposed to basal cell carcinoma (BCC), as well as medulloblastoma and rhabdomyosarcoma. Subsequent studies have demonstrated that nearly all cases of sporadic BCCs are caused by ligand-independent, cell-autonomous Hh pathway activation, typically by loss of Ptc1 function [27] and less frequently by genetic lesions in Smo that render it constitutively active (e.g., the SmoM2 allele which has a W535L mutation) [42, 86]. Activating mutations in Smo can also be found in a subset of medulloblastomas [61], and inactive, truncated alleles of Sufu have been discovered in other cases of this pediatric brain cancer [76].

It is now widely appreciated that Hh pathway activation contributes to the etiology of several cancers, including those of the lung, stomach, pancreas, prostate, and blood [77]. Tumor cells derived from these tissues typically express at least one member of the Hh protein family, and it appears that ligand-dependent pathway activation is required for the progression rather than the onset of these diseases. Although autocrine Hh signaling was initially believed to be the predominant oncogenic mechanism in these cases, subsequent animal model studies suggest that paracrine signaling between the cancer cells and their surrounding stroma maintains a microenvironment favorable for tumor growth [90]. Tumor responses to Hh ligand-expressing stroma could, in principle, represent an alternative mode of oncogenic, paracrine Hh signaling.

Cancer more commonly associated with other oncogenic pathways may also involve Gli-dependent Hh target gene expression, albeit through noncanonical signaling mechanisms. For example, recent findings from several laboratories indicate that tumorigenic Ras signaling can activate Hh transcriptional programs. This crosstalk between the Ras and Hh pathways appears to be Smo-independent and may be mediated through SCL/TAL interrupting locus (SIL), a cytoplasmic protein that uncouples Gli proteins from Sufu-mediated repression [32, 55]. Transforming growth factor-β (TGFβ) signaling can also induce Gli2 expression, leading to Smo-independent Hh target gene expression [22], and phosphoinositol 3-kinase (PI3K), mitogen-activated protein kinase (MAPK), and protein kinase C (PKC) signaling have been reported to modulate Gli activity downstream of Smo [62, 63]. Noncanonical Hh pathway activation can even result from anomalous factors generated by chromosomal translocations; several reports demonstrate that the chimeric transcription factor EWS-FLI1 associated with Ewing's Sarcoma can directly induce *Gli1* transcription and consequently promote Hh target gene expression [5, 94]. In fact, Gli1 may be more critical for tumorigenesis than Gli2 in at least some instances, since genetic loss of Gli1 function significantly reduces the incidence of medulloblastoma in mouse models of NBCCS (39).

Given the variety of mechanisms that can lead to Hh target gene expression and cancer, developing a single pharmacological agent that effectively targets all Hh pathway-related tumors will be a major challenge. Rather, gaining comprehensive small-molecule control of Gli-dependent transcription likely will require several classes of drugs that target multiple activities within the pathway. In the next sections, we summarize current attempts to develop Hh pathway antagonists, a growing pharmacopoeia of diverse structure and function.

## Small-Molecule Inhibitors of Smoothened

The first example of pharmacological blockade of Hh signaling actually pre-dates the discovery of genetic mutations that disrupt this pathway. During the 1950s, lambs exhibiting cyclopia, limb defects, and other congenital abnormalities were observed in ranches located in parts of the Boise, Sawtooth, and Challis National Forests of Idaho [9]. Scientists affiliated with the United States Department of Agriculture (USDA) determined that these birth defects were caused by ingestion of the corn lily *Veratrum californicum* by pregnant ewes [10], and further studies by the USDA Poisonous Plant Research Laboratory isolated the steroid alkaloid cyclopamine (1; Table 11.1) and structurally related compounds as the causative agents [38]. For over 30 years, these one-eyed lambs were featured in developmental biology textbooks as examples of naturally occurring teratogenesis, yet the mechanism of cyclopamine action remained a mystery. Studies of human genetics and knockout mice in the 1990s, however, revealed a role for impaired Shh signaling in holopresencephaly [6, 18, 66], a congenital disorder characterized by cyclopia and other axial defects. These findings suggested that cyclopamine might act as a Hh pathway inhibitor, and subsequent investigations demonstrated that this natural product binds directly to Smo and inhibits its activity [14, 75].

The discovery of cyclopamine as a specific Hh pathway antagonist has led to its extensive use as a basic research tool, allowing the blockade of Hh target gene expression in cultured cells, tissue explants, and whole organisms with temporal precision. Cyclopamine and its derivatives also provided the first pharmacological evidence that Hh pathway activation is required for the proliferation of certain cancers. For example, cyclopamine can inhibit abnormal cell growth associated with the loss of Ptc1 function, and the more potent analog KAAD-cyclopamine (2) can overcome oncogenic Smo mutations in cultured cells [75]. It was further shown that cyclopamine induces tumor regression in murine allograft models of medulloblastoma [7], demonstrating for the first time that sustained Hh pathway activity can be required for cancer cell growth.

Despite these successes, cyclopamine is not without its limitations. While its *in vivo* activity suggests that it inhibits Smo with good selectivity, the natural product can have off-target effects at doses similar to that required for Hh pathway blockade [51, 92]. In fact, it appears that at least some reports of cyclopamine-sensitive, autocrine Hh signaling in cultured tumor cells are confounded by Smo-independent cytotoxicity of the compound [90]. In addition, cyclopamine itself has limited potential as an orally bioavailable therapeutic agent, since its spiroketal functionality readily

**Table 11.1** Hp pathway inhibitors that target Smo

| | Compound | | IC50 (*in vitro*) | Preclinical anti-cancer activity | Clinical trials |
|---|---|---|---|---|---|
| 1 | 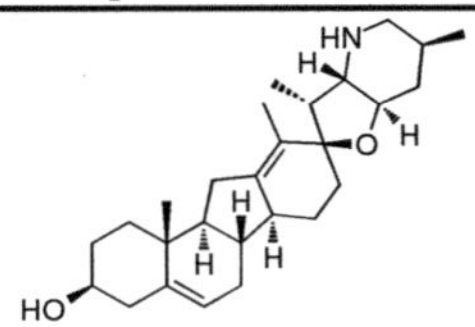 | Cyclopamine | 300 nM | Inhibits growth of several tumor allografts and xenografts in mice | |
| 2 | 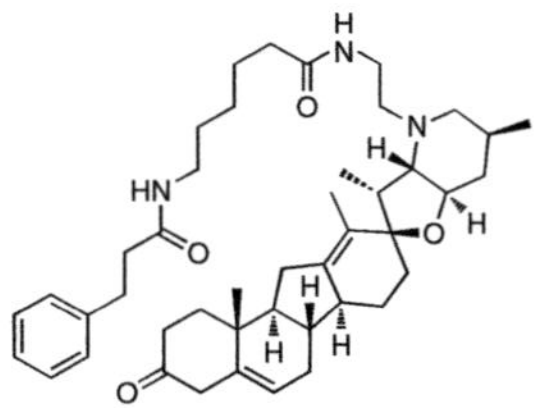 | KAAD-cyclopamine | 20 nM | | |
| 3 | 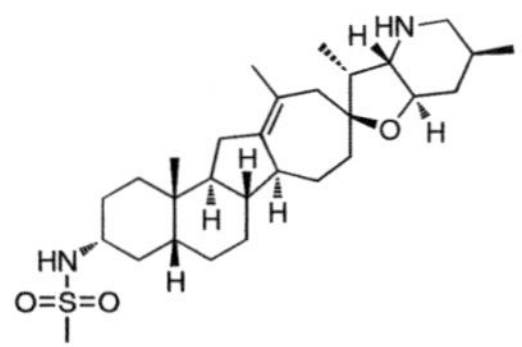 | IPI-926 | 7 nM | Inhibits growth of several tumor xenografts in mice | Currently in Phase I |
| 4 | 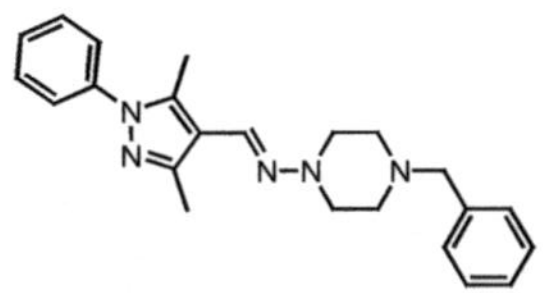 | SANT-1 | 20 nM | | |

| | | | | | |
|---|---|---|---|---|---|
| 5 | 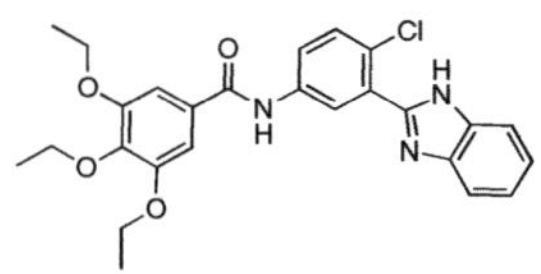 | SANT-2 | 30 nM | | |
| 6 | 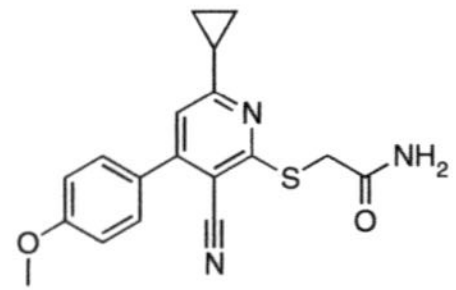 | SANT-3 | 100 nM | | |
| 7 | 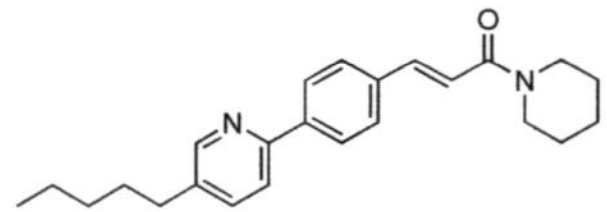 | SANT-4 | 200 nM | | |
| 8 | 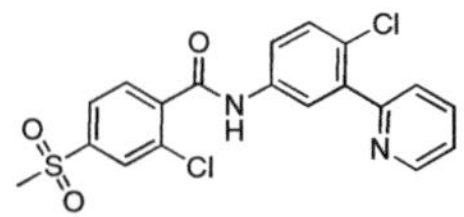 | GDC-0449 | 10 nM | Inhibits growth of medulloblastoma allografts in mice | Currently in Phase II |
| 9 | 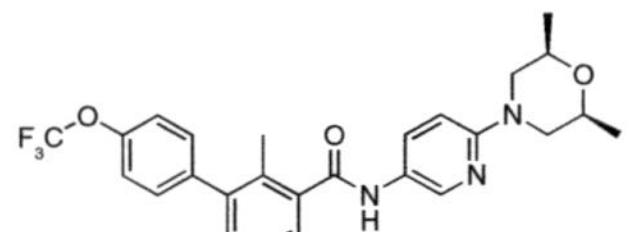 | LDE225 | 0.6 nM | Inhibits growth of medulloblastoma allografts in mice | Currently in Phase I |
| 10 | 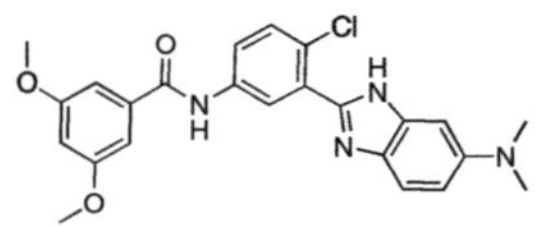 | HhAntag | 30 nM | Inhibits growth of medulloblastomas that arise in $Ptc1^{+/-}$; $p53^{-/-}$ mice and several tumor xenografts | |

(continued)

**Table 11.1** (continued)

| | Compound | | IC50 (*in vitro*) | Preclinical anti-cancer activity | Clinical trials |
|---|---|---|---|---|---|
| 11 | 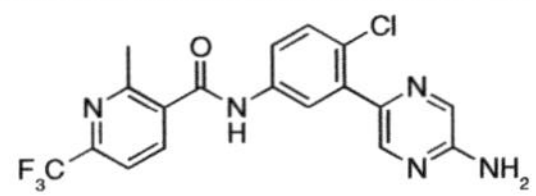 | AstraZenica inhibitor | 10 nM | | |
| 12 | 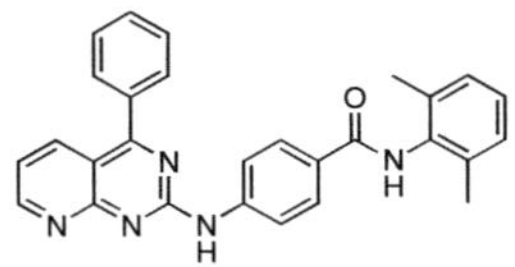 | Exelixis/BMS inhibitor | 4 nM | | |
| 13 | 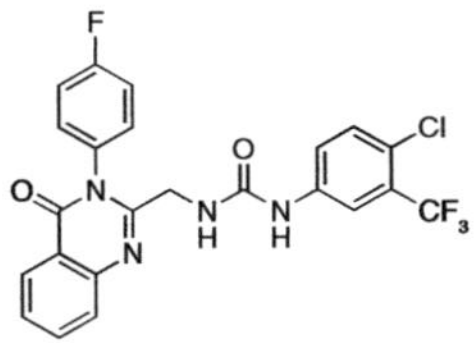 | Compound Z | 70 nM | | |
| 14 | 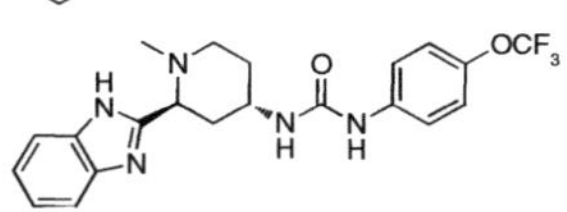 | Pfizer inhibitor | Not reported | | |
| 15 | 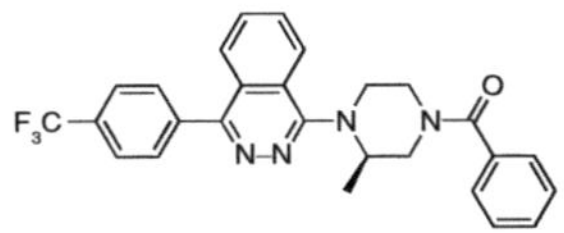 | Amgen inhibitor | 3 nM | Inhibits growth of medulloblastoma allografts in mice | |
| 16 | 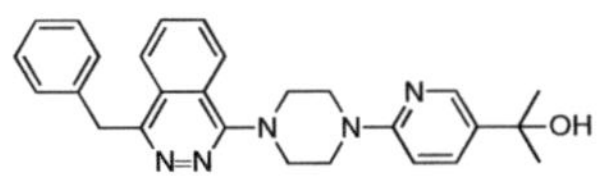 | Novartis inhibitor | 3 nM | Inhibits growth of medulloblastoma allografts in mice | |

| | | | | | |
|---|---|---|---|---|---|
| 17 | 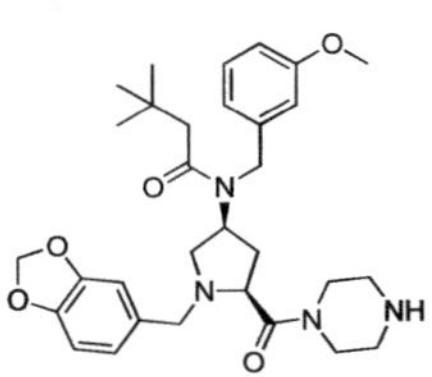 | Cur61414 | 100 nM | Inhibits growth of BCC-like lesions in *Ptc1*$^{+/-}$ mice | Unsuccessful Phase I trial as a topical agent |
| 18 | 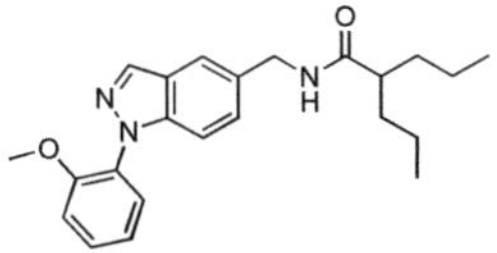 | Merck inhibitor | 5 nM | | |
| 19 | 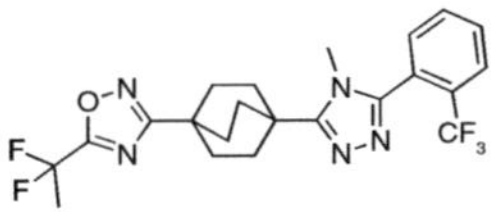 | Merck inhibitor | 300 nM | Inhibits growth of medulloblastoma allografts in mice | |
| 20 | 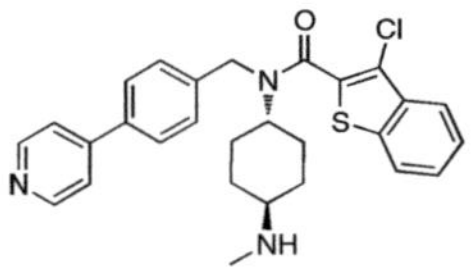 | SAG | 30 nM | | |
| 21 | 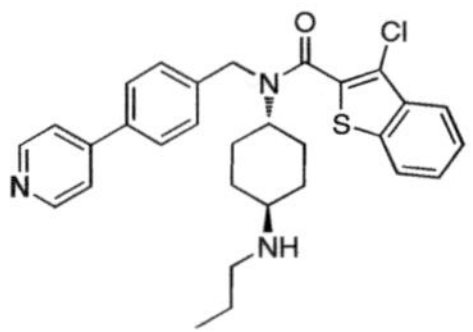 | SANT-75 | 20 nM | | |

undergoes acid-catalyzed ring-opening to yield a functionally inactive derivative [11]. These pharmacological issues combined with the promise of Smo as an anti-cancer target have prompted several research groups in academia and industry to identify new chemical scaffolds capable of inhibiting Smo with high potency and specificity.

One strategy pursued by infinity pharmaceuticals has been the semisynthetic preparation of cyclopamine analogs with improved pharmacokinetic properties and potency [80]. In particular, they modified three structural features in cyclopamine to generate IPI-926 (3): expansion of the acid-sensitive D ring to a seven-membered system, saturation of the B ring, and replacement of the A ring hydroxyl group with a sulfonamide. IPI-926 is orally available and is now being evaluated in phase I clinical trials. Most Smo antagonists, however, have been discovered through high-throughput screens of synthetic libraries using Hh-responsive cell lines and enzymatic reporters of Hh pathway activation. The first synthetic Smo inhibitors to be reported were the SANTs (*S*mo *ant*agonists; 4–7), four compounds with distinct pharmacophores that abrogate Hh signaling with nanomolar potencies and bind to Smo in a manner that is competitive with cyclopamine [15]. The *N*-phenylbenzamide pharmacophore in SANT-2 (5) can be found in the largest class of Smo antagonists known to date, exemplified by GDC-0994 (8), a Genentech compound that was the first Smo inhibitor to progress to Phase II clinical trials [64], and LDE225 (9), a Novartis drug candidate currently in Phase I trials [58]. Additional Smo antagonists containing a *N*-phenylbenzamide core structure include HhAntag (10) [69] and a series of patented compounds from AstraZeneca (11) [26] and Exelixis (12) [3], the latter of which may be related to BMS-833923/XL139, a Phase I drug candidate of undisclosed structure developed through an Exelixis/Bristol Myers Squibb partnership.

Small molecules based upon a phenyl urea scaffold constitute a second class of Smo inhibitors, represented by compound Z (13) developed by Curis and Evotec [12] and a chemical series disclosed by Pfizer (14) [35] that may define structure elements of PF-04449913, a drug candidate currently in Phase I trials. According to the patent literature, these molecules can completely suppress Hh pathway activation in cultured cells at 2-μM doses, although detailed preclinical data is not yet available. A third class of synthetic Smo antagonists is characterized by a pharmacophore composed of piperazine and phthalazine elements that is somewhat reminiscent of SANT-1 (4). Representatives of this inhibitor class developed by Amgen (15) [48] and Novartis (16) [53] can inhibit Hh signaling with single-digit nanomolar potency. Novartis recently initiated a Phase I trial for a second Smo antagonist, LEQ506, although it has not yet been reported whether its structure is related to LDE225, contains the piperazine/phthalazine pharmacophore, or represents yet another class of Smo modulators. In fact, the structural diversity of synthetic Smo ligands is striking, perhaps reflecting the conformational flexibility of this GPCR-like protein and the myriad of inactive states that can be stabilized by small-molecule binding. Smo inhibitors that are structurally distinct from the three classes described above include Cur61414 (17) [85], a drug candidate developed by Curis and one of the first synthetic Smo antagonists tested in animal models, an aryl indazole (18) with nanomolar potency [23], and a family of bicyclooctyltriazoles (19) [4], the latter two being molecular leads reported by Merck. It has even been found that a chemical agonist of Smo called SAG (20) can be converted into Smo antagonists by simple structural modifications, mirroring the structure–activity relationships of certain GPCR ligands [88]. For instance, replacing the methyl group in SAG with a propyl substituent yields SANT75 (21), a potent Hh pathway inhibitor *in vitro* and *in vivo*.

In principle, these structurally distinct inhibitors could selectively stabilize different inactive conformations of Smo, thereby targeting specific aspects of Smo function. Critical steps in Smo homeostasis include protein processing in the endoplasmic reticulum (ER) and Golgi, trafficking to and within the primary cilium, and changes in protein conformation and aggregation state upon Hh ligand-dependent activation [33]. Recent studies are consistent with this model, as there appear to be functional differences between SANT-1, SANT-2, and cyclopamine. For example, the Shh-dependent ciliary accumulation of Smo can be blocked by either SANT-1 or SANT-2, whereas cyclopamine treatment actually increases Smo levels in the primary cilium in the absence of Hh ligand [68]. The SANTs also appear to bind to a different site in Smo than cyclopamine, as the synthetic antagonists do not directly compete with SAG for Smo binding, while the steroid alkaloid does [70]. Furthermore, SANT-1 and SANT-2 but not cyclopamine preferentially reduce cellular levels of post-ER glycosylated Smo [68]. These observations indicate that SANT-1 and SANT-2 perturb a Smo activation step that is upstream of and necessary for its ciliary accumulation, while cyclopamine traps an inactive, ciliary form of Smo to block Hh pathway activation. They also suggest that small-molecule antagonists might differentially counteract oncogenic mutations in Smo, and Hh pathway activation induced by SmoM2 overexpression is in fact several-fold more resistant to cyclopamine than the SANTs [15].

To date, Smo inhibitors have been the primary focus of chemotherapies that target Hh pathway-dependent cancers, and several of the compounds described above have demonstrated efficacy in tumor models. Preclinical studies have typically involved cancers that require Hh target gene expression for their onset and progression, such as BCC and medulloblastoma. For example, cyclopamine, IPI-926, GDC-0449, LDE225, HhAntag, and other lead compounds from Amgen, Novartis, and Merck have been shown to induce the regression of subcutaneous medulloblastoma allografts derived from $Ptc1^{+/-}$, $Ptc1^{+/-}$;$p53^{-/-}$, or $Ptc1^{+/-}$;$Hic1^{+/-}$ mice [4, 7, 48, 53, 58, 64, 69, 80]. LDE225 has been similarly successful in orthotopic tumor models in which medulloblastoma cells are transplanted intracranially, demonstrating that this molecule is capable of passing the blood–brain barrier and inhibiting medulloblastoma growth in an environment similar to that of the native cancer.

More recent investigations have begun to explore the ability of Smo antagonists uncouple Hh pathway-dependent tumor–stroma interactions that promote cancer survival. HhAntag has been found to inhibit the proliferation of xenografted human colon cancer and pancreatic adenocarcinomas in mouse models, tumors that secrete Hh ligands but do not require Smo-dependent Hh signaling for their survival. In these cases, it is believed that Hh target gene expression in the surrounding stroma creates a physiological niche that fosters tumor growth, a process that is blocked by the Smo inhibitor. Hh ligand production by stromal compartments may also contribute to tumorigenesis, as illustrated by the Hh pathway-dependent expansion of Eμ-*Myc*-driven B-cell lymphomas in syngeneic mouse models [24]. Cyclopamine treatment of these mice disrupts paracrine Hh signaling presumably initiated by bone marrow, nodal, and splenic stroma, leading to tumor cell apoptosis and increased survival of the host animals.

Altogether, these preclinical results strongly suggest that Smo inhibitors may be effective chemotherapies for several types of cancers, many of which currently have poor clinical prognoses. The success of these animal model studies has generated considerable enthusiasm for human clinical investigations of Smo antagonists, and nearly 30 clinical trials have been initiated since 2007. A wide variety of therapeutic applications have been proposed, including the use of Smo-targeting compounds as single agents or in combination with other drugs in patients with medulloblastoma, glioblastoma, pancreatic adenocarcinoma, stomach cancer, colorectal carcinoma, breast cancer, hematological malignancies, and generally advanced or refractory solid tumors. Due to the recent inception of these studies, relatively little clinical data is yet available, but results from the first completed Phase I trial underscore the therapeutic potential of Hh pathway-targeting drugs. In this study, patients with metastatic BCC were treated with GDC-0449, with the majority exhibiting objective clinical responses and a few achieving complete regression of the disease [81]. A single patient with advanced, systemic medulloblastoma also responded dramatically to GDC-0449 within 2 months of therapy [91]. At least in this latter case, however, drug-resistant tumors subsequently emerged, resulting in withdrawal of the patient from the trial and eventual death. Genetic analyses of tumor biopsies from the medulloblastoma patient revealed the presence of a mutant Smo allele (D473H) that was not observed in pretreatment tumor samples. Cell-based studies of this mutant protein showed that it is phenotypically normal with respect to Hh response but resistant to binding and inhibition by GDC-0449. Smo D473H is also insensitive to KAAD-cyclopamine, and whether other Smo inhibitors will be similarly ineffective against this or other drug-resistant alleles that arise remains to be determined. Remarkably, a structurally equivalent Smo mutation was observed in one GDC-0449-resistant tumor after prolonged treatment of a mouse medulloblastoma model [91]. Two other GDC-0449-insensitive tumors examined in this preclinical study did not exhibit any Smo mutations, indicating that resistance to Smo inhibitors can arise through multiple mechanisms.

## Small-Molecule Inhibitors that Act Upstream of Smoothened

While Smo is perhaps the most druggable target within the Hh pathway, small molecules that block Hh signaling through other mechanisms could also be useful therapeutic agents (Table 11.2), especially given the emergence of Hh pathway-dependent tumors that are resistant to Smo inhibitors. The most upstream targets within the pathway would be those involved in the biogenesis and secretion of Hh ligands. To date, there have been no reports of compounds that block these steps in Hh signaling, although the feasibility of this approach is strongly suggested by the

**Table 11.2** Hh pathway inhibitors that do not target Smo

| | Compound | | IC50 (*in vitro*) | Preclinical anti-cancer activity | Clinical trials |
|---|---|---|---|---|---|
| 22 | 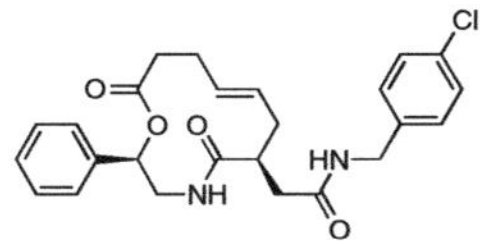 | Robotnikinin | Binds to Shh | 10 μM | |
| 23 | 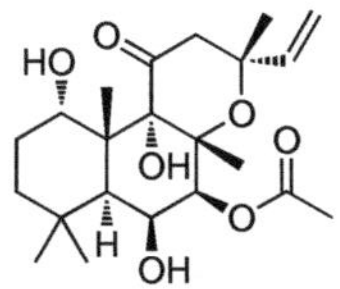 | Forskolin | Indirectly activates PKA and promotes Gli processing | 10 μM | |
| 24 | 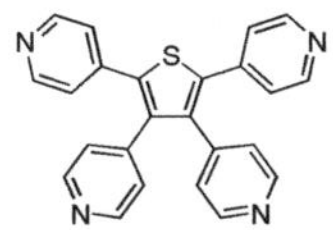 | GANT-58 | Inhibits overexpressed Gli1 and Gli2 | 5–10 μM | Inhibits growth of prostate tumor xenografts in mice |
| 25 | 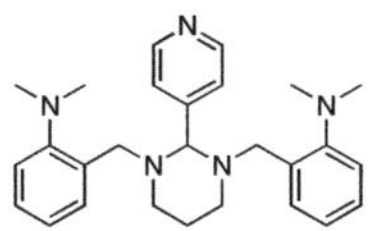 | GANT-61 | Inhibits overexpressed Gli1 and Gli2 | 5–10 μM | Inhibits growth of prostate tumor xenografts in mice |
| 26 | 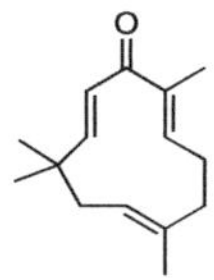 | Zerumbone | Inhibits overxpressed Gli1 and Gli2 | 7 μM for Gli1<br>1 μM for Gli2 | |
| 27 | 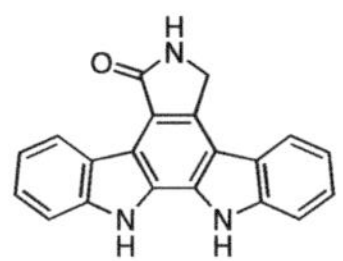 | Staurosporinone | Inhibits overxpressed Gli1 and Gli2 | 2 μM for Gli1<br>3 μM for Gli2 | Inhibits proliferation of a pancreatic cancer cell line |

(continued)

**Table 11.2** (continued)

| | Compound | | IC50 (*in vitro*) | Preclinical anti-cancer activity | Clinical trials |
|---|---|---|---|---|---|
| 28 | 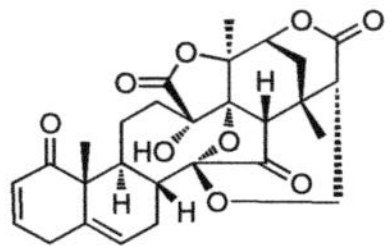 | Physalin B | Inhibits overexpressed Gli1 and Gli2 | 0.6 μM for Gli1<br>2 μM for Gli2 | Inhibits proliferation of a pancreatic cancer cell line |
| 29 | 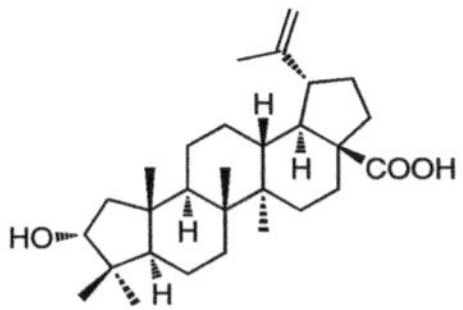 | Betulinic acid | Inhibits overexpressed Gli1 | 30 μM | Inhibits proliferation of several cancer cell lines and tumor xenografts in mice |
| 30 | 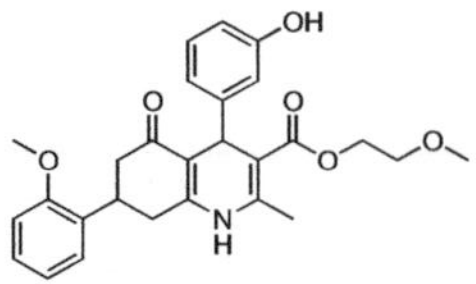 | HPI-1 | Inhibits overexpressed Gli1 and Gli2 | 2 μM | Inhibits proliferation of SmoM2-expressing CGNPs in primary cultures |

| | | | | | |
|---|---|---|---|---|---|
| 31 | 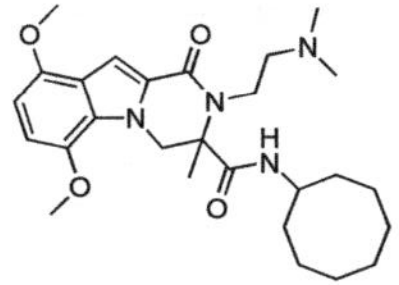 | HPI-2 | Inhibits overexpressed Gli2 | 2 μM | |
| 32 | 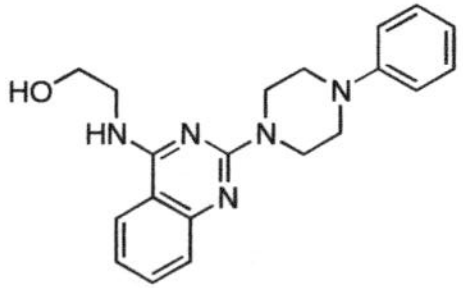 | HPI-3 | Inhibits Hh signaling downstream of Su(fu) | 3 μM | |
| 33 | 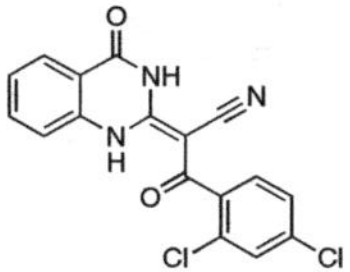 | HPI-4 | Inhibits Hh signaling downstream of Su(fu) | 7 μM | Inhibits proliferation of SmoM2-expressing CGNPs in primary cultures |

recent discovery of synthetic Wnt signaling inhibitors that target Porcupine, an *O*-acyltransferase that palmitoylates members of the Wnt morphogen family [16].

In contrast, small molecules that can bind directly to lipid-modified, secreted Hh ligands and block their signaling capabilities have been identified, challenging the conventional wisdom that ligand–receptor interactions cannot be pharmacologically targeted. In this study, a collection of compounds prepared by diversity-oriented synthesis was immobilized on glass slides and screened for molecules capable of binding fluorescently labeled recombinant Shh. Subsequent lead optimization efforts led to the macrocycle robotnikinin (22), which can inhibit Hh pathway activation induced by Hh ligands but not that caused by the loss of Ptc1 function or activation of Smo by a synthetic agonist [73]. These activities support a model in which robotnikinin prevents Shh from engaging the ligand reception machinery in Hh-responsive cells. However, it is not yet known which portion of the Shh morphogen binds to robotnikinin or which Hh ligand-binding proteins are unable to recognize the Shh/robotnikinin complex. Although the specificity of robitnikinin for Shh versus Ihh and Dhh has not been established, in principle it should be possible to develop robotnikinin-like compounds that discriminate between Hh ligand homologs. Such ligand specificity could be therapeutically advantagenous, as it would allow the pharmacological intervention of diseases associated with a particular Hh ligand, while minimizing perturbations of other Hh pathway-dependent physiology. For example, Smo inhibitors used to treat Shh-dependent tumors in children would adversely and permanently disrupt Ihh-regulated bone development, resulting in dwarfism [40]; a Shh-specific inhibitor, however, might be able to avoid these deleterious side effects.

Inhibition of the Hh pathway upstream of Ptc1 may also be clinically desirable, as recent evidence suggests that Ptc1 can directly promote programmed cell death through the recruitment of a complex-containing DRAL (downregulated in rhabdomyosarcoma LIM-domain protein) and caspase 9 and that this process is inhibited by Hh ligands [52]. Small molecules that reestablish this Ptc1-facilitated apoptosis pathway in tumors, therefore, might be more efficacious than Smo inhibitors against certain cancers, such as those that rely on autocrine Hh signaling.

## Small-Molecule Inhibitors that Act Downstream of Smoothened

Hh pathway inhibitors that act downstream of Smo could be particularly efficacious anti-cancer agents since they can counteract Hh pathway activation induced by drug-resistant Smo mutants and possibly even oncogenic lesions that affect downstream components (Table 11.2). Prior to the advent of high-throughput screening campaigns for Hh pathway inhibitors, the most common pharmacological method for disrupting cytoplasmic Hh signaling was treatment with forskolin (23), a plant-derived agonist of adenylate cyclase that increases cellular levels of cyclic adenosine monophosphate (cAMP) and consequently promotes PKA-dependent Gli repressor formation. While forskolin has been a valuable tool for studying Hh signaling mechanisms, its indirect mode of action limits its potential as a therapeutic agent.

Identifying new small molecules that alter Gli function more directly is an area of increasing biomedical interest and clinical importance, albeit one complicated by the exceptional sensitivity of Smo to pharmacological modulation and the regulation of Gli activity by signaling proteins with pleiotropic functions. Screens for compounds that block Hh ligand-mediated pathway activation have nearly exclusively yielded new Smo antagonists, reflecting the difficulty in discovering inhibitors with novel mechanisms amidst the large number of Smo-targeting pharmacophores. Efforts to find downstream-acting molecules have consequently relied on other means of Hh pathway activation. The first of these studies surveyed a collection of 1,990 compounds from the National Cancer Institute for their ability to block the expression of a Gli-dependent reporter in HEK293 cells transiently transfected with exogenous *Gli1* [43]. Two *G*li *ant*agonists, GANT-58 (24) and GANT-61 (25), were uncovered through this screen, and the compounds were also able to block exogenous Gli2 activity in HEK293 cells and Hh pathway activity in embryonic fibroblasts derived from *Sufu*$^{-/-}$ mice. GANT-58 and GANT-61 further demonstrated anti-proliferative activities toward cultured human pancreatic adenocarcinoma PANC1 and prostate cancer 22Rv1 cells, even inhibiting the growth of 22Rv1 xenografts in athymic nude mice.

GANT-58 and GANT-61 have subsequently been found to block the proliferation of Ewing's sarcoma cell lines and Gli-dependent transcription in several cellular contexts [36]. The precise mechanism by which either compound acts remains unclear, but exogenous Gli1 expressed in GANT-61-treated HEK293 cells exhibits diminished affinity for its DNA recognition sequence [43]. Since this effect is not observed *in vitro*, the action of GANT-61 on Gli1 is likely through an indirect mechanism, possibly involving a posttranslational modification that impedes DNA binding. It is also uncertain whether the GANT-58 or GANT-61 pharmacophores will be appropriate scaffolds for drug development; their simple, symmetrical structures are atypical for pharmacological agents, and the compounds may have target specificities and pharmacokinetic properties that are limiting.

Similar Gli overexpression screens have led to the discovery of several natural products that can inhibit Gli-dependent transcription. A survey of 286 natural substances yielded members of the zerumbone (26), staurosporinone (27), and physalin (28) families as potential Gli antagonists [29], and a subsequent study of *Zizyphus cambodiana* extracts identified three structurally related pentacyclic triterpenes as pathway inhibitors, exemplified by betulinic acid (29) [2]. How these compounds block Gli function has not yet been established either, but activities ascribed previously to molecules in these structural families may provide some clues. For example, molecules containing the staurosporinone pharmacophore are well known to promiscuously inhibit kinases by competitively blocking adenosine triphosphate binding [71]. Physalins have been reported to abrogate phorbol ester-induced activation of the nuclear factor-κB pathway and therefore may perturb Gli function through the inhibition of one or more PKC isoforms [21]. Betulinic acid is most commonly known for its pro-apoptotic activity in multiple cancer cell lines and murine tumor models [54], although how these mechanisms might relate to Gli function is unclear.

In spite of these unanswered questions, the screens described above demonstrate the feasibility of pharmacologically modulating Gli activity. The mechanisms by which the Gli antagonists act are likely to be divergent, and it is almost certain that several cellular targets remain untapped. In addition, Hh target gene expression induced by exogenous Gli1 or Gli2 escapes at least some regulatory processes that would otherwise control the endogenous transcription factors. To search for Hh pathway inhibitors in manner that surveys downstream signaling more comprehensively, our laboratory established an NIH-3T3 cell-based assay in which a stably transfected Gli-dependent reporter is activated by high doses of SAG. Under these conditions, Smo antagonists such as cyclopamine are largely ineffectual while forksolin and other downstream-acting compounds retain their inhibitory activities. A collection of 122,755 synthetic compounds was screened for molecules capable of blocking Gli-dependent transcription in SAG-treated cells. Four *H*h pathway inhibitors (HPI-1 through HPI-4; 30–33) were discovered in this study, each based upon a unique molecular scaffold.

Consistent with their differing structures, the four Gli antagonists appear to act through distinct mechanisms. HPI-1 (30) and HPI-2 (31) can overcome Hh pathway activation induced by Gli1 or Gli2 overexpression, with the latter exhibiting selectivity for the Gli2 isoform [31]. HPI-3 (32) and HPI-4 (33) are less effective against exogenous Gli transcription factors, but their potent activities in $Sufu^{-/-}$ fibroblasts indicate that they can block endogenous Gli function. As with the other Gli antagonists, the cellular targets of the HPIs await further study, but their effects on Gli processing provides some mechanistic insights. HPI-1 and HPI-4 uncouple Shh signaling from Gli2 repressor formation, whereas Shh-induced Gli2 stabilization is maintained in cells treated with either HPI-2 or HPI-3. Most strikingly, HPI-4 perturbs primary cilia structure, and it is possible that this compound disrupts Gli activator formation by targeting a ciliary transport protein. Some of the HPIs also exhibit activities against murine SmoM2-expressing cerebellar granule neuron precursors (CGNPs), which give rise to medulloblastoma-like tumors. In contrast to cyclopamine, HPI-1 and HPI-4 can effectively inhibit SmoM2 CGNP proliferation *in vitro*, coincident with their ability to block the expression *Gli1*, *N-Myc*, and other Hh target genes in these cells. Surprisingly, HPI-2 and HPI-3 exhibit neither activity, yet both are able to prevent Hh pathway activation in SmoM2-expressing NIH-3T3 cells. This unexpected result indicates that Hh pathway inhibitors can have tissue-specific activities, a finding that could have important clinical implications.

## Conclusions and Future Directions

Since the connection between Hh pathway activation and oncogenesis was first recognized nearly 15 years ago, there have been significant advances toward the development of chemotherapies that specifically target this signaling mechanism. In this relatively brief period, Smo has been identified as a druggable target within the Hh pathway, potent and specific inhibitors of Smo have been developed, and the anti-cancer activities of these compounds have been validated in preclinical models and now human subjects. The dramatic responses of some clinical trial participants to Smo antagonists may herald new targeted therapies for cancer patients, including those battling metastatic BCCs, medulloblastoma, and other Hh pathway-dependent tumors that currently lack effective treatments.

This new-found optimism is tempered somewhat, however, by the appearance of drug-insensitive tumors in these initial clinical trials, which focused on cancers caused by ligand-independent pathway activation. The subset of these cancers that acquire drug-resistant Smo mutations could be ameliorated by new classes of Smo inhibitors, but murine models suggest that the majority of tumors that gain resistance to Smo antagonists will have genetic lesions in downstream effectors or parallel oncogenic pathways. Whether drug resistance will emerge in Hh ligand-dependent tumors and the mechanisms by which it occurs remains to be seen. Cancers that proliferate in response to Hh ligands, either by autocrine or stroma-to-tumor paracrine signaling, could acquire resistance as these genomically unstable cells are subjected to pharmacological pressure. Cancers that require tumor-to-stroma paracrine Hh signaling to maintain a tumorigenic microenvironment, however, might be more likely to maintain their drug sensitivity since the Hh-responsive stroma cells presumably will be less subject to drug-mediated selection pressure and have normal genomic mutation rates. In any case, choosing the most appropriate chemotherapy for each cancer patient will depend on the mode of Hh pathway activation, the susceptibility of the tumor cells to drug-resistance mechanisms, and the patient's need for either systemic or tissue-selective Hh pathway inhibition.

Realizing this level of personalized medicine will require a structurally and mechanistically diverse ensemble of Hh pathway-targeting drugs. While our growing collection of Smo antagonists has been a logical and promising starting point, new Hh pathway inhibitors need to be developed. Small molecules that block the biogenesis or trafficking of Hh ligands, as well as robotnikinin-like compounds that directly target these secreted morphogens, would constitute valuable additions to our anti-cancer arsenal. Discovering drugs that more directly perturb Gli function is a particularly important area of future research, since in principle, Gli antagonists would be more broadly applicable to cancers that require Hh target gene expression, including those in which Gli activation is achieved through noncanonical signaling mechanisms. Achieving these goals will undoubtedly be challenging, involving new high-throughput small-molecule screens, extensive medicinal chemistry, and the identification of cellular targets for the resulting chemical leads. Yet if these efforts are successful, Hh pathway inhibitors will be known not only as the "monster-generating" compounds in our developmental textbooks, but also as life-saving drugs in our fight against cancer.

## References


1. Allen BL, Tenzen T, McMahon AP (2007) The Hedgehog-binding proteins Gas1 and Cdo cooperate to positively regulate Shh signaling during mouse development. Genes Dev 21: 1244–1257
2. Arai M, Tateno C, Hosoya T, Koyano T, Kowithayakorn T, Ishibashi M (2008) Hedgehog/GLI-mediated transcriptional inhibitors from Zizyphus cambodiana. Bioorg Med Chem 16: 9420–9424

3. Bahceci S, Bajjalieh W, Chen J, Epshteyn S, Forsyth TP, Huynh TP, Kim BG, Lehy JW, Lee MS, Lewis GL, *et al.* (2008) Inhibitors of the Hedgehog pathway. WO/2008/112913, 18 Sept 2008
4. Balkovec JM, Thieringer R, Waddell ST (2008) Triazole derivatives which are Smo antagonists. WO/2008/130552, 30 Oct 2008
5. Beauchamp E, Bulut G, Abaan O, Chen K, Merchant A, Matsui W, Endo Y, Rubin JS, Toretsky J, Uren A (2009) GLII is a direct transcriptional target of EWS-FLI1 oncoprotein. J Biol Chem 284: 9074–9082
6. Belloni E, Muenke M, Roessler E, Traverso G, Siegel-Bartelt J, Frumkin A, Mitchell HF, Donis-Keller H, Helms C, Hing AV, *et al.* (1996) Identification of Sonic hedgehog as a candidate gene responsible for holoprosencephaly. Nat Genet 14: 353–356
7. Berman D, Karhadkar S, Hallahan A, Pritchard J, Eberhart C, Watkins D, Chen J, Cooper M, Taipale J, Olson J, *et al.* (2002) Medulloblastoma growth inhibition by hedgehog pathway blockade. Science 297: 1559–1561
8. Berman DM, Karhadkar SS, Maitra A, Montes De Oca R, Gerstenblith MR, Briggs K, Parker AR, Shimada Y, Eshleman JR, Watkins DN, *et al.* (2003) Widespread requirement for Hedgehog ligand stimulation in growth of digestive tract tumours. Nature 425: 846–851
9. Binns W, Thacker EJ, James LF, Huffman WT (1959) A congenital cyclopiantype malformation in lambs. J Am Vet Med Assoc 134: 180–183
10. Binns W, James LF, Shupe JL, Everett G (1963) A congenital cyclopian-type malformation in lambs induced by maternal ingestion of a range plant, Veratrum Californicum. Am J Vet Res 24: 1164–1175
11. Brown D (1978) Structure-activity relation of steroidal amine teratogens. In effects of poisonous plants on livestock. Keeler RF, Van Kampen KR, James LF, (eds) New York, Academic Press, p 409
12. Brunton SA, Stibbard JHA, Rubin LL, Kruse LI, Guicherit OM, Boyd EA, Price S (2008) Potent inhibitors of the hedgehog signaling pathway. J Med Chem 51: 1108–1110
13. Buglino JA, Resh MD (2008) Hhat is a palmitoylacyltransferase with specificity for N-palmitoylation of Sonic Hedgehog. J Biol Chem 283: 22076–22088
14. Chen J, Taipale J, Cooper M, Beachy P (2002) Inhibition of Hedgehog signaling by direct binding of cyclopamine to Smoothened. Genes Dev 16: 2743–2748
15. Chen J, Taipale J, Young K, Maiti T, Beachy P (2002) Small molecule modulation of Smoothened activity. Proc Natl Acad Sci USA 99: 14071–14076
16. Chen B, Dodge M, Tang W, Lu J, Ma Z, Fan C, Wei S, Hao W, Kilgore J, Williams N, *et al.* (2009) Small molecule-mediated disruption of Wnt-dependent signaling in tissue regeneration and cancer. Nat Chem Biol 5: 100–107
17. Cheng SY, Bishop JM (2002) Suppressor of Fused represses Gli-mediated transcription by recruiting the SAP18-mSin3 corepressor complex. Proc Natl Acad Sci USA 99: 5442–5447
18. Chiang C, Litingtung Y, Lee E, Young KE, Corden JL, Westphal H, Beachy PA (1996) Cyclopia and defective axial patterning in mice lacking Sonic hedgehog gene function. Nature 383: 407–413
19. Chuang PT, McMahon AP (1999) Vertebrate Hedgehog signalling modulated by induction of a Hedgehog-binding protein. Nature 397: 617–621
20. Corbit KC, Aanstad P, Singla V, Norman AR, Stainier DY, Reiter JF (2005) Vertebrate Smoothened functions at the primary cilium. Nature 437: 1018–1021
21. Damu AG, Kuo PC, Su CR, Kuo TH, Chen TH, Bastow KF, Lee KH, Wu TS (2007) Isolation, structures, and structure - cytotoxic activity relationships of withanolides and physalins from Physalis angulata. J Nat Prod 70: 1146–1152
22. Dennler S, André J, Verrecchia F, Mauviel A (2009) Cloning of the human GLI2 Promoter: transcriptional activation by transforming growth factor-beta via SMAD3/beta-catenin cooperation. J Biol Chem 284: 31523–31531
23. Dessole G, Branca D, Ferrigno F, Kinzel O, Muraglia E, Palumbi M, Rowley M, Serafini S, Steinkuhler C, Jones P (2009) Discovery of N-[(1-aryl-1H-indazol-5-yl)methyl]amides

derivatives as smoothened antagonists for inhibition of the hedgehog pathway. Bioorg Med Chem Lett 19: 4191–4195
24. Dierks C, Grbic J, Zirlik K, Beigi R, Englund NP, Guo GR, Veelken H, Engelhardt M, Mertelsmann R, Kelleher JF, *et al.* (2007) Essential role of stromally induced hedgehog signaling in B-cell malignancies. Nat Med 13: 944–951
25. Ding Q, Fukami S, Meng X, Nishizaki Y, Zhang X, Sasaki H, Dlugosz A, Nakafuku M, Hui C (1999) Mouse suppressor of fused is a negative regulator of sonic hedgehog signaling and alters the subcellular distribution of Gli1. Curr Biol 9: 1119–1122
26. Fauber B, Hird A, Janetka J, Russell DJ, Bin Y (2009) Phenylcarboxamide derivatives as inhibitors and effectors of the Hedgehog pathway. WO/2009/030952, 3 Dec 2009
27. Gailani MR, Stahle-Backdahl M, Leffell DJ, Glynn M, Zaphiropoulos PG, Pressman C, Unden AB, Dean M, Brash DE, Bale AE, *et al.* (1996) The role of the human homologue of Drosophila patched in sporadic basal cell carcinomas. Nat Genet 14: 78–81
28. Haycraft CJ, Banizs B, Aydin-Son Y, Zhang Q, Michaud EJ, Yoder BK (2005) Gli2 and Gli3 localize to cilia and require the intraflagellar transport protein polaris for processing and function. PLoS Genet 1: e53
29. Hosoya T, Arai M, Koyano T, Kowithayakorn T, Ishibashi M (2008) Naturally occurring small-molecule inhibitors of hedgehog/GLI-mediated transcription. Chembiochem 9: 1082–1092
30. Humke EW, Dorn KV, Milenkovic L, Scott MP, Rohatgi R (2010) The output of Hedgehog signaling is controlled by the dynamic association between Suppressor of Fused and the Gli proteins. Genes Dev 24: 670–682
31. Hyman J, Firestone A, Heine V, Zhao Y, Ocasio C, Han K, Sun M, Rack P, Sinha S, Wu J, *et al.* (2009) Small-molecule inhibitors reveal multiple strategies for Hedgehog pathway blockade. Proc Natl Acad Sci USA 106: 14132–14137
32. Ji Z, Mei FC, Xie J, Cheng X (2007) Oncogenic KRAS activates hedgehog signaling pathway in pancreatic cancer cells. J Biol Chem 282: 14048–14055
33. Jiang J, Hui C-C (2008) Hedgehog signaling in development and cancer. Dev Cell 15: 801–812
34. Johnson RL, Rothman AL, Xie J, Goodrich LV, Bare JW, Bonifas JM, Quinn AG, Myers RM, Cox DR, Epstein EH, *et al.* (1996) Human homolog of patched, a candidate gene for the basal cell nevus syndrome. Science 272: 1668–1671
35. Jones CS, La Greca S, Li Q, Munchhof MJ, Reiter LA (2009) Benzimidazole derivatives. WO/2009/004427; 1 Aug 2009
36. Joo J, Christensen L, Warner K, States L, Kang H-G, Vo K, Lawlor ER, May WA (2009) GLI1 is a central mediator of EWS/FLI1 signaling in Ewing tumors. PLoS One 4: e7608
37. Karhadkar SS, Bova GS, Abdallah N, Dhara S, Gardner D, Maitra A, Isaacs JT, Berman DM, Beachy PA (2004) Hedgehog signalling in prostate regeneration, neoplasia and metastasis. Nature 431: 707–712
38. Keeler RF, Binns W (1966) Teratogenic compounds of Veratrum californicum (Durand). II. Production of ovine fetal cyclopia by fractions and alkaloid preparations. Can J Biochem 44: 829–838
39. Kimura H, Stephen D, Joyner A, Curran T (2005) Glil is important for medulloblastoma formation in Ptc1 +/– mice. Oncogene 24: 4026–4036
40. Kimura H, Ng JM, Curran T (2008) Transient inhibition of the Hedgehog pathway in young mice causes permanent defects in bone structure. Cancer Cell 13: 249–260
41. Kogerman P, Grimm T, Kogerman L, Krause D, Unden AB, Sandstedt B, Toftgard R, Zaphiropoulos PG. (1999) Mammalian suppressor-of-fused modulates nuclear-cytoplasmic shuttling of Gli-1. Nat Cell Biol 1: 312–319
42. Lam CW, Xie J, To KF, Ng HK, Lee KC, Yuen NW, Lim PL, Chan LY, Tong SF, McCormick F (1999) A frequent activated smoothened mutation in sporadic basal cell carcinomas. Oncogene 18: 833–836
43. Lauth M, Bergstrom A, Shimokawa T, Toftgard R (2007) Inhibition of GLI-mediated transcription and tumor cell growth by small-molecule antagonists. Proc Natl Acad Sci USA 104: 8455–8460

44. Le H, Kleinerman R, Lerman OZ, Brown D, Galiano R, Gurtner GC, Warren SM, Levine JP, Saadeh PB (2008) Hedgehog signaling is essential for normal wound healing. Wound Repair Regen 16: 768–773
45. Levy V, Lindon C, Harfe BD, Morgan BA (2005) Distinct stem cell populations regenerate the follicle and interfollicular epidermis. Dev Cell 9: 855–861
46. Litingtung Y, Dahn RD, Li Y, Fallon JF, Chiang C (2002) Shh and Gli3 are dispensable for limb skeleton formation but regulate digit number and identity. Nature 418: 979–983
47. Liu A, Wang B, Niswander LA (2005) Mouse intraflagellar transport proteins regulate both the activator and repressor functions of Gli transcription factors. Development 132: 3103–3111
48. Lucas BS, Aaron W, An S, Austin RJ, Brown M, Chan H, Chong A, Hungate R, Huang T, Jiang B, *et al.* (2010) Design of 1-piperazinyl-4-arylphthalazines as potent Smoothened antagonists. Bioorg Med Chem Lett 20: 3618–3622
49. Ma Y, Erkner A, Gong R, Yao S, Taipale J, Basler K, Beachy PA (2002) Hedgehog-mediated patterning of the mammalian embryo requires transporter-like function of dispatched. Cell 111: 63–75
50. Marigo V, Davey RA, Zuo Y, Cunningham JM, Tabin CJ (1996) Biochemical evidence that patched is the Hedgehog receptor. Nature 384: 176–179
51. Mich JK, Blaser H, Thomas NA, Firestone AJ, Yelon D, Raz E, Chen JK (2009) Germ cell migration in zebrafish is cyclopamine-sensitive but Smoothened-independent. Dev Biol 328: 342–354
52. Mille F, Thibert C, Fombonne J, Rama N, Guix C, Hayashi H, Corset V, Reed JC, Mehlen P (2009) The Patched dependence receptor triggers apoptosis through a DRAL-caspase-9 complex. Nat Cell Biol 11: 739–746
53. Miller-Moslin K, Peukert S, Jain R, McEwan M, Karki R, Llamas L, Yusuff N, He F, Li Y, Sun Y, *et al.* (2009) 1-amino-4-benzylphthalazines as orally bioavailable smoothened antagonists with antitumor activity. J Med Chem 52: 3954–3968
54. Mullauer FB, Kessler JH, Medema JP (2010) Betulinic acid, a natural compound with potent anticancer effects. Anticancer Drugs 21, 215–227
55. Nolan-Stevaux O, Lau J, Truitt ML, Chu GC, Hebrok M, Fernández-Zapico ME, Hanahan D (2009) GLI1 is regulated through Smoothened-independent mechanisms in neoplastic pancreatic ducts and mediates PDAC cell survival and transformation. Genes Dev 23: 24–36
56. Ochoa B, Syn WK, Delgado I, Karaca GF, Jung Y, Wang J, Zubiaga AM, Fresnedo O, Omenetti A, Zdanowicz M, *et al.* (2010) Hedgehog signaling is critical for normal liver regeneration after partial hepatectomy in mice. Hepatology 51: 1712–1723
57. Pan Y, Bai CB, Joyner AL, Wang B (2006) Sonic hedgehog signaling regulates Gli2 transcriptional activity by suppressing its processing and degradation. Mol Cell Biol 26: 3365–3377
58. Pan S, Wu X, Jiang J, Gao W, Wan Y, Cheng D (2010) Discovery of NVP-LDE225, a Potent and Selective Smoothened Antagonist. ACS Med Chem 1 (3): 130–134
59. Panakova D, Sprong H, Marois E, Thiele C, Eaton S (2005) Lipoprotein particles are required for Hedgehog and Wingless signalling. Nature 435: 58–65
60. Porter JA, Young KE, Beachy PA (1996) Cholesterol modification of hedgehog signaling proteins in animal development. Science 274: 255–259
61. Reifenberger J, Wolter M, Weber RG, Megahed M, Ruzicka T, Lichter P, Reifenberger G (1998) Missense mutations in SMOH in sporadic basal cell carcinomas of the skin and primitive neuroectodermal tumors of the central nervous system. Cancer Res 58: 1798–1803
62. Riobo NA, Haines GM, Emerson CP (2006) Protein kinase C-delta and mitogen-activated protein/extracellular signal-regulated kinase-1 control GLI activation in hedgehog signaling. Cancer Res 66: 839–845
63. Riobo NA, Lu K, Ai X, Haines GM, Emerson CP (2006) Phosphoinositide 3-kinase and Akt are essential for Sonic Hedgehog signaling. Proc Natl Acad Sci USA 103: 4505–4510
64. Robarge KD, Brunton SA, Castanedo GM, Cui Y, Dina MS, Goldsmith R, Gould SE, Guichert O, Gunzner JL, Halladay J, *et al.* (2009) GDC-0449-a potent inhibitor of the hedgehog pathway. Bioorg Med Chem Lett 19: 5576–5581

65. Roelink H, Augsburger A, Heemskerk J, Korzh V, Norlin S, Ruiz i Altaba A, Tanabe Y, Placzek M, Edlund T, Jessell TM, *et al.* (1994) Floor plate and motor neuron induction by vhh-1, a vertebrate homolog of hedgehog expressed by the notochord. Cell 76: 761–775
66. Roessler E, Belloni E, Gaudenz K, Jay P, Berta P, Scherer SW, Tsui LC, Muenke M (1996) Mutations in the human Sonic Hedgehog gene cause holoprosencephaly. Nat Genet 14: 357–360
67. Rohatgi R, Milenkovic L, Scott MP (2007) Patched1 regulates hedgehog signaling at the primary cilium. Science 317: 372–376
68. Rohatgi R, Milenkovic L, Corcoran RB, Scott MP (2009) Hedgehog signal transduction by Smoothened: pharmacologic evidence for a 2-step activation process. Proc Natl Acad Sci USA 106: 3196–3201
69. Romer JT, Kimura H, Magdaleno S, Sasai K, Fuller C, Baines H, Connelly M, Stewart CF, Gould S, Rubin LL, *et al.* (2004) Suppression of the Shh pathway using a small molecule inhibitor eliminates medulloblastoma in Ptc1(+/–)p53(–/–) mice. Cancer Cell 6: 229–240
70. Rominger CM, Bee W-LT, Copeland RA, Davenport EA, Gilmartin A, Gontarek R, Hornberger KR, Kallal LA, Lai Z, Lawrie K, *et al.* (2009) Evidence for allosteric interactions of antagonist binding to the smoothened receptor. J Pharmacol Exp Ther 329: 995–1005
71. Ruegg UT, Burgess GM (1989) Staurosporine, K-252 and UCN-01: potent but nonspecific inhibitors of protein kinases. Trends Pharmacol Sci 10: 218–220
72. St-Jacques B, Dassule HR, Karavanova I, Botchkarev VA, Li J, Danielian PS, McMahon JA, Lewis PM, Paus R, McMahon AP (1998) Sonic hedgehog signaling is essential for hair development. Curr Biol 8: 1058–1068
73. Stanton B, Peng L, Maloof N, Nakai K, Wang X, Duffner J, Taveras K, Hyman J, Lee S, Koehler A, *et al.* (2009) A small molecule that binds Hedgehog and blocks its signaling in human cells. Nat Chem Biol 5: 154–156
74. Svard J, Heby-Henricson K, Persson-Lek M, Rozell B, Lauth M, Bergstrom A, Ericson J, Toftgard R, Teglund S (2006) Genetic elimination of Suppressor of fused reveals an essential repressor function in the mammalian Hedgehog signaling pathway. Dev Cell 10: 187–197
75. Taipale J, Chen J, Cooper M, Wang B, Mann R, Milenkovic L, Scott M, Beachy P (2000) Effects of oncogenic mutations in Smoothened and Patched can be reversed by cyclopamine. Nature 406: 1005–1009
76. Taylor MD, Liu L, Raffel C, Hui CC, Mainprize TG, Zhang X, Agatep R, Chiappa S, Gao L, Lowrance A, *et al.* (2002) Mutations in SUFU predispose to medulloblastoma. Nat Genet 31: 306–310
77. Teglund S, Toftgard R (2010) Hedgehog beyond medulloblastoma and basal cell carcinoma. Biochim Biophys Acta 1805: 181–208
78. Tenzen T, Allen BL, Cole F, Kang JS, Krauss RS, McMahon AP (2006) The cell surface membrane proteins Cdo and Boc are components and targets of the Hedgehog signaling pathway and feedback network in mice. Dev Cell 10: 647–656
79. The I, Bellaiche Y, Perrimon N (1999) Hedgehog movement is regulated through tout velu-dependent synthesis of a heparan sulfate proteoglycan. Mol Cell 4: 633–639
80. Tremblay M, Lescarbeau A, Grogan M, Tan E, Lin G, Austad B, Yu L, Behnke M, Nair S, Hagel M, *et al.* (2009) Discovery of a potent and orally active hedgehog pathway antagonist (IPI-926). J Med Chem 52: 4400–4418
81. Von Hoff DD, LoRusso PM, Rudin CM, Reddy JC, Yauch RL, Tibes R, Weiss GJ, Borad MJ, Hann CL, Brahmer JR, *et al.* (2009) Inhibition of the hedgehog pathway in advanced basal-cell carcinoma. N Engl J Med 361: 1164–1172
82. Vorechovský I, Tingby O, Hartman M, Strömberg B, Nister M, Collins VP, Toftgård R (1997) Somatic mutations in the human homologue of *Drosophila* patched in primitive neuroectodermal tumours. Oncogene 15: 361–366
83. Wang B, Fallon JF, Beachy PA (2000) Hedgehog-regulated processing of Gli3 produces an anterior/posterior repressor gradient in the developing vertebrate limb. Cell 100: 423–434
84. Watkins DN, Berman DM, Burkholder SG, Wang B, Beachy PA, Baylin SB (2003) Hedgehog signalling within airway epithelial progenitors and in small-cell lung cancer. Nature 422: 313–317

85. Williams JA, Guicherit OM, Zaharian BI, Xu Y, Chai L, Wichterle H, Kon C, Gatchalian C, Porter JA, Rubin LL, *et al.* (2003) Identification of a small molecule inhibitor of the hedgehog signaling pathway: effects on basal cell carcinoma-like lesions. Proc Natl Acad Sci USA 100: 4616–4621
86. Xie J, Murone M, Luoh SM, Ryan A, Gu Q, Zhang C, Bonifas JM, Lam CW, Hynes M, Goddard A, *et al.* (1998) Activating Smoothened mutations in sporadic basal-cell carcinoma. Nature 391: 90–92
87. Yang Y, Guillot P, Boyd Y, Lyon MF, McMahon AP (1998) Evidence that preaxial polydactyly in the Doublefoot mutant is due to ectopic Indian Hedgehog signaling. Development 125: 3123–3132
88. Yang H, Xiang J, Wang N, Zhao Y, Hyman J, Li S, Jiang J, Chen J, Yang Z, Lin S (2009) Converse conformational control of smoothened activity by structurally related small molecules. J Biol Chem 284: 20876–20884
89. Yao S, Lum L, Beachy P (2006) The ihog cell-surface proteins bind Hedgehog and mediate pathway activation. Cell 125: 343–357
90. Yauch RL, Gould SE, Scales SJ, Tang T, Tian H, Ahn CP, Marshall D, Fu L, Januario T, Kallop D, *et al.* (2008) A paracrine requirement for hedgehog signalling in cancer. Nature 455: 406–410
91. Yauch RL, Dijkgraaf GJP, Alicke B, Januario T, Ahn CP, Holcomb T, Pujara K, Stinson J, Callahan CA, Tang T, *et al.* (2009) Smoothened mutation confers resistance to a Hedgehog pathway inhibitor in medulloblastoma. Science 326: 572–574
92. Zhang X, Harrington N, Moraes RC, Wu MF, Hilsenbeck SG, Lewis MT (2009) Cyclopamine inhibition of human breast cancer cell growth independent of Smoothened (Smo). Breast Cancer Res Treat 115: 505–521
93. Zhao C, Chen A, Jamieson CH, Fereshteh M, Abrahamsson A, Blum J, Kwon HY, Kim J, Chute JP, Rizzieri D, *et al.* (2009) Hedgehog signalling is essential for maintenance of cancer stem cells in myeloid leukaemia. Nature 458: 776–779
94. Zwerner JP, Joo J, Warner KL, Christensen L, Hu-Lieskovan S, Triche TJ, May WA (2008) The EWS/FLI1 oncogenic transcription factor deregulates GLI1. Oncogene 27: 3282–3291

# Chapter 12
# Cyclopamine and Its Derivatives for Cancer Therapeutics

**Martin Tremblay and Karen McGovern**

## Introduction

In the 1950s, researchers from the United States Department of Agriculture (USDA) investigated cases of congenital cyclopia in sheep grazing in high mountain ranges in central Idaho. After nearly a decade of research, steroidal alkaloids present in the corn lily plant (*Veratrum californicum*) were found responsible for the induction of cyclopic-type craniofacial birth defects that occurred when *Veratrum* was ingested by pregnant sheep on day 14 of gestation [7]. Jervine and cyclopamine were two important teratogenic compounds isolated from *Veratrum californicum* (Fig. 12.1, compound **2** and **6**, respectively), while numerous nonteratogenic but toxic *Veratrum* alkaloids were also present such as veratramine and muldamine [42–46]. Of note, the maternal ewes do not suffer ill effects from ingestion of the plants or cyclopamine [7], with birth defects being confined to a specific window of time during fetal development [103]. Because of its steroidal structure, cyclopamine was originally proposed to antagonize putative hormones involved in regulation of specific genes [42]. The pharmacology of cyclopamine remained dormant for nearly 30 years until genetic studies revealed that mutations in the Hedgehog pathway impacted development.

The role of the hedgehog pathway in development was first discovered in the fruit fly Drosophila by Nüsslein-Volhard and Wieschaus [72] and was ultimately recognized by a Nobel Prize in 1995. Their groundbreaking mutational analysis of genes in Drosophila that control segmentation and polarity elucidated a pathway that, when mutated, resulted in larvae with spiculated cuticles on their skin, resembling the spines of a hedgehog. Subsequent identification of the specific gene products revealed a unique signaling pathway with related orthologues in vertebrate organisms [37]. Mutations generated in the Hh pathway in vertebrates resulted in animals with cyclopic features [14]. These findings were substantiated in humans,

K. McGovern (✉)
Infinity Pharmaceuticals, Inc., 780 Memorial Drive, Cambridge, MA 02139, USA
e-mail: Karen.McGovern@infi.com

J. Xie (ed.), *Hedgehog Signaling Activation in Human Cancer and its Clinical Implications*, 187
DOI 10.1007/978-1-4419-8435-7_12, 

General carbon framework of steroid

General carbon framework of C-Nor D-Homo steroid

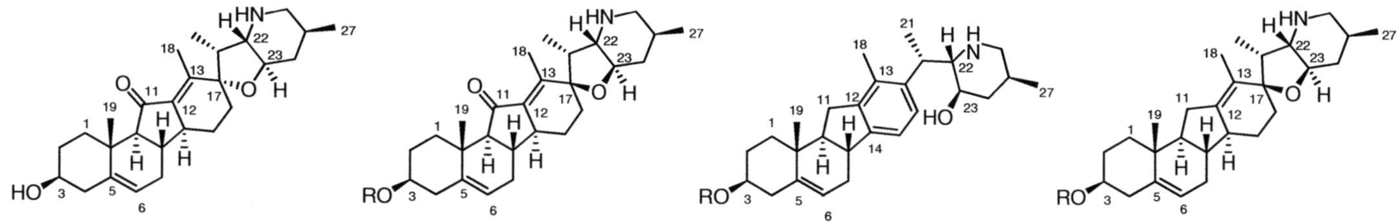


**1** Jervine (Originally proposed structure) **2** jervine (revised structure, R = H)
**3** pseudojervine (R = Glu)
**4** veratramine (R = H)
**5** veratrosine (R = Glu)
**6** cyclopamine (R = H)
**7** cycloposine (R = Glu)

**Fig. 12.1** General structure of steroid and C-nor D-homo steroids. Chemical structures of a subset of *Veratrum* alkaloids

where mutations in Shh were linked to holoprosencepaly, which can include cyclopic features [5, 80]. Thus, all inhibitors of the Hh pathway, whether derived from cyclopamine or not, would be expected to impact embryogenesis.

Studies of the Hh pathway in multiple developmental systems have identified key signaling proteins and genes that are regulated in response to ligand activation. Important links between Drosophila genetics and vertebrate biology led to breakthroughs in our understanding of the Hh pathway. One such link, as described above, was the discovery of the plant-derived alkaloid cyclopamine that is found in *Veratrum californicum*. Cyclopamine was subsequently found to antagonize the Hh pathway [15, 32] and to exert its inhibitory effects by binding to Smoothened (Smo) [12, 88]. The natural product cyclopamine, while not active against Drosophila Smo, has served as a powerful tool to help understand the role of the Hh pathway in many aspects of mammalian physiology and disease. While the entire book describes how the Hh pathway is involved in development and cancer, this chapter will review several aspects of the chemistry, pharmacology, and therapeutic potential of cyclopamine and its derivatives.

## Cyclopamine, a Natural Steroidal Alkaloid from *Veratrum* Species: Extraction, Isolation, and Structure Elucidation

As early as in the 1870s, a number of alkaloids with the C-nor D-homo steroid skeleton have been isolated from plants of the lily family such as *Veratrum* species. Over the years, a subset of alkaloids shown in Fig. 12.1 has triggered interest because of either their relative abundance and/or their pharmacological effects. Jervine was first isolated from *Veratrum album* [108, 109]. Since this pioneering work by Wright, several extraction and isolation methods for jervine have been described [82, 83] and its structure determination has been debated over several decades. Originally, a number of proposed structures placed jervine in the category of regular steroids constructed with a 6-6-6-5 tetracyclic framework [34, 35, 106]. However, careful exploration of the chemical reactivity of the C–D ring α/β-unsaturated ketone and degradation studies provided evidence that "this alkaloid does not have a normal steroid nucleus" [22]. The structure as depicted in **1** was then proposed, but there continued to be uncertainty concerning the C17 configuration [107]. Further characterization work [54] and X-ray crystallographic data of related alkaloids [79] subsequently showed that the configuration at the C17 position was inverted in the original assignment and that structure **2** was indeed jervine's accurate molecular structure. The structure elucidation of jervine served as a cornerstone to the discovery of other *Veratrum* alkaloids.

A jervine congener, cyclopamine **6** (11-deoxojervine), was isolated independently from *Veratrum glandiforum* (aka *Veratrum album*) collected in the Tokachi district of Hokkaido in Japan [65] and from *Veratrum californicum* collected in the western US [44, 46]. As was the case originally for jervine, the C17 configuration was wrongly assigned initially but corrected when more data became available on jervine.

Early after a reliable isolation method was published for cyclopamine by the USDA research group, they also reported the isolation of the glycosylated form of cyclopamine called cycloposine **7** [47]. Several years after pioneering work by Masamune and Keeler, improved methods to extract and isolate cyclopamine were published [36, 50, 73, 86], and it is clear that *Veratrum californicum* is rich in cyclopamine from all of these reports.

## Synthetic Chemistry of Cyclopamine

The amazing structural constitution of jervine and cyclopamine classified these plant natural products as C-nor D-homo steroids. Both compounds have 27 carbons arranged in four carbocycles (A–D rings) and dimethyloctahydrofuro[3,2-b] pyridine (E–F rings), the latter being spiro-connected to the D-ring at the C17 position. Stereochemical complexity in a form of ten stereogenic centers, two of which are quaternary centers, can be found in jervine and cyclopamine. Two polar functions (C3-alcohol and basic piperidine nitrogen) are on the two opposite sides of the lipophilic steroidal skeleton, which is unsaturated at two positions (B-ring and D-ring). Jervine is only different from cyclopamine by the presence of the C11-keto function, which is conjugated with the alkene on the D-ring. Relative to conventional steroids, the preparation of C-nor D-homo steroids has received far less attention from the synthetic community [11]. Nevertheless, a few strategies were exploited to access small quantities of jervine and cyclopamine (Fig. 12.2).

The rearrangement of regular steroids to C-nor D-homo system discovered by Hirschmann [31] provided the basis for the semisynthesis of jervine **2** from hecogenin **8** [67] and more recently of cyclopamine **6** from dehydroepiandrosterone **9** [24]. One of the first successful approaches to jervine started with hecogenin **8** that was converted to intermediate **11** by a six step sequence [69]. Alternatively, the same intermediate could be synthesized from Hagemann ester **10** by a series of annulation reactions [38]. Conversion of the tetracyclic intermediate **11** to jervine involves addition of the E–F ring system in multiple steps [55, 67]. Cyclopamine can be obtained from jervine through a Wolff–Kishner reduction of the C11-ketone [87]. For nearly three decades following these reports, the synthetic community paid little attention to the synthesis of in *Veratrum* alkaloids. The elucidation of the mode of action of cyclopamine coupled with its therapeutic potential revamped interest for this wonderful but complex molecule. Nowadays, many synthetic groups have active research programs aimed at the synthesis of cyclopamine. Recently, a semisynthetic approach to cyclopamine from dehydroepiandrosterone was described [24]. While this approach utilized essentially the same rearrangement pioneered by Hirschmann (*vide infra*), this key transformation was performed on the advanced intermediate **12** bearing the extremely challenging E-ring with the correct configuration at C17 (Fig. 12.2). This approach is very powerful because the spiro-γ-lactone in intermediate **12** can be readily installed and ultimately serves as a versatile building block for the construction of the piperidine ring.

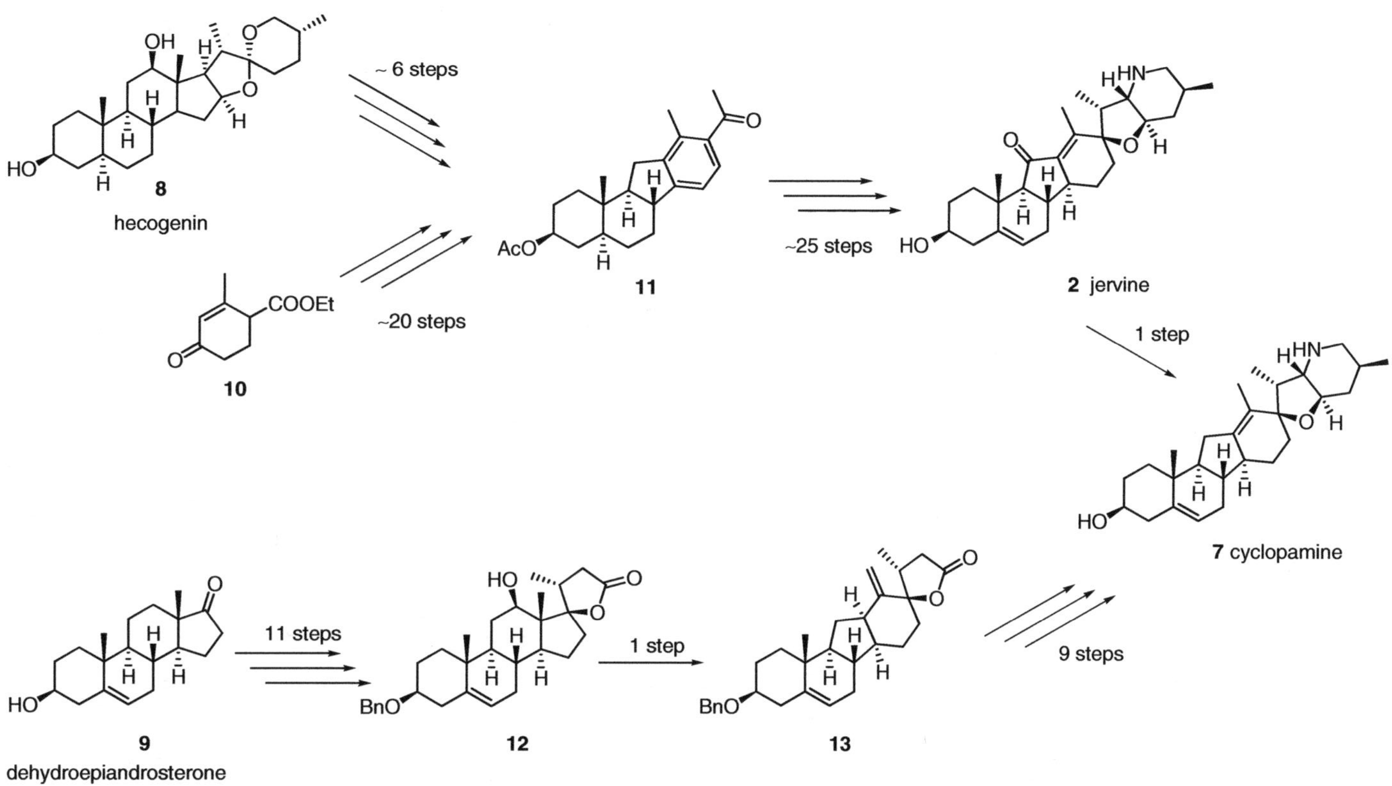


**Fig. 12.2** Synthetic approaches to jervine and cyclopamine through rearrangement of C/D ring of conventional steroids to C-nor D-homo steroids or by total synthesis

## Evaluation of Cyclopamine as Drug Lead: Its Drug Properties

With the emerging role of the Hh pathway in disease in the late 1990s, the need for selective modulators to test their therapeutic potential became apparent. The discovery of small molecule modulators of the Hh pathway has been comprehensively reviewed elsewhere [62, 77, 95] and the optimization of cyclopamine analogs is the focus of this book chapter. By virtue of its role in target validation of Smo and its availability from nature, cyclopamine is a very interesting starting point for the discovery of modulators of the Hh pathway. Cyclopamine can be readily isolated in high yield from *Veratrum californicum* [36, 50, 73, 86]. This plant species is naturally occurring in the western US, particularly in Utah and Idaho [49], where it was once the target for eradication due to the harmful effects on livestock. Alternatives to field cultivation can already be envisioned for *Veratrum* plant species. Successfully applied to taxol, a therapeutic compound isolated from the yew tree [52], a plant cell culture technique has been reported to generate green plants from embryonic calli of *Veratrum californicum* [61]. Most importantly, these green plants produced veratramine and traces of cyclopamine when grown in suspension media in the presence of naphthalene acetic acid. To complement the latter approach, alternative production platforms such as metabolic engineering are emerging [56] and may be applicable to *Veratrum* alkaloids. Toward that end, important steps in the biosynthetic pathway of cyclopamine have been elucidated [39, 40], which may supplement the development of in vitro techniques to produce cyclopamine. In addition to biomass production and processing, cyclopamine can be obtained by semisynthetic approaches. As one example, jervine can be converted to cyclopamine via Wolff–Kishner reduction [87] (Fig. 12.2). Another example is the single step conversion of cycloposine **7** to cyclopamine **6** by deglycosylation methods [36]. These one step transformations from other C-nor D-homo steroids provide additional source of starting materials.

While cyclopamine represents a very interesting starting point for the discovery of modulators of the Hh pathway, its clinical development was hampered mainly by poor pharmaceutical properties and suboptimal potency. Cyclopamine has low water solubility, which initially impeded the development of formulations for administration to animals. A variety of cyclopamine formulations, including ethanol [3, 50], triolein/EtOH (4:1, v/v) [6, 89], dimethylformamide [98], and dimethylsulfoxide [96], have recently been reviewed [59]. The addition of high concentrations of complexing agents such as 2-hydroxypropyl-β-cyclodextrin resulted in better aqueous formulations of cyclopamine [75, 99]. Alternatively, decreasing the concentration of 2-hydroxypropyl-β-cyclodextrin and thereby the viscosity can be achieved by using the hydrochloride salt of cyclopamine to generate chemically and physically stable formulations [21, 92]. In addition to the hydrochloride salt, water soluble tartrate salts of cyclopamine have recently been described in the literature [103, 110].

In addition to its poor aqueous solubility, cyclopamine is acid labile and readily converts to veratramine (**4**, Fig. 12.1) [48] as well as other isomeric products [104]. Although structurally related to cyclopamine, veratramine does not act as a Smo antagonist but affects several receptors [70, 90] and causes hemolysis [2]. This liability may not be problematic for oral administration of cyclopamine in preclinical species where the stomach pH may not promote conversion, but in humans the stomach pH ranges between 1.5 and 3 [20] and gastrointestinal transit time is typically longer (20–30 h) than other laboratory animals [41].

In some of the early studies with cyclopamine in sheep, observation of the teratogenic effects required daily oral (p.o.) administration of 2–3 g/animal (average weight of animal ~50 kg; 40–60 mg/kg) [42, 45, 50]. This need for high and frequent doses could be due to poor pharmacokinetic properties and/or potency; some recent studies have shed some light on both aspects. In sheep, a short elimination half-life (1.1 ± 1 h) was measured when cyclopamine tartrate salt (1.6 mg/kg) was administered intravenously (i.v.) [103]. In rodents, the oral (p.o.) bioavailability of cyclopamine is modest (33% relative to intraperitoneal (i.p. administration [59]) and the poor pharmaceutical properties are likely responsible for this observation.

Many cellular assays have been developed to evaluate the potency of Hh pathway inhibitors. Cyclopamine inhibits Hh-dependent processes including (a) HNF3β-induction in chick embryo neural plate assay ($IC_{50}$ of 10 nM; [33]), (b) Gli-reporter assay ($IC_{50}$ of 300 nM; [12]), and (c) C3H10T1/2 differentiation ($IC_{50}$ of 300–400 nM [18]). While the teratogenic effect of cyclopamine is associated with on-target activity on Smo and the Hh pathway, there is evidence that cyclopamine may not be entirely selective. High concentrations of cyclopamine (10 μM) have shown cytotoxic effect on cancer cells that do not express Smo, indicating that the observed effect was off-target. This phenomenon is not unique to cyclopamine since another small molecule Hh antagonist (Cur61614) that is not structurally related to cyclopamine has shown the same behavior [114]. The use of cancer cells grown in culture for the evaluation of Hh pathway antagonists has been extremely controversial. There have been reports in the literature revealing that cancer cell lines grown in culture lose their dependence on the Hh signaling pathway for growth [85, 101]. For this reason, in vitro assessment of a therapeutic window related to treatment with cyclopamine needs to be interpreted carefully. Recent studies have provided some insights into the toxicity of cyclopamine in vivo [59]. The observed toxicity is dependent on the route of administration and could be circumvented by infusion of the drug. This mode of administration delivers cyclopamine at steady state concentrations, whereas bolus intraperitoneal (i.p.) and oral (p.o.) dosing resulted in high and transient plasma peak concentrations that lead to severe dystonia and lethargy, respectively. The relatively rapid onset of the observed toxicity (between 2 and 6 h postdose for i.p. and p.o., respectively) suggests that off-target mechanisms are likely the cause. However, it is unclear if this toxicity is due to the parent drug or a metabolite of cyclopamine. Indeed, high plasma concentration of cyclopamine (20 μM) are well tolerated by the animals dosed i.p., whereas toxicity is observed in orally dosed animals with relatively low (2 μM) plasma concentration of cyclopamine. In summary, derivatives of cyclopamine with improved solubility, stability, and potency are highly desirable to address some of the issues outlined above.

## Medicinal Chemistry of Cyclopamine Analogs

Well before the elucidation of cyclopamine's biological target, scientists have been intrigued by the structure–activity relationship (SAR) of naturally occurring jervine [9, 10] and cyclopamine, [48], in particular with respect to their teratogenic potential. However, the determination of biological effects required gram quantities of material which obviously hampered the ability to obtain SAR data. Moreover, the teratogenic potential measured in vivo encompasses both pharmacodynamic and pharmacokinetic properties of the molecules, which makes it difficult to draw conclusive SAR. With that caveat in mind, given that both jervine **2** and cyclopamine **6** are active teratogens, one may conclude that the 11-keto group is not necessary for this activity. However, this functional group plays a role in increasing the stability of D/E-ring system by reducing acid-mediated opening of the ether bridge [9]. Interestingly, cyclopamine-4-en-3-one **14** [66] (Fig. 12.3) was found to be at least twofold more bioactive than jervine as demonstrated independently by its teratogenity in hamsters [9, 23] and its inhibitory effect of Shh signaling in chick neural plate [33]. Also shown in these two assays, the reduction of C5–C6 and C12–C13 double bonds produced a weaker teratogen tetrahydrojervine **15** [68], whereas substantial activity was retained for 12β,13α-dihydrojervine **16**. Cyclo-posine **7** bearing a glucosyl group at the C3 position has also shown teratogenic activity [45], but its effects on the Shh signaling in the chick neural plate was modest [33]. It is plausible that the glycosyl group of cycloposine could be hydrolyzed to liberate cyclopamine in vivo [103], thus explaining the difference between these two results. In addition to alterations of the steroidal skeleton and substitution at the C3 position, the piperidine

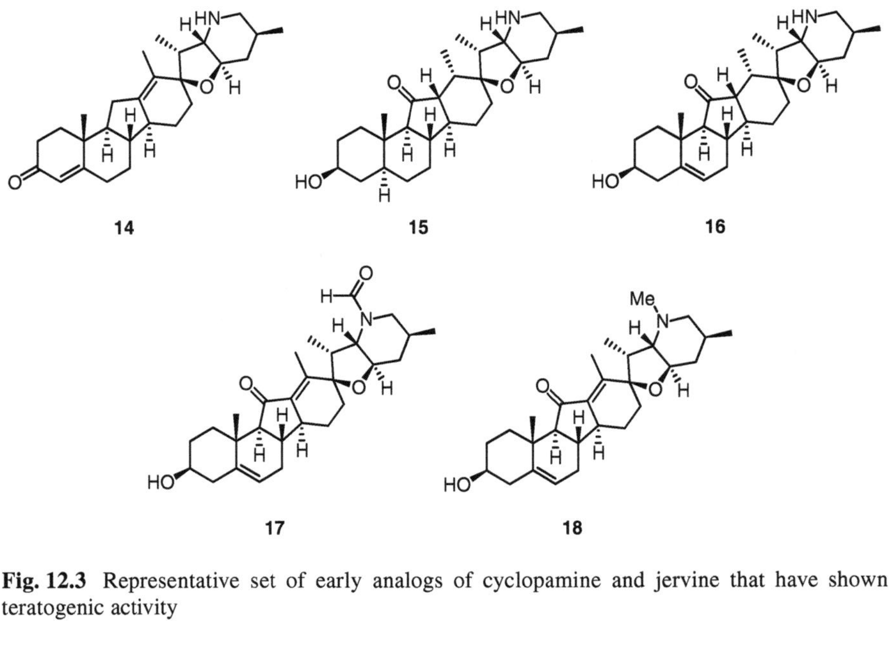

14 15 16

17 18

**Fig. 12.3** Representative set of early analogs of cyclopamine and jervine that have shown teratogenic activity

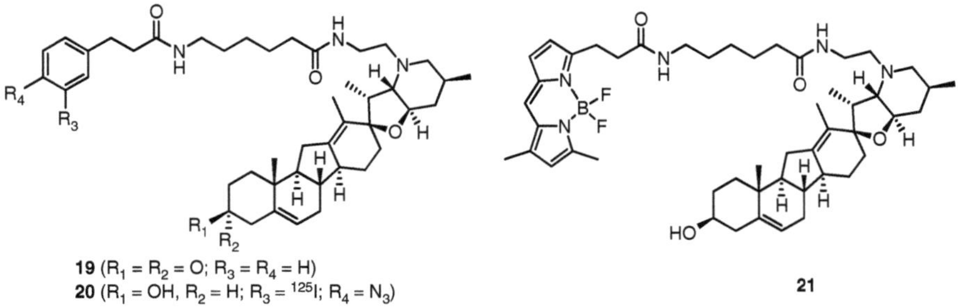


**Fig. 12.4** Semisynthetic analogs and probes of cyclopamine based on the natural steroid skeleton

nitrogen was also subjected to modifications [10]. Notably, *N*-formyl (**17**) and *N*-methyl (**18**) jervine analogs were determined to be active teratogens in hamsters, while quaternization of the nitrogen or bulkier alkyl substituents almost completely eliminates the biological activity. In summary, early studies around the structure–teratogenicity relationships of jervine and cyclopamine revealed that these structures were amenable to changes while retaining the biological response. However, improvement of the drug-like properties of these compounds (*vide infra*) would require further investigations.

A synthetic analog of cyclopamine named KAAD-cyclopamine **19** ($IC_{50}$ = 20 nM) was determined to be one order of magnitude more potent than cyclopamine ($IC_{50}$ = 300 nM) in the Shh-Light Gli-reporter assay [13] (Fig. 12.4). This compound played a significant role in establishing Smo as the target of cyclopamine. A radiolabeled and photoaffinity derivative of KAAD-cyclopamine (**20**) ($IC_{50}$ = 150 nM, Shh-Light) was used in the target identification studies [12] and BODIPY derivative **21** ($IC_{50}$ = 150 nM, Shh-Light) is commonly used to assess binding affinity of Smo antagonists. A number of related analogs have also been reported in the patent literature [4]. These studies showed that the potency of cyclopamine could be improved through chemical modifications.

A diversity-oriented approach was designed to generate a focused library of carbohydrate cyclopamine conjugates with improved aqueous solubility [113]. As an example, compound **22** bearing L-rhamnose moiety has better aqueous solubility and similar anticancer activity ($IC_{50}$ = 33 μM) relative to cyclopamine in A549 lung cancer cell line. Although no binding data on Smo were reported for these compounds, it is notable that polar solubilizing groups on the piperidine nitrogen are tolerated while KAAD-cyclopamine **20** bears highly lipophilic side-chain at the same position.

Aiming to develop a targeted therapy for prostate cancer and perhaps move away from potential on-target effects of cyclopamine on the normal stem cell niche, prodrug derivative **23** was designed and synthesized (Fig. 12.5) [53]. In this approach, a peptide carrier susceptible to selective hydrolysis by the active form of prostate-specific antigen (PSA) was grafted to cyclopamine. While the prodrug itself had no significant activity, cyclopamine was indeed released in presence of PSA and biological response was observed in DU-145 prostate cell line. The same concept,

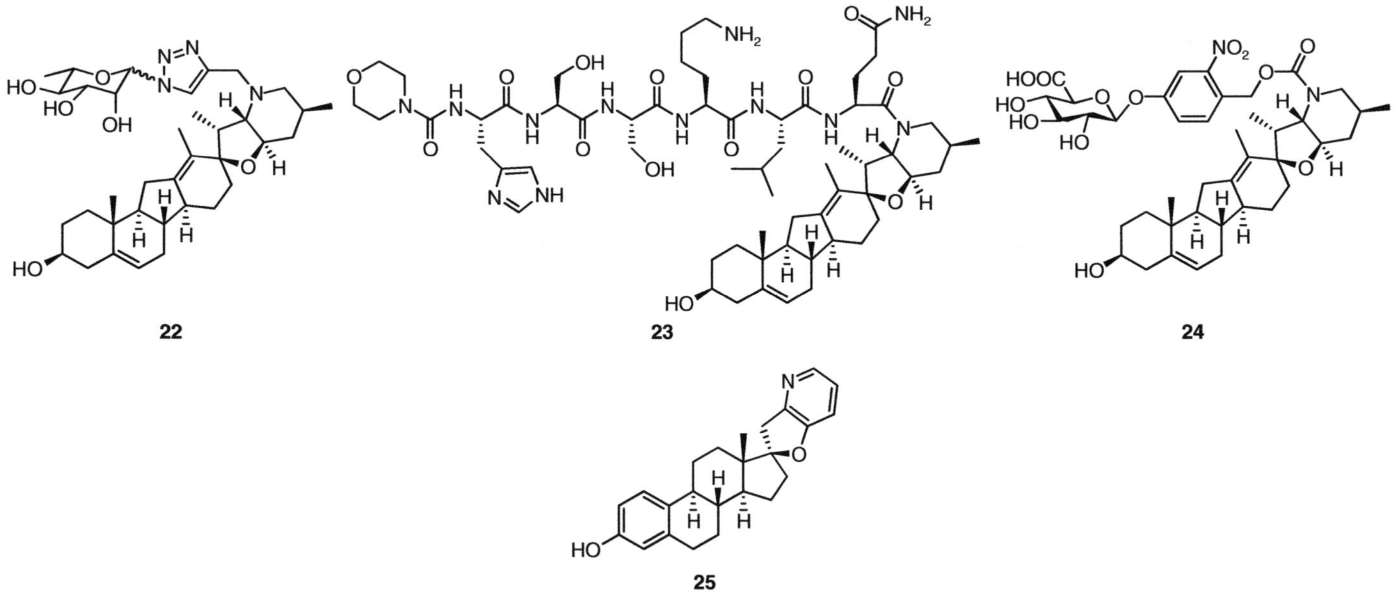


Fig. 12.5 Semisynthetic analogs of cyclopamine based on natural steroid skeletons

with potentially broader scope, led to the design and synthesis of the carbamate cyclopamine analog **24**. Carbamate prodrug **24** was not active by itself, but recapitulated cyclopamine's growth inhibition in glioma U87 cell line when exposed to β-glucuronidase, which is typically present in high levels in the tumor vicinity [26]. These two distinct approaches demonstrated that cyclopamine prodrugs offer the opportunity to selectively deliver cyclopamine to tumor environments.

Readily available steroidal synthetic starting points provided novel cyclopamine analogs that were evaluated for their effect on the Hh pathway [105]. For instance, estrone derivative **25**, which displays a simplified E/F ring system, was shown to inhibit Shh-induced proliferation of mouse granule neuron precursors as well as Shh-Light2 cells at 10 μM. This study revealed that simplification of the core structure of cyclopamine could result in analogs with desirable properties, but more work needs to be done to understand the consequence of these simplifications on the binding to Smo. As a complementary approach, the synthesis of cyclopamine from dehydroepiandrosterone (Fig. 12.2) could deliver very interesting analogs, particularly on the E/F rings, which would be otherwise challenging to obtain from the natural product [28, 29].

## Discovery and Development of IPI-926, a Semisynthetic Cyclopamine Derivative in Clinical Trials

The first strategy to improve the pharmaceutical properties of cyclopamine was to address the acid lability. The approach was to alter the influence of the D-ring allylic ether on the cleavage of the spirotetrahydrofuran E-ring, with minimal change to the D-ring geometry. It was originally hypothesized that the oxygen of the allylic ether could direct a chemo- and stereo-selective cyclopropanation of the D-ring double bond. Serendipitously, it was found that the D-ring could be expanded synthetically by cyclopropanation and subsequent acid-catalyzed rearrangement. The resulting 7-membered ring cyclopamine analogs, exemplified by compound **26** (Fig. 12.6), were found to have increased chemical stability [93]. Reminiscent of the case of cyclopamine (*vide infra*), the change in oxidation state from homoallylic alcohol to the α/β-unsaturated ketone in the A/B-ring system brings significant improvement of potency and, in this case, solubility. Also, the structure–activity relationships for N-substitution on compound **26** and cyclopamine closely track, suggesting that expansion of the D-ring did not cause major changes in the binding mode. Compound **26** is equipotent to cyclopamine, yet has improved aqueous solubility and stability relative to cyclopamine. The α/β-unsaturated ketone system found in compound **26** is a very common and structurally important functionality in endogenous steroid hormones. However, this functionality on compound **26** was found to be readily metabolized to the corresponding saturated alcohol and glucuronide conjugate after oral administration in cynomolgus monkeys [64]. The conversion of the half-chair A-ring system of compound **26** to a *cis*-ring fusion system provided a remarkable improvement (approximately tenfold) in potency of

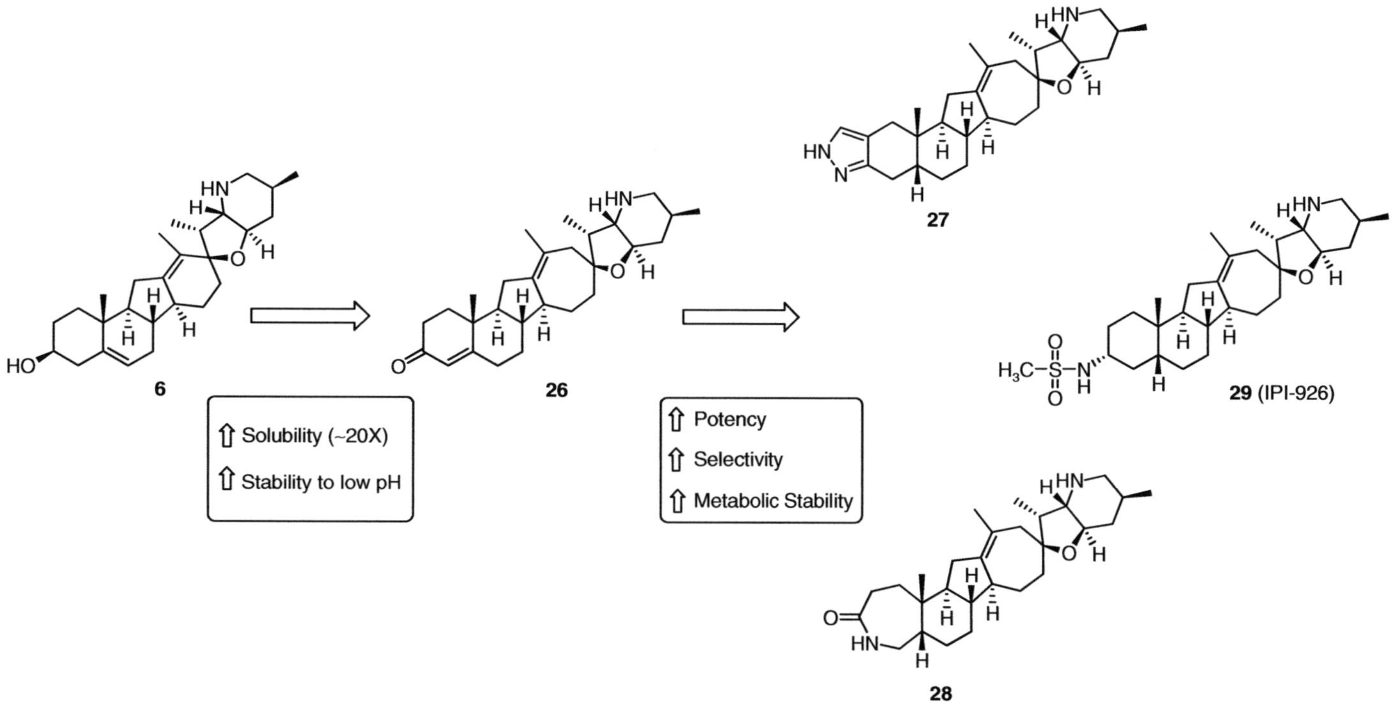


**Fig. 12.6** Semisynthetic analogs of cyclopamine based on homologated D-ring skeleton. Progression toward the discovery of **29** (IPI-926)

these D-homo cyclopamine derivatives. Further modifications of the A-ring generated three new series of analogs pyrazole **27**, lactam **28**, and sulfonamide **29** with significantly greater potency and less susceptibility to metabolism than the first generation compound **26**. One key discriminating factor between the three series of analogs is their pharmacokinetic profiles. Although all three series displayed good exposure when administered orally to multiple species, the sulfonamide **29** showed a significant increase in plasma half-life due to low clearance and high tissue distribution (Table 12.1). These properties translated into greater tumor-free intervals following treatment and more robust efficacy than the two other lead compounds when studied in an Hh-dependent B837Tx tumor model, which is described in more detail below [94].

To support its clinical development, capabilities for robust large-scale production of compound **29** were established. First, sourcing of starting material and large-scale extraction and isolation of cyclopamine *Veratrum californicum* was developed and optimized. A robust process was developed to produce multiple kilograms of advanced intermediates (e.g., compound **26**) in fixed equipment. Likewise, the original conversion of cyclopamine to IPI-926 utilized multiple steps that involved potential throughput-limiting purification steps [93, 94].

**Table 12.1** In vitro and in vivo profiling of cyclopamine analogs **27–29**[a]

| *Potency* | **6** (cyclopamine) | **27** | **28** | **29** (IPI-926) |
|---|---|---|---|---|
| C3H10T1/2 (EC50) | 300–400 nM | 10–20 nM | 30–40 nM | 7–15 nM |
| Smo Binding (IC50) | 114 nM | 6 nM | 6 nM | 1–2 nM |
| *DMPK* | **6** (cyclopamine) | **27** | **28** | **29** (IPI-926) |
| Range $F_{PO}$ (mouse, rat, dog, cynomolgus) | 33%[b] | 30–80% | >90% | 50–100% |
| Range $t_{½}$ (mouse, rat, dog, cynomolgus) | 4 h[b] | 1–7 h | 3–7 h | 8–24 h |
| Range Vd (mouse, rat, dog, cynomolgus) | – | 6–13 L/kg | 2–5 L/kg | 9–30 L/kg |
| *Efficacy Ptc1$^{+/-}$/Hic1$^{+/-}$ B837Tx allograft mouse model* | **6** (cyclopamine) (40 mg/kg p.o.) | **27** (80 mg/kg p.o.) | **28** (30 mg/kg p.o.) | **29** (IPI-926) (40 mg/kg p.o.) |
| Tumor-free interval 21 days posttreatment | Not observed | 19 days | 10 days | 21 days |
| Recurrence rate posttreatment | 100% | 40% | 30% | 0% |

[a] From Tremblay et al. [94]
[b] From Lipinski et al. [59]. Species: Female C57BL/6J mice at 12–16 weeks of age

Following optimization, IPI-926 drug substance is produced without the involvement of chromatography [1]. Finally, IPI-926 drug product is a solid dosage form intended for oral administration, which was developed to support dose-escalation in the clinic. In a single molecule, IPI-926 (**29**) encompasses the structural features and pharmaceutical properties that overcome many of the deficiencies identified for cyclopamine – namely, solubility, stability, pharmacokinetic profile, and in vivo potency.

## Preclinical Pharmacology of IPI-926

The activity of cyclopamine and analogs (e.g., **27–29**) were assessed in an assay that measures activation and inhibition of Hh pathway-dependent cellular differentiation. This assay is a Smo-mediated differentiation of the murine mesenchymal cell line C3H10T1/2; it has been demonstrated that cells exposed to either Hh ligands or oxysterols will differentiate to osteoblasts and that cyclopamine will inhibit this differentiation [18, 71, 100]. Cyclopamine inhibits the Smo-dependent differentiation of C3H10T1/2 with an $IC_{50}$ of 300 nM. Derivatives of cyclopamine (**27–29**) are much more active than cyclopamine. A Smo binding competition assay was conducted with BODIPY-cyclopamine, as previously described above [12], to confirm that the inhibition of the Hh pathway is through targeting Smo. While cyclopamine blocks the binding of BODIPY-cyclopamine with an $EC_{50}$ of 114 nM, the derivatives of cyclopamine bind more tightly (Table 12.1). IPI-926 (**29**) has an $IC_{50}$ between 7 and 15 nM in the differentiation assay and an $EC_{50}$ of ~2 nM in the Smo binding assay. The inhibition of Smo-driven differentiation correlates with inhibition of expression of Gli-1 regulated genes such as Gli1 and Ptch1. Similar inhibition by IPI-926 of Hh pathway gene expression is detected in the human mesenchymal cell line HEPM.

As described in previous chapters, some tumor types have mutations in Hh pathway members that lead to constitutive activity of the pathway, such as loss of Ptch function or activation of Smo. These types of mutations are found in BCC and some medulloblastoma. The activity of IPI-926 was investigated in a mouse model of medulloblastoma with the Hh pathway constitutively active due to heterozygous $Ptch1^{+/-}$. This mouse model is also heterozygous for Hic-1, the gene for *H*ypermethylated *I*n *C*ancer [8]. The B837Tx cell line was derived from a medulloblastoma in this mouse and passaged as a subcutaneous allograft in NOD/SCID mice. A single oral dose of IPI-926 leads to a dose-dependent decrease in Gli-1 in a dose-dependent manner (Fig. 12.7). A subsequent study determined the pharmacokinetic (PK) and pharmacodynamic (PD) relationship in this model. A single oral dose of either 4 or 40 mg/kg was administered to B837Tx tumor-bearing mice. Levels of Gli1 expression in the tumor and drug levels in the tumor and plasma were assessed at multiple time points postdose, taken over 7 days. The plasma drug concentration profile indicates exposure over many days after both low and high doses of IPI-926 (Fig. 12.8). Gli1 expression correlated well with tumor drug levels,

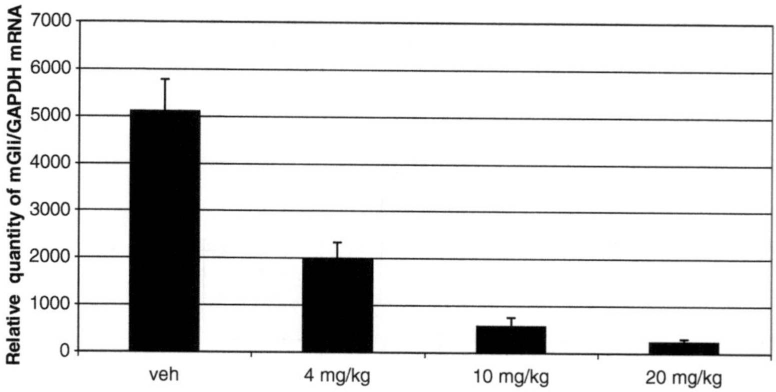


**Fig. 12.7** Inhibition of Gli-1 expression at 8 h in B837Tx tumors following single oral administration of IPI-926

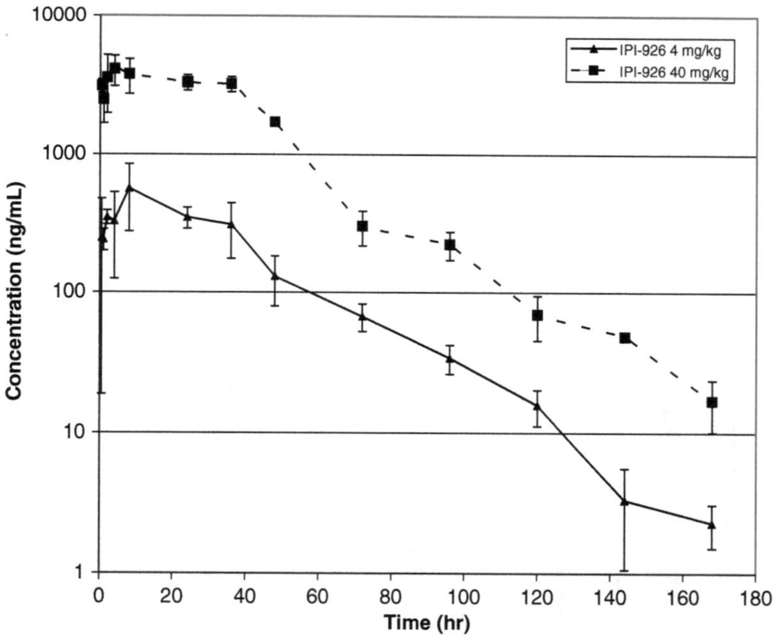


**Fig. 12.8** Plasma concentration-time profile of IPI-926 following 4 and 40 mg/kg oral administration of IPI-926 to NOD/SCID mice bearing a B837Tx tumor

and a single dose of 40 mg/kg IPI-926 led to inhibition of Gli1 in the tumor beyond 6 days (Figs. 12.9 and 12.10) [78].

The antitumor activity of IPI-926 was also assessed in the medulloblastoma allograft. IPI-926 was administered orally to tumor-bearing animals daily at 4, 10, or 20 mg/kg for 21 days. During treatment, tumors regressed to undetectable levels in animals administered 10 or 20 mg/kg (Fig. 12.11), with the 4 mg/kg group showing

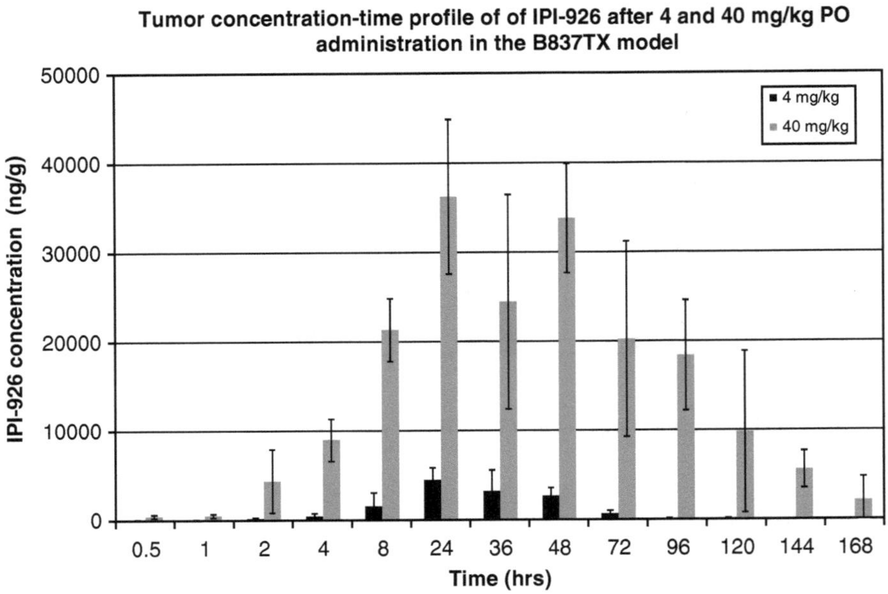


**Fig. 12.9** Tumor levels of IPI-926 following single-dose administration of 4 and 40 mg/kg in B837Tx tumor-bearing mice

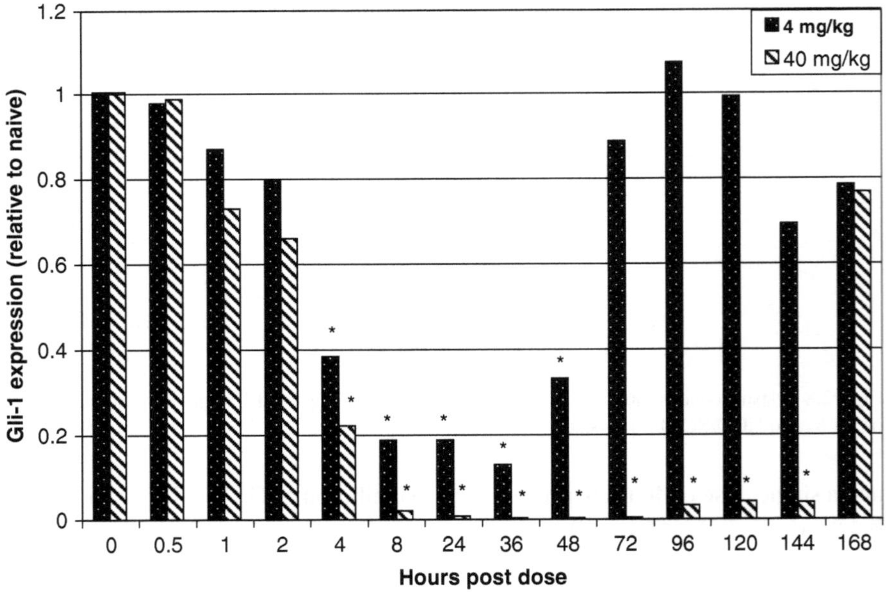


**Fig. 12.10** Gli-1 mRNA expression in B837Tx tumors following administration of a single oral dose of IPI-926 (4 and 40 mg/kg); $p < 0.05$

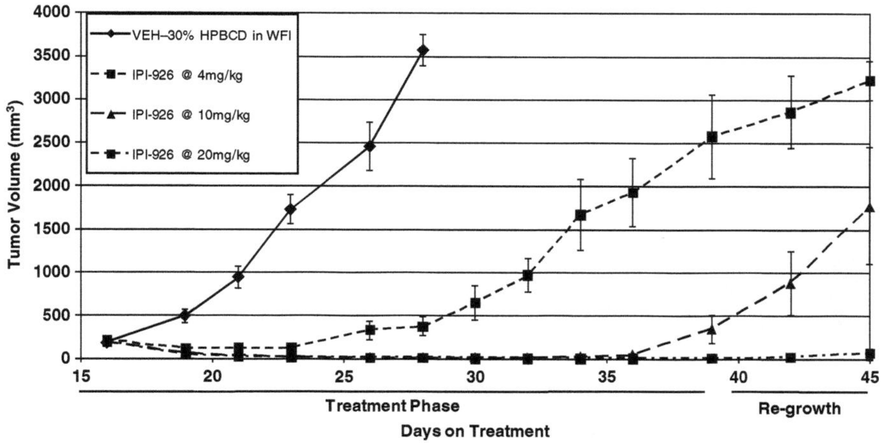


**Fig. 12.11** Inhibition of B837Tx tumor growth with daily oral administration of IPI-926; $p<0.05$

50% tumor growth inhibition. In a separate study, daily administration of 40 mg/kg led to regression with no regrowth of the tumors posttreatment, thereby demonstrating that IPI-926 is active in inhibiting growth of Ptch mutant driven tumors [78].

The activity of IPI-926 was also determined in nongenetic tumors. These tumors are driven by malignant activation of the Hh pathway in a Hh ligand-dependent manner. Examples of this type of tumor include pancreatic, ovarian, prostate, and breast ([111] and also see Chaps. 7, 10, and 11). Utilizing a sensitive immunohistochemistry (IHC) method for detection of Sonic hedgehog (SHh) ligand, a high percentage of tumors of various types were found to express high levels of SHh (Fig. 12.12). In many tumor types, Hh signaling occurs with tumors secreting Hh ligand and the surrounding stroma cells responding. Blocking signaling with a Smo antagonist inhibits the Hh gene expression in the stroma and can lead to tumor growth inhibition. This has been detected in multiple tumor models with IPI-926 and other Smo inhibitors (e.g., Fig. 12.13; [112]).

An improved understanding of the mechanism of action of Hh pathway inhibition in ligand-driven tumors has been described in studies with a transgenic model of pancreatic cancer. This model is driven by activated Kras and loss of p53 [KPC] [30] and recapitulates the human disease, from PanIN lesions to the development of pancreatic adenocarcinoma (PDA) and liver metastases. The PDA tumors are highly desmoplastic, as are human PDAs, with an abundance of stromal cells producing collagen and fibronectin among the tumor cells. PDAs are not very susceptible to chemotherapy, including the standard of care, gemcitabine, in both humans and in the KPC model. It was demonstrated that treatment with IPI-926 decreased the stromal content and increased the microvascular density of the KPC tumors [74]. The effect of IPI-926 on the stroma enabled delivery of the chemotherapeutic gemcitabine to the tumor, leading to tumor growth inhibition and a doubling of the median survival of these mice. The IPI-926 plus gemcitabine-treated mice also had

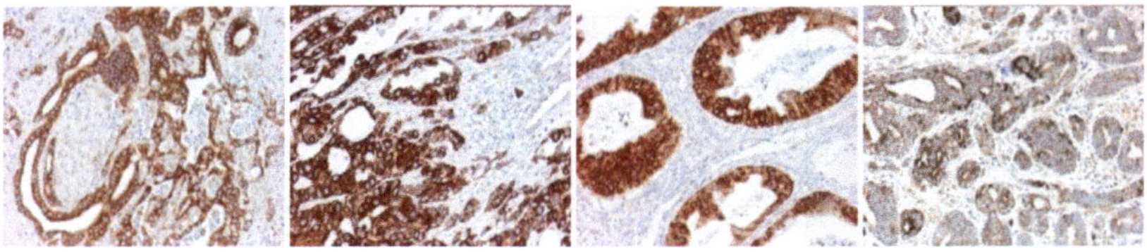


| Cancer Type | Total # of Samples | SHh Positive | SHh Negative | Percent Positive |
|---|---|---|---|---|
| Pancreatic | 92 | 65 | 27 | 71% |
| Colon | 69 | 58 | 11 | 84% |
| Ovarian | 68 | 32 | 36 | 47% |
| Prostate | 73 | 56 | 17 | 77% |

**Fig. 12.12** Sonic Hedgehog is highly expressed in multiple tumor types. IHC for SHh shown in *top*; staining in tumor microarrays summarized in *bottom*

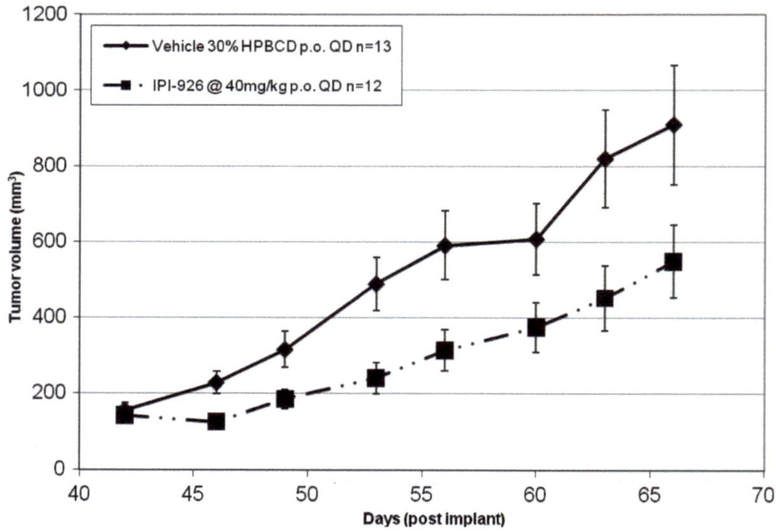


**Fig. 12.13** Activity of IPI-926 in the pancreatic cancer xenograft model BxPC3, 40 mg/kg daily p.o.; $p<0.05$

a lower incidence of liver metastases. These data provide rationale for evaluation of the combined treatment of chemotherapy and IPI-926 in pancreatic cancer.

The potential for a role for Hh pathway in tumor progenitor cells has been suggested in multiple indications (reviewed in [97]), in both hematological malignancies, including multiple myeloma [76], acute lymphocytic leukemia [58] and chronic myeloid leukemia [17] and in solid tumors, such as glioma [3, 19] breast cancer [60] and pancreatic cancer [57]. It is believed that tumor progenitor cells are resistant to chemotherapy and therefore suspected to be responsible for disease relapse following treatment with conventional therapeutic agents. To address the role of Hh pathway in a model of minimal residual disease (MRD) postchemotherapy, IPI-926 was testing in a primary tumor model of small cell lung cancer (SCLC). Clinically, SCLC responds well to chemotherapy but then relapses within months with no further response to therapy (ref). A chemotherapy-sensitive primary tumor-derived xenograft model, LX22 was utilized to address whether IPI-926 would have an effect on time to relapse postchemotherapy [27]. The LX22 model responds well to chemotherapy and regresses, then regrows, resembling a clinical "complete response." Inhibition of Hh pathway alone in established tumors does not affect tumor growth. However, treatment with IPI-926 after SCLC tumors regress with chemotherapy leads to a significant delay in time to tumor reoccurrence (Fig. 12.14) [91]. Recently, we have determined, along with collaborators, that IPI-926 is active in additional models of minimal residual disease. Similar activity as seen in LX22 of IPI-926 in delaying tumor relapse postchemotherapy was detected in multiple primary ovarian tumor xenografts [25]. In addition, IPI-926 is also active posttumor reduction with targeted therapy, such as tyrosine kinase inhibitors (TKIs). Tumor reduction occurs in a cell line xenograft model of non-small cell lung cancer (NSCLC), NCI-H1650, with treatment with the TKI gefitinib. Maintenance therapy with

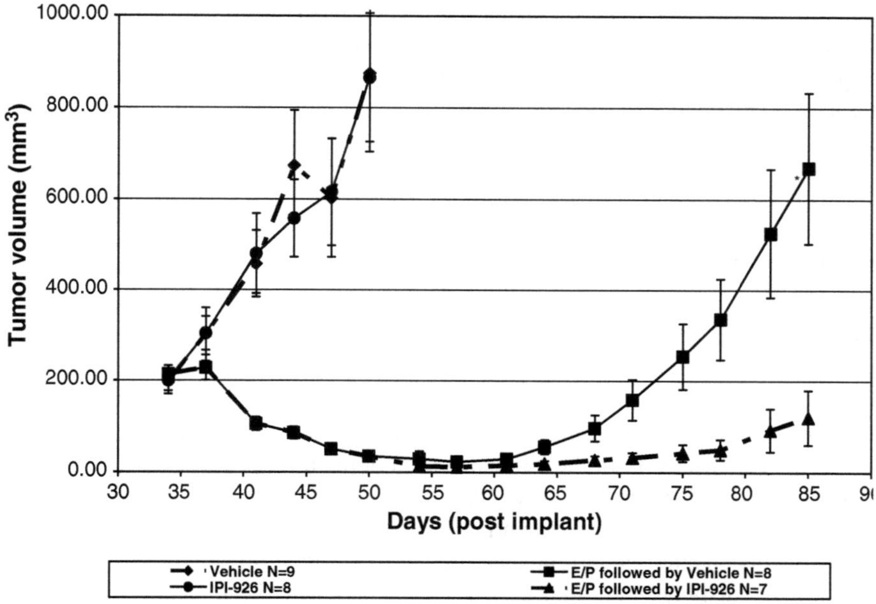


**Fig. 12.14** Inhibition of the Hh pathway with IPI-926 delays LX22 SCLC growth following etoposide/carboplatin treatment. Etoposide: 12 mg/kg i.v. three consecutive doses; Carboplatin: 60 mg/kg i.v. once per week. IPI-926 40 mg/kg p.o. QD; $p = 0.0101$

IPI-926 post-gefitinib significantly delays time to tumor regrowth [63]. Finally, multiple primary tumor xenograft models of head and neck squamous cell carcinoma (HNSCC) that regress with cetuximab treatment are sensitive to IPI-926 posttumor reduction, with a delay in time to tumor regrowth [51]. These data suggests an important role for the Hedgehog pathway in tumor relapse, potentially through a cancer stem cell and provides rationale for evaluation of a Smo inhibitor such as IPI-926 in the minimal residual disease setting in the clinic.

## Clinical Application

The role of the Hedgehog pathway in the preclinical settings described above suggests a broad potential for Smo antagonists in multiple clinical settings. The most straight-forward is in tumors driven by activating mutations in key Hh pathway members, such as in basal cell carcinoma and some medulloblastoma. Clinical proof of concept has been demonstrated in these indications [81, 102]. In addition, targeting the tumor microenvironment with a Hedgehog inhibitor could affect drug delivery to desmoplastic tumors such as pancreatic cancer and the studies in the KPC mouse model support evaluation of chemotherapy combined with IPI-926 in pancreatic cancer. The minimal residual disease setting is another example where the Hedgehog pathway plays a key role in relapse of tumors posttumor reduction with chemotherapy. While the precise mechanism is still being elucidated, whether Hh-dependent tumor initiating cells or a higher dependence on the microenvironment for tumor regrowth, the cells responsible for regrowth posttumor reduction are dependent on the Hh pathway and inhibited by IPI-926. There are potentially many uses for a Hh antagonist in the treatment of cancer and the trials with multiple Smo inhibitors will provide more insight into the role of Hh in these clinical settings.

## Conclusion

Plant-derived natural products continue to play an important role in drug discovery for many therapeutic areas [84]. In oncology, several drugs that have greatly impacted the life of cancer patients are plant derived such as taxol, camptothecin, combrestatin, podophyllotoxin, and vinca alkaloids such as vinblastine and vincristine [16]. The discovery of the mode of action of plant-derived cyclopamine and the importance of this target in cancer has attracted the attention of many researchers to investigate the therapeutic potential of cyclopamine and its analogs. IPI-926, the focus of this chapter, is currently in clinical trials. Preclinical studies have shown that IPI-926 displays antitumor activity in both Hh ligand-independent and ligand-dependent models of malignant activation of the Hh pathway. Clinical evaluation of the first semisynthetic analog of cyclopamine will determine whether this approach will prove beneficial in an array of different clinical settings.

**Acknowledgments** Several data on IPI-926 summarized in this chapter were the fruit of the entire Infinity Hedgehog team who are specially acknowledged. Outstanding contributions made by Dr. Phillip A. Beachy and Dr. James K. Chen served as the starting point for Infinity's program. In addition, we gracefully acknowledged long-standing collaborators of Infinity Pharmaceuticals, Inc. who had a major impact on the work presented in this chapter: Dr. Neil Watkins (Monash Institute of Medical Research), Dr. David A. Tuveson (Cancer Research UK), Dr. André B. Charette (University of Montreal) and Mr. Steve Monsen (USFS retiree). Finally, the authors thank Drs. Margaret A. Read, Alfredo C. Castro, and Vito J. Palombella for critical review of the manuscript.

## References

1. Austad B, Behnke ML, Castro AC, Charette AB, Grogan MJ, Janardanannair S, Lescarbeau A, Peluso S, Tremblay M (2009) Cyclopamine analogs. US 2009/0012109 A1, 8 Jan 2009
2. Badria FA, McChesney JD, Halim AF, Zaghloul AM, El Sayed KA (1995) Time course and inhibition of stavaroside K, veratramine and cevine-induced hemolysis by other pregnane glycosides and *Veratrum* alkaloids. Pharmazie 50:421–423
3. Bar EE, Chaudhry A, Lin A, Fan X, Schreck K, Matsui W, Piccirillo S, Vescovi AL, Dimeco F, Olivi A et al (2007) Cyclopamine-mediated hedgehog pathway inhibition depletes stem-like cancer cells in glioblastoma. Stem Cells 25:2524–2533
4. Beachy PA (2001) Synthesis, compositions and uses of steroidal alkaloids as regulators of the hedgehog pathway. WO2001/027135, 19 April 2001
5. Belloni E, Muenke M, Roessler E, Traverso G, Siegel-Bartelt J, Frumkin A, Mitchell HF, Donis-Keller H, Helms C, Hing AV et al (1996) Identification of *Sonic hedgehog* as candidate gene responsible for holoprosencephaly. Nat Genet 14:353–356
6. Berman DM, Karhadkar SS, Hallahan AR, Pritchard JI, Eberhart CG, Watkins DN, Chen JK, Cooper MK, Taipale J, Olson JM et al (2002) Medulloblastoma growth inhibition by hedgehog pathway blockade. Science 297:1559–1561
7. Binns W, James LF, Shupe JL, Everett G (1963) A congenital cyclopian-type malformation in lambs induced by maternal ingestion of a range plant, *Veratrum californicum*. Am J Vet Res 24:1164–1174
8. Briggs KJ, Corcoran-Schwartz IM, Zhang W, Harcke T, Devereux WL, Baylin SB, Eberhart CG, Watkins DN (2008) Cooperation between the Hic1 and Ptc1 tumor suppressors in medulloblastoma. Genes Dev 22:770–785
9. Brown D, Keeler RF (1978) Structure-activity relation of steroid teratogens. 1 Jervine ring system. J Agric Food Chem 26:561–563
10. Brown D, Keeler RF (1978) Structure-activity relation of steroid teratogens. 2. N-Substituted jervines. J Agric Food Chem 26:564–566
11. Brown E, Ragault M (1979) The syntheses of C-Nor D-Homo Steroids. Tetrahedron 35:911–927
12. Chen JK, Taipale J, Cooper MK, Beachy PA (2002) Inhibition of hedgehog signaling by direct binding of cyclopamine to Smoothened. Genes Dev 16:2743–2748
13. Chen JK, Taipale J, Young KE, Maiti T, Beachy PA (2002) Small molecule modulation of Smoothened activity. Proc Natl Acad Sci U S A 99:14071–14076
14. Chiang C, Litingtung Y, Lee E, Young KE, Corden JL, Westphal H, Beachy PA (1996) Cyclopia and defective axial patterning in mice lacking *Sonic hedgehog gene* function. Nature 383:407–413
15. Cooper MK, Porter JA, Young KE, Beachy PA (1998) Teratogen-mediated inhibition of target tissue response to Shh signaling. Science 280:1603–1607
16. Cragg GM, Kingston DGI, Newman DJ (eds) (2005) Anticancer agents from natural products. CRC Taylor & Francis, Boca Raton

17. Dierks C, Beigi R, Guo G-R, Zirlik Z, Stegert MR, Manley P, Trussell C, Schmitt-Graeff A, Landwerlin K, Veelken H et al (2008) Expansion of Bcr-Abl-positive leukemic stem cells is dependent on hedgehog pathway activation. Cancer Cell 14:238–249
18. Dwyer JR, Sever N, Carlson M, Nelson SF, Beachy PA, Parhami F (2007) Oxysterols are novel activators of the hedgehog signaling pathway in pluripotent mesenchymal cells. J Biol Chem 282:8959–8968
19. Ehtesham M, Sarangi A, Valadez JG, Chanthaphaychith S, Becher MW, Abel TW, Thompson RC, Cooper MK (2007) Ligand-dependent activation of the hedgehog pathway in glioma progenitor cells. Oncogene 26:5752–5761
20. Fallingborg J (1999) Intraluminal pH of the human gastrointestinal tract. Dan Med Bull 46:183–196
21. Feldmann G, Dhara S, Fendrich V, Bedja D, Beaty R, Mullendore M, Karikari C, Alvarez H, Iacobuzio-Donahue C, Jimeno A et al (2007) Blockade of hedgehog signaling inhibits pancreatic cancer invasion and metastases: a new paradigm for combination therapy in solid cancers. Cancer Res 67:2187–2196
22. Fried J, Wintersteiner O, Moore M, Iselin BM, Klingsberg A (1951) The structure of jervine. II. Degradation to perhydrobenzofluorene derivatives. J Am Chem Soc 73:2970–2971
23. Gaffield W, Keeler RF (1996) Craniofacial malformations induced in hamsters by steroidal alkaloids. J Nat Toxins 5:25–38
24. Giannis A, Heretsch P, Sarli V, Stoessel A (2009) Synthesis of cyclopamine using a biomimetic and diastereoselective approach. Angew Chem Int Ed 48:7911–7914
25. Growdon W, McCann C, Foster R, Curley M, Ross R, Proctor J, Borger D, MacDougall J, Rueda B (2010) Hedgehog pathway inhibition impedes ovarian cancer xenograft growth in the adjuvant and maintenance settings. American Society of Clinical Oncology, Chicago, 4–8 June 2010
26. Hamon F, Renoux B, Chadeneau C, Muller J-M, Papot S (2010) Study of a cyclopamine glucuronide prodrug for the selective chemotherapy of glioblastoma. Eur J Med Chem 45:1678–1682
27. Hann CL, Daniel VC, Sugar EA, Dobromilskaya I, Murphy SC, Cope L, Lin X, Hierman JS, Wilburn DL, Watkins DN et al (2008) Therapeutic efficacy of ABT-737, a selective inhibitor of BCL-2, in small cell lung cancer. Cancer Res 68:2321–2328
28. Heretsch P, Rabe S, Giannis A (2009) Synthesis of all diastereomers of the piperidine-alkaloid substructure of cyclopamine. Org Lett 11:5410–5412
29. Heretsch P, Tzagkaroulaki L, Giannis A (2010) Cyclopamine and hedgehog signaling: chemistry, biology, medical perspectives. Angew Chem Int Ed 49:3418–3427
30. Hingorani SR, Wang L, Multani AS, Combs C, Deramaudt TB, Hruban RH, Rustgi AL, Chang S, Tuveson DA (2005) $Trp53^{R172H}$ and $Kras^{G12D}$ cooperate to promote chromosomal instability and widely metastatic pancreatic ductal adenocarcinoma in mice. Cancer Cell 7:469–483
31. Hiskey CF, Hirschmann R, Wendler NL (1953) Rearrangement of the steroid C-D rings. Synthesis of an 11-oxo-Delta 13(17a)-C-nor-D-homosteroid. J Am Chem Soc 75:5135–5136
32. Incardona JP, Gaffield W, Kapur RP, Roelink H (1998) The teratogenic *Veratrum* alkaloid cyclopamine inhibits Sonic hedgehog signal transduction. Development 125:3553–3562
33. Incardona JP, Gaffield W, Lange Y, Cooney A, Pentchev PG, Liu S, Watson JA, Kapur RP, Roelink H (2000) Cyclopamine inhibition of sonic hedgehog signal transduction is not mediated through effects on cholesterol transport. Dev Biol 224:440–452
34. Jacobs WA, Craig LC (1943) Veratrine alkaloids. XVI. The formulation of jervine. J Biol Chem 148:51–55
35. Jacobs WA, Sato Y (1948) Veratrine alkaloids. XXVIII. Structure of jervine. J Biol Chem 175:57–65
36. Jayatilake GS, Richheimer SL, Mann DA (2010) Isolation of cyclopamine from *Veratrum* plant. WO 2010/002970 A2, 7 Jan 2010
37. Jiang J, Hui CC (2008) Hedgehog signaling in development and cancer. Dev Cell 15:801–812
38. Johnson WS, Cox JH, Graham DW, Whitlock HW Jr (1967) The total synthesis of 17-acetyl-5α-etiojerva-12,14,16-tien-3β-ol. J Am Chem Soc 89:4524–4526

39. Kaneko K, Mitsuhashi H, Hirayama K, Ohmori S (1970) 11-Deoxojervine as a precursor for jervine biosynthesis in *Veratrum grandiflorum*. Phytochemistry 9:2497–2501
40. Kaneko K, Mitsuhashi H, Hirayama K, Yoshida N (1970) Biosynthesis of C-nor-D-homosteroidal alkaloids from acetate-1-14C, cholesterol-4-14C and cholesterol-26-14C in *Veratrum grandiflorum*. Phytochemistry 9:2489–2495
41. Kararli TL (1995) Comparison of the gastrointestinal anatomy, physiology, and biochemistry of humans and commonly used laboratory animals. Biopharm Drug Dispos 16:351–380
42. Keeler RF, Binns W (1966) Teratogenic compounds of *Veratrum californicum* (Durand). II. Production of ovine fetal cyclopia by fractions and alkaloid preparations. Can J Biochem 44:829–838
43. Keeler RF, Binns W (1966) Teratogenic compounds of *Veratrum californicum* (Durand). I. Preparation and characterization of fractions and alkaloids for biologic testing. Can J Biochem 44:819–828
44. Keeler RF (1968) Teratogenic compounds of *Veratrum californicum* (Durand)-IV. First isolation of veratramine and alkaloid Q and a reliable method for isolation of cyclopamine. Phytochemistry 7:303–306
45. Keeler RF, Binns W (1968) Teratogenic compounds of *Veratrum californicum* (Durand)-V. Comparison of cyclopian effects of steroidal alkaloids from the plant and structurally related compounds from other sources. Teratology 1:5–10
46. Keeler RF (1969) Teratogenic compounds of *Veratrum californicum* (Durand)-VI. The structure of cyclopamine. Phytochemistry 8:223–225
47. Keeler RF (1969) Teratogenic compounds from *Veratrum californicum* (Durand) VII. The structure of glycosidic alkaloid cycloposine. Steroids 13:579–588
48. Keeler RF (1970) Teratogenic compounds in *Veratrum californicum* (Durand) IX. Structure-activity relation. Teratology 3:169–174
49. Keeler RF, Binns W (1971) Teratogenic compounds of *Veratrum californicum*. XII. Teratogenic compounds of *Veratrum californicum* as a function of plant part, stage, and site of growth. Phytochemistry 10:1765–1769
50. Keeler RF, Baker DC (1989) Oral, osmotic minipump, and intramuscular administration to sheep of the *Veratrum* alkaloid cyclopamine. Proc Soc Exp Biol Med 192:153–156
51. Keysar S, Anderson R, Takimoto S, Tan AC, Deyneko I, McGovern K, Ross RW, Song J, Jimeno A (2010) Combined hedgehog and epidermal growth factor receptor (EGFR) inhibition in a direct patient tumor model (DPTM) of head and neck squamous cell cancer (HNSCC). American Society of Clinical Oncology, Chicago, 4–10 June 2010
52. Kingston DGI (2005) Taxol and its analogs. In: Cragg GM, Kingston DGI, Newman DJ (eds) Anticancer agents from natural products. CRC Taylor & Francis, Boca Raton, pp 89–122
53. Kumar SK, Roy I, Anchoori RK, Fazli S, Maitra A, Beachy PA, Khan SR (2008) Targeted inhibition of hedgehog signaling by cyclopamine prodrugs for advanced prostate cancer. Bioorg Med Chem 16:2764–2768
54. Kupchan SM, Suffness MI (1968) *Veratrum* alkaloids. LIV. Stereochemistry of jervine and related alkaloids. J Am Chem Soc 90:2730–2731
55. Kutney JP, Cable J, Gladstone WAF, Hanssen HW, Nair GV, Torupka EJ, Warnock WDC (1975) Synthetic studies in the *Veratrum* alkaloid series. II. The total synthesis of verarine, veratramine, jervine, and veratrobasine. Can J Chem 53:1796–1817
56. Leonard E, Runguphan W, O'Connor S, Prather KJ (2009) Opportunities for metabolic engineering to facilitate alkaloid production. Nat Chem Biol 5:292–300
57. Li C, Heidt DG, Dalerba P, Burant CF, Zhang L, Adsay V, Wicha M, Clarke MF, Simeone DM (2007) Identification of pancreatic cancer stem cells. Cancer Res 67:1030–1037
58. Lin TL, Wang Q, Brown P, Peacock C, Merchant AA, Brennan S, Jones E, McGovern K, Watkins DN, Sakamoto K et al (2010) Self-renewal of acute lymphocytic leukemia cells is limited by the Hedgehog pathway inhibitors cyclopamine and IPI-926. PLoS One 5:e15262

59. Lipinski RJ, Hutson PR, Hannam PW, Nydza RJ, Washington IM, Moore RW, Girdaukas GG, Peterson RE, Bushman W (2008) Dose- and route-dependent teratogenicity, toxicity, and pharmacokinetic profiles of the hedgehog signaling antagonist cyclopamine in the mouse. Toxicol Sci 104:189–197
60. Liu S, Dontu G, Mantle I, Patel S, Ahn N, Jackson K, Suri P, Wicha M (2006) Hedgehog signaling and Bmi-1 regulate self-renewal of normal and malignant human mammary stem cells. Cancer Res 66:6063–6071
61. Ma R, Ritala A, Oksman-Caldentey K-M, Rischer H (2006) Development of in vitro techniques for the important medicinal plant *Veratrum californicum*. Planta Med 72:1142–1148
62. Mahindroo N, Punchihewa C, Fujii N (2009) Hedgehog-Gli signaling pathway inhibitors as anticancer agents. J Med Chem 52:3829–3845
63. Mandley E, White K, Faia K, Travaglione V, Read M, McGovern K, MacDougall JR, Fritz C (2010) The Hh inhibitor IPI-926 inhibits relapse of a non-small cell lung cancer xenograft model following treatment with an EGFR targeted tyrosine kinase inhibitor. American Association for Cancer Research, Washington, 17–21 April 2010
64. Manna JD, Alvarez-Diez TM, Grogan MJ, Porter JR, Tremblay MR, Castro AC, Sydor JR (2008) Metabolite identification of IPI-609, a novel and potent inhibitor of the hedgehog pathway, in different species. ASMS Conference, Denver, 1–5 June 2008
65. Masamune T, Mori Y, Takasugi M, Murai A, Ohuchi S, Sato N, Katsui N (1965) 11-Deoxojervine, a new alkaloid from *Veratrum* species. Bull Chem Soc Jpn 38:1374–1378
66. Masamune T, Sato N, Kobayashi K, Yamazaki I, Mori Y (1967) Syntheses and NMR spectra of 22,27-imino-17,23-oxidojervane derivatives. Tetrahedron 23:1591–1612
67. Masamune T, Takasugi M, Murai A, Kobayashi K (1967) The synthesis of jervine and related alkaloids. J Am Chem Soc 89:4521–4523
68. Masamune T, Murai A, Ono H, Orito K, Suginome H (1969) Revision of the $C_{12}$ configuration of dihydrojervine. Tetrahedron Lett 10:255–258
69. Mitsuhashi H, Shibata K (1964) Synthesis of a nitrogen-free derivative of veratramine. Studies on C-nor D-homosteroids. III. Tetrahedron Lett 5:2281–2283
70. Nagata R, Izumi K (1991) Veratramine-induced behavior associated with serotonergic hyperfunction in mice. J Pharmacol 55:129
71. Nakamura T, Aikawa T, Iwamoto-Enomoto M, Iwamoto M, Higuchi Y, Pacifici M, Kinto N, Yamaguchi A, Noji S, Kurisu K et al (1997) Induction of osteogenic differentiation by hedgehog proteins. Biochem Biophys Res Commun 237:465–469
72. Nüsslein-Volhard C, Wieschaus E (1980) Mutations affecting segment number and polarity in Drosophila. Nature 287:795–801
73. Oatis JE Jr, Brunsfeld P, Rushing JW, Moeller PD, Bearden DW, Gallien TN IV, Cooper G IV (2008) Isolation, purification, and full NMR assignments of cyclopamine from *Veratrum californicum*. Chem Cent J 2:12–18
74. Olive KP, Jacobetz MA, Davidson CJ, Gopinathan A, McIntyre D, Honess D, Madhu B, Goldgraben MA, Caldwell ME, Allard D et al (2009) Inhibition of hedgehog signaling enhances delivery of chemotherapy in a mouse model of pancreatic cancer. Science 324:1457–1461
75. Palma V, Lim Daniel A, Dahmane N, Sanchez P, Brionne Thomas C, Herzberg Claudia D, Gitton Y, Carleton A, Alvarez-Buylla A, Ruiz i Altaba A (2005) Sonic hedgehog controls stem cell behavior in the postnatal and adult brain. Development 132:335–344
76. Peacock CD, Wang Q, Gesell GS, Corcoran-Schwartz IM, Jones E, Kim J, Devereux WL, Rhodes JT, Huff CA, Beachy PA et al (2007) Hedgehog signaling maintains a tumor stem cell compartment in multiple myeloma. Proc Natl Acad Sci U S A 104:4048–4053
77. Peukert S, Miller-Moslin K (2009) Hedgehog signaling pathway inhibitors as cancer therapeutics. Annu Rep Med Chem 44:323–337
78. Pink M, Proctor J, Briggs K, MacDougall J, Whitebread N, Tremblay MR, Grogan M, Palombella V, Castro A, Adams J et al (2008) Activity of IPI-926, a potent HH pathway inhibitor, in a novel model of medulloblastoma derived from Ptch/HIC$^{+/-}$ mice. American Association for Cancer Research, San Diego, 12–16 April 2008

79. Reeke GN Jr, Vincent RL, Lipscomb WN (1968) Molecular structure of veratrobasine. J Am Chem Soc 90:1663–1664
80. Roessler E, Belloni E, Gaudenz K, Jay P, Berta P, Scherer SW, Tsui L-C, Muenke M (1996) Mutations in the human *Sonic Hedgehog* gene cause holoprosencephaly. Nat Genet 14:357–360
81. Rudin CM, Hann CL, Laterra J, Yauch RL, Callahan CA, Fu L, Holcomb T, Stinson J, Gould SE, Coleman B et al (2009) Treatment of medulloblastoma with hedgehog pathway inhibitor GDC-0449. N Engl J Med 361:1173–1178
82. Saito K, Suginome H, Takaoka M (1934) On the alkaloids of white hellebore I. Isolation of constituent alkaloids. Bull Chem Soc Jpn 9:15–23
83. Saito K, Suginome H (1935) On the alkaloids of white hellebore II. Isolation of alkaloids from the so-called resinous matters. Bull Chem Soc Jpn 11:168–171
84. Saklani A, Kutty SK (2008) Plant-derived compounds in clinical trials. Drug Discov Today 13:161–171
85. Sasai K, Romer JT, Lee Y, Finkelstein D, Fuller C, McKinnon PJ, Curran T (2006) Shh pathway activity is down-regulated in cultured medulloblastoma cells: implications for preclinical studies. Cancer Res 66:4215–4222
86. Splinter S, Kadali S (2010) Methods for obtaining cyclopamine. WO2010/000070A1, 7 Jan 2010
87. Suginome H, Yonekuma N, Masamune T (1980) The transformation of jervine into 18-functional C-nor-D-homosteroids. Bull Chem Soc Jpn 53:210–213
88. Taipale J, Chen JK, Cooper MK, Wang B, Mann RK, Milenkovic L, Scotts MP, Beachy PA (2000) Effects of oncogenic mutations in Smoothened and Patched can be reversed by cyclopamine. Nature 406:1005–1009
89. Thayer SP, diMagliano MP, Heiser PW, Nielsen CM, Roberts DJ, Lauwers GY, Qi YP, Gysin S, Fernandez-del Castillo C, Yajnik V et al (2003) Hedgehog is an early and late mediator of pancreatic cancer tumorigenesis. Nature 425:851–856
90. Thron CD, McCann FV (1998) Pharmacological tests of the mechanism of the periodic rhythm caused by veratramine in the sinoatrial node of the guinea pig. Gen Pharmacol 32:81–89
91. Travaglione V, Peacock CD, MacDougall J, McGovern K, Cushing J, Yu L-C, Trudeau M, Palombella V, Adams J, Hierman J et al (2008) A novel Hh pathway inhibitor, IPI-926, delays recurrence post-chemotherapy in a primary human SCLC xenograft model. Annual Meeting of AACR, San Diego, 2008
92. Tremblay MR, McGovern K, Nevalainen M, Nair S, Porter JR, Behnke M, Yu L-C, Grenier L, Campbell M, Cushing J et al (2007) Synthesis of novel, chemically stable D-homo-cyclopamine analogs via a cyclopropanation/ring-expansion sequence. Gordon Research Conference on Natural Products, Tilton, 22–27 July 2007
93. Tremblay MR, Nevalainen M, Nair SJ, Porter JR, Castro AC, Behnke ML, Yu L-C, Hagel M, White K, Faia K et al (2008) Semi-synthetic cyclopamine analogues as potent and orally bioavailable hedgehog pathway antagonists. J Med Chem 51:6646–6649
94. Tremblay MR, Lescarbeau A, Grogan MJ, Tan E, Lin G, Austad BC, Yu L-C, Behnke ML, Nair SJ, Hagel M et al (2009) Discovery of a potent and orally active hedgehog pathway antagonist (IPI-926). J Med Chem 52:4400–4418
95. Tremblay MR, Nesler M, Weatherhead R, Castro AC (2009) Recent patents for Hedgehog pathway inhibitors for the treatment of malignancy. Expert Opin Ther Pat 19: 1039–1056
96. Trowbridge JJ, Scott MP, Bhatia M (2006) Hedgehog modulates cell cycle regulators in stem cells to control hematopoietic regeneration. Proc Natl Acad Sci U S A 103:14134–14139
97. Tung DC, Chao KS (2007) Targeting hedgehog in cancer stem cells: how a paradigm shift can improve treatment response. Future Oncol 3:569–574
98. Turner TT, Bang HJ, Attipoe SA, Johnston DS, Tomsig JL (2006) Sonic hedgehog pathway inhibition alters epididymal function as assessed by the development of sperm motility. J Androl 27:225–232

99. Van Den Brink GR, Hardwick JCH, Tytgat GNJ, Brink MA, Ten Kate FJ, Van Deventer SJH, Peppelenbosch MP (2001) Sonic hedgehog regulates gastric gland morphogenesis in man and mouse. Gastroenterology 121:317–328
100. van der Horst G, Farih-Sips H, Lowik CWGM, Karperien M (2003) Hedgehog stimulates only osteoblastic differentiation of undifferentiated KS483 cells. Bone 33:899–910
101. Vestergaard J, Pedersen Mikkel W, Pedersen N, Ensinger C, Tumer Z, Tommerup N, Poulsen Hans S, Larsen Lars A (2006) Hedgehog signaling in small-cell lung cancer: frequent in vivo but a rare event in vitro. Lung Cancer 52:281–290
102. Von Hoff DD, LoRusso PM, Rudin CM, Reddy JC, Yauch RL, Tibes R, Weiss GJ, Borad MJ, Hann CL, Brahmer JR et al (2009) Inhibition of the hedgehog pathway in advanced basal-cell carcinoma. N Engl J Med 361:1164–1172
103. Welch KD, Panter KE, Lee ST, Gardner DR, Stegelmeier BL, Cook D (2009) Cyclopamine-induced synophthalmia in sheep: defining a critical window and toxicokinetic evaluation. J Appl Toxicol 29:414–421
104. Wilson SR, Strand MF, Krapp A, Rise F, Petersen D, Krauss S (2010) Hedgehog antagonist cyclopamine isomerizes to less potent forms when acidified. J Pharm Biomed Anal 52:707–713
105. Winkler JD, Isaacs A, Holderbaum L, Tatard V, Dahmane N (2009) Design and synthesis of inhibitors of hedgehog signaling based on the alkaloid cyclopamine. Org Lett 11:2824–2827
106. Wintersteiner O, Moore M, Fried J, Iselin B (1951) The structure of jervine: diacetyl-7-ketojervine. Proc Natl Acad Sci U S A 37:333–340
107. Wintersteiner O, Moore M (1964) Jervine. XIII. C/D Ring cis-trans equilibria in tetrahydrojervine derivatives with open ring E. Tetrahedron 20:1947–1954
108. Wright CRA, Luff AP (1878) XLVII – The alkaloids of *Veratrums*. Part II. The alkaloids of *Veratrum album*. J Chem Soc 33:338
109. Wright CRA (1879) XLVIII – The alkaloids of *Veratrums*. Part III. The Alkaloids of *Veratrum viride*. J Chem Soc 35:421–430
110. Xie J, Garrossian M (2009) Cyclopamine tartrate salt for cancer treatment. WO2009099625, 5 Feb 2009
111. Yang L, Xie G, Fan Q, Xie J (2010) Activation of the hedgehog-signaling pathway in human cancer and the clinical implications. Oncogene 29:469–481
112. Yauch RL, Gould SE, Scales SJ, Tang T, Tian H, Ahn CP, Marshall D, Fu L, Januario T, Kallop D et al (2008) A paracrine requirement for hedgehog signaling in cancer. Nature 455:406–410
113. Zhang J, Garrossian M, Gardner D, Garrossian A, Chang Y-T, Kim YK, Chang CWT (2008) Synthesis and anticancer activity studies of cyclopamine derivatives. Bioorg Med Chem Lett 18:1359–1363
114. Zhang X, Harrington N, Moraes RC, Wu M-F, Hilsenbeck SG, Lewis MT (2009) Cyclopamine inhibition of human breast cancer cell growth independent of Smoothened (Smo). Breast Cancer Res Treat 115:505–521

# Index




DOI 10.1007/978-1-4419-8435-7, 

MIX
Papier aus verantwortungsvollen Quellen
Paper from responsible sources
FSC® C105338

If you have any concerns about our products,
you can contact us on
**ProductSafety@springernature.com**

In case Publisher is established outside the EU,
the EU authorized representative is:
**Springer Nature Customer Service Center GmbH**
**Europaplatz 3, 69115 Heidelberg, Germany**

Printed by Libri Plureos GmbH
in Hamburg, Germany